Petroleum Science and Technology

Petroleum Science and Technology

Petroleum Generation, Accumulation and Prospecting

Muhammad Abdul Quddus

CRC Press
Taylor & Francis Group
Boca Raton London New York

CRC Press is an imprint of the
Taylor & Francis Group, an **informa** business

First edition published 2021
by CRC Press
6000 Broken Sound Parkway NW, Suite 300, Boca Raton, FL 33487-2742

and by CRC Press
2 Park Square, Milton Park, Abingdon, Oxon, OX14 4RN

© 2021 Taylor & Francis Group, LLC
CRC Press is an imprint of Taylor & Francis Group, LLC

ISBN: 978-0-367-50322-2 (hbk)
ISBN: 978-0-367-50441-0 (pbk)
ISBN: 978-1-003-04948-7 (ebk)

Typeset in Times
by Deanta Global Publishing Services Chennai India

*The book is dedicated to my evening class
students who inspired me to write it.*

Contents

Preface

A number of books are available covering all aspects of petroleum technology, and many more are expected to be on the market in the future. One book triggers the need for another on the same subject. The need for new books will never be exhausted. Every book written on the same subject adds a different color, taste and shows different aspects of the subject matter. One book may be written to target a specific group and the other may be written for the general reader. This book, *Petroleum Science and Technology*, has been written with a specific purpose. It introduces almost half of the 'upstream' sector of the petroleum industry. The author has taught petroleum technology to morning/evening program graduate students for 17 years (2000–2016). The evening students comprise both genders and are of various age groups. Most of them are gainfully employed in diversified industries, services and in the teaching profession. Accordingly they have diversified educational backgrounds. One student may have studied one subject and another may have not learned the other subject in their undergraduate science program. The students have learned English as a second language in different educational environments. The students have different levels of English proficiency in listening, writing and comprehension. This applies universally for those who have not studied English as a first language. This book aims to resolve their language difficulties along with providing sufficient knowledge to the petroleum engineer, technologist and geologist. It is hoped that the book equally serves the purpose of English-speaking persons and that they will find it useful. Evening students have their own specific expectations and aspirations by attending the petroleum technology course at the cost of their extra efforts, time and money. To meet the expectations and requirements of such a diversified group of students seems to be a challenging assignment. The proper choice of words and sentences is necessary for meaningful communication, according to the skill and understanding of each individual student. There is hardly any book available in the market which completely fulfills the requirements of such diversified students seeking petroleum knowledge. The non-availability of such a comprehensive yet concise book in a single volume has been pointed out by the students themselves during my teaching experience, and I felt the same. Additionally, with diversified experience in the petroleum field coupled with teaching diversified students in the departments of petroleum technology and applied chemistry, it is expected that the author is well placed to meet the expectations of students and readers of the book. As already stated, the book has been written to address the expectations of readers across the globe who have different levels of English proficiency in reading. Additionally, new entrants in the petroleum technology sector will find it useful as the major activities of upstream petroleum are available to them.

The enthusiasm and interest shown by the students during the lectures have inspired me to write this book. I owe this book to my students. Although the book is written focusing on the need of students, it can be equally useful for any person associated with the industry who wants to obtain technical/scientific information on petroleum.

The data described in the book are taken from the references cited at the end of the book. Most of the tables and figures of the book have appeared in more than one place in the literature. If any work has been left without acknowledgment or there is copyright omission, it is regretted, and I am ready to take remedial measures. The book's text has occasionally used the unpublished work, notes, discussion and suggestions of senior and junior colleagues as well as professionals from industry. The book is a review of the published material.

It is debated that petroleum is a depleting commodity and becoming costlier day by day or that it is not environmentally friendly. It is also suspected that since the reserves of hydrocarbon are finite, it may soon be exhausted due to rapid production and consumption. Whatever the reason, society is looking for its substitute. What are the other options for energy requirements? Hardly any real substitute for natural hydrocarbon exists in terms of economy and ease of utilization. The scientific communities are working to find out such material, but the natural hydrocarbon is likely to stay on the scene for quite a long time.

Acknowledgments

I am indebted and thankful to Dr Akhlaq Ahmed, Pro Vice Chancellor, Dr Majeedullah Qadri, Chairman, Department of Petroleum Technology and Dr Fasihullah Khan, Chairman, Department of Applied Chemistry, University of Karachi, for inducting me into the teaching profession and the cooperation of all the members of the two departments throughout my teaching tenure and writing this book. I am thankful to Mr. Muhammad Munirullah, Graphic Designer, Department of Geology, for helping me in the drawing of figures and typing of the book content. I would like to thank Gagandeep Singh, Senior Editor at Taylor & Francis/CRC Press for his persistence, encouragement and guidance, for making the book proposal a reality and bringing the draft proposal up to the publisher standard. I must humbly acknowledge the authors of the literature cited in the reference pertaining to the book's subject.

Author

Muhammad Abdul Quddus, PhD, has been serving the petroleum sector and R&D organizations, both national and multinational, for more than 40 years and has worked in various capacities in the laboratory, office, field and plant; he has also engaged in teaching petroleum technology as a visiting professor for 17 years. He earned BSc (Hons) and MSc degrees along with a PhD with thesis title 'Oxidation of Asphalt'. As a result of constant research, he has published nine international and twelve national papers, obtained one patent, presented five papers in conferences and prepared six technical reports. He has also visited the USA, Canada and Indonesia for short courses in petroleum technology and teacher training.

Dr. Quddus is now retired, having earlier worked in the Hydrocarbon Development Institute of Pakistan, Pakistan Council of Scientific & Industrial Research, Arabian American Oil Company Saudi Arabia and Karachi University. He thought to share his knowledge and experience with readers of the book. He feels that a simple treatment of the subject is needed among diversified readers, students, industrial professionals and academics in addition to the advanced treatment of the matter which is available in abundance.

Introduction

Petroleum is a necessity of society and contributes to the survival and advancement of society. It is meeting energy demands and providing useful petrochemical-based consumer products.

The whole petroleum industry has been divided into several 'streams' or sectors.

- The 'upstream' sector generally refers to the recovery of hydrocarbon raw material from underground sources to be used in the 'downstream' industry for making consumer products. Upstream activity in the oil/gas sector involves surveying and exploration for oil/gas prospecting, drilling operations, production technology and raw petroleum (oil/gas) treatment at the surface to make the fluid safe and harmless. The sector also includes the study of the origins, generation, migration and accumulation of petroleum in subsurface source rock, carrier-bed and reservoir rock under suitable geological conditions.
- The 'mid-stream' sector is the transportation and storage of oil/gas. It links the upstream and downstream sectors.
- The 'downstream' sector involves the processing and refining of the raw oil/gas to produce finished products and deliver them to the consumer through marketing companies.
- The 'chemical-stream', popularly known as 'petrochemicals', is the fourth sector. The sector is already well developed and has a long history but is receiving renewed attention. Significant innovation in petrochemicals is coming up. It involves producing a wide variety of materials from common consumable products to advance quality industrial materials. The chemical-stream is expected to revolutionize the use of petrochemicals in commodities. The best feed stocks for the chemical-stream are the gaseous and lighter petroleum hydrocarbons. As the petrochemical industry develops further, the lighter hydrocarbons including the gases will become more precious than oil. Oil is fast depleting compared to gas and is also replaceable by alternate fuels. Moreover new gas reserves are fast coming up. It is anticipated that gas has a better future than oil.

The upstream petroleum sector is further divided broadly into four groups. The first group is petroleum origins, generation, migration and accumulation under suitable subsurface geological conditions. The second group deals with petroleum oil and gas prospecting by geological, geophysical and geochemical techniques. The third is exploration by drilling and well logging methods and the extraction of fluid (oil/gas) from underground strata by drilling and employing different production methods. The fourth group of activities includes the characterization of subsurface reservoir fluid and surface treatment to make the raw oil/gas a saleable commodity.

Drilling operations are first employed for the exploration of oil/gas; if the fluid is found in commercial quantity the drill borehole is converted into a production well.

Originally it was planned that the entire above-stated description of the upstream sector be reported in a single volume of the book, as it may be convenient to buyers and students. But it became too voluminous. After consultation across the concerned personnel it was decided to bifurcate into two books. The title of the present first book is *Petroleum Science and Technology: Petroleum Generation, Accumulation and Prospecting.* This volume comprises 41 figures and 11 tables. Similarly the proposed Volume II will be titled *Petroleum Science and Technology: Exploration, Production and Treatment.*

Chapters 2 to 8 of the first book describe the natural hydrocarbon geology along with applicable aspects of physics, chemistry, biology, environment science, mathematics and engineering/technology. The chapters begin with basic principles and then build up to give a full conception of the subject. The introductory knowledge then can be applied to advanced topics and higher level courses to be taken up by the student in the future. The principles stated in the book are general and broadly applicable. No specific or targeted example is taken up. The narration gives the overall concepts of the subject rather than in-depth study. The conclusion given at the end of each chapter is written in such a way as to invoke further study among readers.

A brief description of each chapter is summarized as follows:

- Chapter 1. The 'Petroleum Pre-Period'. The chapter is included to see what was before organic matter in the universe. It describes the origin and evolution of the universe and inorganic matter.
- Chapter 2. 'Petroleum Origin and Generation'. The chapter describes the evolution of organic matter and its preservation in sedimentary rock for conversion to petroleum. Further the chapter describes the origin and generation of petroleum under subsurface conditions.
- Chapter 3. The petroleum system consists of the source rock, oil/gas migration path and reservoir rock. This is described in the chapter along with petroleum migration and accumulation.
- Chapter 4. This chapter is a continuation of Chapters 2 and 3. A brief description of the igneous and metamorphic rocks is given. Mechanically (clastic sandstone and limestone), organically (calcite) and chemically (halite and gypsum) formed sedimentary rock and remains of dead organisms as fossilized sediments are discussed. Physical features, structural deformation features, stratigraphic sedimentary structural features, bio-stratigraphy and magneto-stratigraphy have been mentioned. Methods for the relative/absolute age determination of rock, geological mapping, topography of sea bottom, formation of the seven continents, sea floor spreading, formation of ocean ridges and trenches, collision of crusts, earthquake, volcanic activity, rock subduction, plate tectonics and aerial/satellite surveying are described.
- Chapters 5, 6 and 7. Geophysical surveying, that is gravity and magnetic, reflection and refraction seismic and electrical methods for prospecting for oil/gas, is described.

- Chapter 8. Explanations and broader applications of geochemical principles along with geochemical surveying for the evaluation of kerogen and asphalt using geochemical analytical techniques are discussed in the chapter.

The key points of the subject are well explained with appropriate black/white figures and data presented in tabular form. The main feature of the book is that the basic information of the subject is described in the volume, so the reader is able to get full information without looking at other books. However a need for referring to other literature remains for enhanced and diversified knowledge of the subject. The subject matter of the book is dealt with by a theoretical treatment as well as giving applied information. The book has been made attractive to the reader.

There are mainly three basic 'unit systems', namely the International System of Units (SI) based on the meter, kilogram, second and kelvin; the metric (cgs) physical units based on the centimeter, gram, second and centigrade; and the traditional (FPS) foot, pound, second and Fahrenheit system, being used in industry and literature for the expression of a particular measurement or parameter. Other units are derived from these basic systems. Multifarious units of a single parameter reported in the literature are considered as an inconvenience and confusing to the user in industry as well as in academics. For example calorific value is reported in many units. This renders the comparison difficult between the two values reported in different units. The units used for gases are more diversified than those used for solids and liquids. SI is the preferred system.

The use of data in scientific dissertation is normal; it improves the quality and substance. The data in the book have been kept to a workable minimum. In doing so it is expected that it not only saves time but also helps quick understanding and better retention of the subject content.

1 Petroleum Pre-Period

1.1 INTRODUCTION

What is the origin of petroleum? Such and related questions are often asked and most probably answered properly. Most agree that organic matter is the precursor of petroleum. How did organic matter evolve? Inorganic lifeless matter prevailed before organic matter and living organisms. What is matter, and how it is generated and held together? Or simply how did the whole universe evolve, and what are its constituents? It will be interesting to seek answers to these questions. A brief account of the different events of universe creation and evolution that ultimately led to the advent of life on earth and the present stage of the universe is given below.

1.2 CREATION OF THE UNIVERSE

What is the nature of universe? Is it a single universe or multiple universes? The commonly defined present universe includes the entire cosmos from the smallest particle on earth to the largest galaxy and black hole in the universe with enormous size. The expanse of the universe is trillions of light years across. Cosmos and universe are interchangeable words. The scientific study of the universe in terms of its origin, development and future possibilities is known as cosmology. The universe observable even by the largest telescope is much smaller than the entire universe. The universe is in an 'excellent ordered' system, created by nature. Most probably the universe originated from a super dense pointed matter that expanded rapidly. The question may arise, from where did the super dense matter come into being? Many theories and hypotheses explain the creation of the universe and super dense matter. Some of them are stated below.

1.2.1 Study State Theory

The 'study state' theory is of the view that the universe existed always similar to the present form with alteration on a minor level. It is observed that a 'star system' within a galaxy collapses (death) with the release of radiation along with fundamental particles, atoms, elements, ions and molecules and highly condensed star residue. Again, these newly created radiations along with fundamental particles, atoms, elements, ions and molecules aggregated and contracted to form (birth) a new star system. The natural cycles of star birth/death or creation/destruction are continuing in the universe and a steady state is maintained in the universe.

1.2.2 Multiple Universes

Multiple universes are a hypothetical possibility. It includes this universe and any other parallel universes. The collapse of one universe into a super-giant point mass is followed by another giant expansion and creation of a new universe.

1.2.3 DARK ENERGY AND DARK MATTER

The fundamental units of the universe consist of 'dark energy' and 'dark matter'. Dark energy and dark matter are inter-convertible under suitable conditions of natural forces. The dark energy is responsible for the expansion and development of the universe by pushing the space matter and galaxies away from each other. It is a repulsive force against the inward gravitational pull or contraction. Dark matter is highly condensed, of high density and with great inward gravitational pull. It emits no light and is not visible. Dark matter is detected only by its capacity for the gravitational pull of the other visible object in space.

1.2.4 PULSATING THEORY

The theory states that the universe is expanding and contracting alternatively. That is to say that the universe is pulsating. The current expansion of the universe will stop at a certain point in time by the gravitational pull of the total mass of the universe. The stoppage of expansion triggers the contraction of the universe. After contraction to a certain point (super dense point matter), an explosion (Big Bang) again occurs and the universe again starts to expand and develop.

1.2.5 BIG BANG THEORY

According to the 'Big Bang' theory, the universe was created about 14.7 billion years ago, as a result of a mass explosion. The Big Bang was an explosion, with a high speed equal to the velocity of light at a very high temperature, of a small point of super dense spherical mass, containing probably an enormous amount of neutron particles. The neutron particles disintegrated into the universe (boundless) with a high temperature and with the velocity of light (186,000 miles/sec). From where did the neutron sphere come into being and form this super dense spherical ball? Now it is speculated that very high energy was concentrated in the small sphere in wave form or radiation!

If it is supposed that the explosion was a forward reaction for universe creation, then a reversible reaction is possible. The universe at a certain time can coalesce or contract into its original small ball point by natural forces.

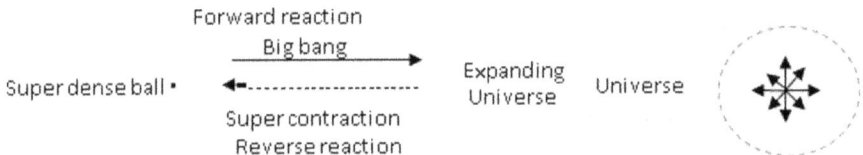

The assumption assured the sustainability of the universe. The explosion was the beginning time (t_0) of the universe. The universe evolved to the present stage through various eras of evolutions and activities.

1.2.6 EINSTEIN'S GENERAL THEORY OF RELATIVITY

Einstein's general theory does not explain the origin of the universe but defines the shape of the universe. The theory deals with space and time in the universe. The theory of relativity is related to the effect of the gravitational force on celestial bodies of the universe. The theory states 'gravity pulling in one direction is equivalent to acceleration in the opposite direction' that creates curvature in space.

1.3   RADIATION ERA

The Big Bang explosion occurred at very a high temperature that produced a tremendous amount of massless radiation. The radiation spread all around, forming an expanding radiation universe. The temperature dropped first rapidly and thereafter gradually decreased. In these conditions the disintegrated radiation coalesced and converted to particles of protons, neutrons and mesons, etc. These particles are called 'heavy particles' since they have appreciable mass. The interaction of radiation with heavy particles resulted in the emergence of lighter fundamental particles, namely electrons, neutrinos and positrons. The lighter fundamental particles were almost massless. An electron has a mass of one-thousandth of the lightest atom, the hydrogen atom. All these light and heavy particles are known as 'fundamental particles' and are considered as the subdivision and basic units of atoms and all matter. At present about 200 fundamental particles have been discovered and are considered as the initial form of matter.

Together all the fundamental particles and associated radiation constituted the expanding fireball of the expanding universe. But the radiation was dominant because small numbers of fundamental particles were produced at that time. The period is called the 'radiation era'. The radiation is carried by massless particles called 'photons'. The expanding fireball contained high-energy photons. Photons are quanta of radiation. A quantum is a naturally fixed quantity (quantum physics). Photons possess the dual properties of wave (radiation) and matter. With the emergence of fundamental particles, the stage is set for the creation of real matter (atom and molecule).

So much happened within a small period of time (a few hundred seconds) after the Big Bang. In any explosion or chemical reaction, the initial period is associated with maximum material conversion and enthalpy changes, so a similar situation occurred with the Big Bang explosion of dense mass as well. The initial Big Bang explosion activities were slowed down with decreasing temperature and expanding universe with increasing time. With diminishing temperatures, the radiation converted to fundamental particles. The concentration and emergence of fundamental particles started to build up at the cost of diminishing radiation, the dominance of the radiation era beginning to end with the passage of time.

1.4   FOUR NATURAL BASIC FORCES AND UNIVERSE

All matter is composed of atoms and subatomic fundamental particles protons, neutrons, electrons, etc. The atom consists of an inner nucleus surrounded by the orbiting electrons. The nucleus of an atom contains protons, neutrons and some other particles. The protons, neutrons and electrons, etc., were thought to be fundamental particles (units) of atoms but are now considered as divisible by modern physicists.

The fundamental particles have sub-structures, sub-units and sub-particles. When the fundamental particles collide with each other and with radiation at a very high speed, subdivisions of fundamental particles take place. The collision of nucleus particles (protons, neutrons) results in the generation of smaller sub-particles known as quarks. In the same collision with electrons, leptons are produced. Quarks are sub-particles of protons and leptons are sub-particles of electrons. The sub-particles quarks and leptons behave like wave having energy and force. The energy/force is expressed in 'electron volts', which is defined as the amount of energy that an electron gains under an electric field of 1 volt. An electromagnetic field can provide thousands of millions of electron volts of energy to particles. The photons are the ultimate particles of electromagnetic radiation. The electromagnetic field is carried by these waves like photon particles. The force, energy and particles are interrelated and indistinguishable under certain conditions. Modern physics considered that the quarks and leptons are the building blocks of fundamental particles. The quarks and leptons need some sort of force to bind themselves and produce fundamental particles leading to the formation of atoms and matter. The force-carrying particles are grouped according to the strength of the force and the particles with which they interact. There are four such basic forces associated with the fundamental interaction of matter and the evolution of the universe. The four basic natural forces along with the unification of all the forces are described below.

1.4.1   Gravitational Force between Particles with Appreciable Mass

Gravitational force is a universal force. Every kind of matter produces gravitational force according to its mass and prevailing energy. Gravity is the weakest of all the four natural basic forces; it is difficult to detect but on the universal scale it is the most dominant force. Gravity force is an attractive force between two bodies and acts over large distances. The weak gravitational force between two massive bodies can produce significant pull, for example between the sun and earth. Gravitational force helps the earth to circulate in an orbit around the sun. Gravity causes the aggregation of small particles (matter) into a large compact mass. Gravitational force held the small particles to form the galaxy. Stars are formed by the aggregation of small particles under gravitational force. The force of gravity between two bodies is supposed to be carried by sub-particles known as 'gravitons', a neutral particle with virtually negligible mass. The gravitational force between two bodies is the exchange of graviton sub-particles.

1.4.2   Electromagnetic Forces between Particles Having Electric/Magnetic Charge

The electromagnetic force is much stronger than the gravitational force and acts between two oppositely charged bodies, for example between negatively charged electrons and positively charged protons or quarks. The force between two oppositely charged bodies is attractive, whereas that between two same-charged bodies is repulsive. The electromagnetic force is significant on the smaller scale of the matter that is particles, atoms and molecules, but on a larger scale there is little net resultant electromagnetic force. Large bodies such as the sun and earth contain an equal number of opposite charges, so they cancel each other out and there is no net electromagnetic force. The electromagnetic forces between the negatively charged electrons and positively charged protons in the nucleus of the atom help the electron to orbit the nucleus. The electromagnetic force is the exchange of massless sub-particles called 'photons' between two oppositely charged masses. The electromagnetic force binds fundamental oppositely charged particles to form atoms and between atoms to form molecules and compounds.

1.4.3 STRONG NUCLEAR FORCES BETWEEN QUARKS

Strong nuclear force is exhibited in the nucleus of the atom. The force binds the same-charged quarks together as protons neutrons, mesons, etc., in the nucleus of the atom in spite of the repulsive force between the same-charged particles. Strong nuclear force is carried by a massless sub-particle called a 'gluon' through interaction with quarks.

1.4.4 WEAK NUCLEAR FORCES BETWEEN QUARKS AND LEPTONS

Weak force acts only between quarks and leptons. The force is responsible for radioactivity and nuclear reactions of fission and fusion. Fission and fusion are the main reactions for the formation, sustainment and destruction of stars in the galaxy. Weak nuclear force is carried by sub-particles known as 'bosons'.

1.4.5 UNIFICATION OF THE FOUR FORCES

All the above described four basic forces of nature are used by the physicist to explain the creation and evolution of the particles, atoms, molecules, matter and the universe as a whole. Scientists are now developing 'unified theory' that states that all the four forces originated from a single unified source. Unified theory is the combination of the 'theory of relativity' and 'quantum physics'. Quantum physics deals with the forces operating at the sub-atomic and atomic levels.

1.5 MATTER ERA, EVOLUTION OF INORGANIC

Matter is a substance that has some weight and occupies a space. Inorganic matter is lifeless non-biogenic matter. Sub-fundamental (quark, lepton, etc.) and fundamental particles (proton, electron, etc.) collided and interacted among themselves in the presence of radiation (photons) in the expanding universe, resulting in the emergence of atoms. For example, the collision of a proton with a neutron resulted in the formation of a nucleus of a 'hydrogen atom' which was surrounded by an orbiting electron. Another example is the formation of the helium atom. Helium was formed by the integration of two protons and two neutrons, forming a nucleus orbited by two electrons. Helium is equivalent to the fusion of two hydrogen atoms, a nuclear reaction.

The formation of atoms happened at the cost of fundamental particles and radiations. With time, the universe expanded and the temperature dropped. The integration of fundamental particles became possible during their collision and collusion. The process multiplied to generate enormous quantities of hydrogen atoms. At this time the universe consisted of radiation, hydrogen and helium atoms. The emergence and increasing number of hydrogen and helium atoms correspond to the reduction of radiation. About 1 billion years after the Big Bang, matter in atomic form was dominant in the universe over radiation.

An atom is the smallest unit of matter. Atoms cannot survive independently. They are converted to a stable neutral element or molecule by the sharing of electrons either with the same kind of atom or a different atom. Elements and molecules are the basic building blocks of all matter. Only lighter elements or molecules were formed in the universe at the initial stage. Heavier elements were generated in the core of stars, where more severe conditions of temperature and pressure exist, conducive for nuclear reactions. A nuclear reaction is accompanied by the fusion or fission of atoms and the formation of other lower and higher atomic weight atoms along with the release of enormous amounts of radiation.

The evolution of matter leads to the development of the universe. The universe is a vast territory with an infinite border, and consists of billions of galaxies and billions of stars and galactic (related to galaxies) and celestial (visible in the sky) bodies. The formation of galaxies, stars and solar systems is a continuation of matter evolution.

1.5.1 FORMATION OF GALAXIES

A galaxy is a huge collection of stars containing lighter and heavy elements, molecules, atoms, ions, gases and particles. A countless number (billions) of galaxies were formed in the universe. It is believed that about 140 billion such galaxies exist in the universe. All galaxies vary greatly in shape and size.

Galaxies are formed as a result of gravitational attraction among different types of matter (atoms, elements, molecules, ions and radiation). A galaxy body is not fully compacted, rather it is diffuse and of immense width and length of several million light years. A galaxy exists as a separate unit because of inter-gravitational forces among constituent stars. The space between different galaxies is filled with highly scattered and distant molecules of lighter gases. The Milky Way is the galaxy in which our solar system is located. It includes millions of clusters of stars and nebulae. The diameter of the Milky Way is about one hundred thousand light years.

1.5.2 FORMATION OF BLACK/DARK HOLES

In addition to stars and nebulae materials consisting of particles, atoms, elements, molecules and ions, the galaxy is inhibited with other bodies known as 'dark holes' and 'radiation dust'. Dark holes are not holes, but super dense, optically opaque, solid residue of a dead (coalesced) star. It is super dense condensed matter packed in a small volume. The black hole is a remnant of a star. A black hole is highly packed and condensed mass, which is ten times bigger than the sun.

A black hole is created after a tremendous explosion (supernova) of a dying star and subsequent coalescing and condensation processes. The dark holes possess immense gravitational pull and attract every object passing nearby to them, even radiation and light. Ordinarily dark holes are not recognizable. The black holes are identified by the behaviors of neighboring galaxy bodies that experience great gravitational attraction. Earlier it was believed that nothing is radiated from the black hole. The attracted matter builds up on the surface of the black hole and heats up to radiate X-rays into space.

1.5.3 Formation of Radiation Dust

Radiation dust is originated from many sources existing in space. One of the main causes of radiation dust is the emission of dying stars. Emission contains fundamental particles, atoms, molecules, ions, gases, light elements and electromagnetic radiation.

1.5.4 Formation of Star Systems

The temperature of the universe further reduced and the density increased. The conditions paved the way for the formation of stars within the galaxy. Billions of star systems exist in a galaxy. Stars are extremely large and include revolving cosmic bodies, planets, meteorites, comets and satellites. The positions of stars are relatively fixed within a galaxy. A star is formed due to the gravitational contraction of the prevailing atoms, elements, ions and molecules. The stars produce large amounts of light and radiation and new elements through nuclear fusion and fission reactions.

The sun is the star which is closest to our earth. A star that glows with beautiful colors is known as a 'nebula'. The glow is due to radiating smaller particles (atoms, elements, molecules, ions) forming a 'nebula cloud'. Nebulae cloud and radiation dust may be used synonymously. A specific group of stars form a 'constellation'; huge numbers of constellations exist in the sky.

1.5.5 Star Cycle Evolution & Extinction

A star is a nuclear reactor. Nuclear reactions among hydrogen, helium and lithium atoms formed heavier elements, and the release (loss) of nuclear radiations. Thus a matter deficient environment (gap) was created inside the star. To balance the loss and fill the gap, the ball contracted. The repetition of nuclear reaction, loss of mass, creation of gap and contraction of the ball formed the stable compact solid mass of the star.

However, if the conditions of nuclear reactions are not stabilized, the process of nuclear reaction, loss of atomic mass as radiation, reducing gap and subsequent contraction of the matter continue. A stage arrives where tremendous pressure and temperature build up. The system cannot compensate the loss of matter and gap. The star explodes with tremendous force. The explosion is known as a 'supernova'. The material of the star disintegrates into the universe from where it came earlier to build the star, now a 'dead star'. The disintegrated element matter initiates once again the formation of a new young star elsewhere. Younger stars are rich in lighter elements since they are dominant in the emitted radiation. The older stars contain a greater number of heavier elements, because lighter elements were converted to heavier elements. Most of the elements were produced in the star by nuclear reaction, and disintegrated into the universe by subsequent explosions.

1.5.6 FORMATION OF THE SUN AND SOLAR SYSTEM

The sun is a huge, glowing star at the center of the solar system. The sun is one star among an enormous number of stars in the Milky Way galaxy. The sun provides light, heat and other forms of energy in the solar system. Without the sun there would be no life on earth.

The solar system is a group of celestial bodies located in a particular galaxy known as the 'Milky Way'. The solar system consists of eight planets, Mercury, Venus, earth, Mars, Jupiter, Saturn, Uranus and Neptune, and their moons, dwarf planets, satellites, asteroids, meteoroids and some other smaller celestial bodies. Each planet of the solar system has similarities and variations. Mercury, Venus, earth and Mars are lighter planets. The other four planets, Jupiter, Saturn, Uranus and Neptune are bigger and heavier. Mercury, the innermost and first after the sun, possesses density similar to earth but has no atmosphere. Next is Venus, earth's inner neighbor, and its atmosphere mostly contains thick carbon dioxide and nitrogen gases. The next outer earth neighbor is Mars. Its atmosphere is covered with cloud and dust with some water molecules.

The formation of the solar system occurred about 5 billion years ago (bya). The size and weight of the solar system are minor compared to a galaxy, where billions of such star systems exist. In comparison to the size of the universe, the solar system is very small, almost a point. The volume and weight of the sun are 1.30 million times and 0.33 million times greater than the volume and weight of earth, respectively. One can imagine, what the status (size and weight) of earth would be compared to the galaxy or universe. It is close to or tends to zero ($\rightarrow 0$).

1.6 FORMATION OF ELEMENTS, NUCLEAR REACTION

A chemical element is a substance that cannot be broken into a smaller chemical substance, except in a nuclear reaction. The element is the simplest form of matter; the elements are produced by bonding among atoms of the same types. About 115 elements exist in the universe. Eighty-three elements occur naturally, and the rest are produced by artificial nuclear reaction. A nuclear reaction differs from a chemical reaction. A chemical reaction is the sharing of outer orbital electrons of the reactant atoms. The nucleus of the atom remains intact in a chemical reaction. Nuclear reactions are the fusion (combination) or fission (disintegration) of atomic nuclei. The change in nucleus composition (neutron and proton) generates new atoms with the release of large amounts of radiation. The fission reaction produces lower atomic mass atoms and fusion generates heavier atomic mass atoms. All nuclear reactions are exothermic, whereas chemical reactions are exothermic or endothermic.

All atomic nuclei contain protons (p) and neutrons (n). According to classical physics, protons (positive charge) would repel each other and could not form a nucleus. Neutrons are neutral. Nuclear physics is governed by the principles of quantum mechanics.

An example of the atomic symbol used in nuclear reactions is given for an atom M, as follows:

$$^A_Z M$$

Where

A = Atomic weight/mass (proton + neutron)

Z = atomic number (number of orbital electrons or number of protons in nucleus)

Any chemical reaction, including a nuclear reaction, depends on the number of factors prevailing in the environment. All elements are produced in the universe through nuclear reactions. In the first few minutes after the Big Bang, during the radiation era, two light, stable gas elements, hydrogen and helium, were produced by the integration of fundamental particles. Hydrogen (^1_1H) was produced by the combination of one proton and one election. A helium atom consists of two protons, two neutrons and two electrons (^4_2H). The majority of elements were produced during the 'star cycle' in a galaxy. From the initial stage of star formation till the final stage of 'supernova' explosion, is a long duration of time, billions of years. In the duration, different conditions of temperature, density and radiation occur, so that several different kinds of elements are produced. The diffuse mass of radiations, atoms and ions in the universe progressively turned into a compact, dense, hotter spherical mass, creating ideal conditions for nuclear reactions to take place. Hydrogen and helium atoms were the first to be created; therefore, these are to be first consumed by hydrogen and helium nuclear reactions. Two main nuclear reactions, nuclear fission and nuclear fusion, are responsible for the generation of the elements.

1.6.1 Nuclear Fission

The breaking of a nucleus (atom) into a smaller nucleus (atom) is a 'nuclear fission reaction'. Atoms with an atomic number greater than 83 are unstable and are isotopes. An unstable nucleus can be broken into a smaller nucleus and smaller atoms by bombardment with neutrons or other particles. Besides neutrons, α-particles (^4_2He) and protons (p)are also used for the bombardment of a fissionable nucleus. For example, the nucleus of a heavy uranium atom, $^{235}_{92}U$, when bombarded with high-energy neutrons, undergoes many fission reactions. One of the fission reactions is as follows:

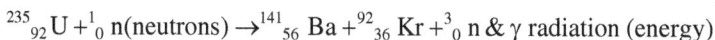

$$^{235}_{92}U + ^1_0 n(\text{neutrons}) \rightarrow ^{141}_{56}Ba + ^{92}_{36}Kr + ^3_0 n \text{ \& } \gamma \text{ radiation (energy)}$$

The reaction generates two new elements having lower mass numbers than the original uranium, along with a lot of kinetic energy and radiation. The reaction also generates free neutrons that trigger further nuclear chain reactions. The fission of uranium-235 generates about 81×10^9 kJ/kg (35×10^9 btu/lb) of energy in a nuclear reaction and the coal on combustion releases about 32×10^3 kJ/kg (14×10^3 btu/lb) of energy. The heating value of uranium fission is 2.5 million times more than the heating value of coal.

1.6.2 NUCLEAR FUSION

Nuclear fusion is the reverse of nuclear fission and leads to the formation of heavy atomic nuclei. Here, two lighter atomic nuclei combine together to form bigger atomic nuclei and atomic particles and radiation along with the release of vast amounts of energy. The mass number of reactant nuclei is greater than that of the product nuclei. The difference in the mass weight between the reactants and product nuclei is released as enormous energy. In nuclear fusion, very large amounts of energy are released compared to fission reactions. Nuclear fusion is also called a 'thermo nuclear reaction'.

Nuclear fusion reaction is the principal source of energy in a star and in the sun. The energy is released when the fusion of hydrogen nuclei occurs, leading to the formation of helium nuclei. From the surface to the core of the sun/star, there is great variation of temperature and density. In the varying conditions of temperature, density and pressure, a variety of nuclear reactions occur from the surface to the core of the sun, producing lighter as well as heavy atoms. A few examples are:

- Two hydrogen nuclei fuse together to form deuterium and gamma rays.
  $$^1_1H + ^1_1H = ^2_1H \text{ (deuterium)} + \text{gamma radiation}$$
- One deuterium and one hydrogen nuclei combine to form one tritium nucleus.
  $$^1_1H + ^1_1H = ^3_1H \text{ (tritium)} + \text{gamma ray}$$
- One deuterium and one tritium nucleus join to form a bigger helium atom.
  $$^2_1H + ^3_1H = ^4_2He \text{ (helium)} + ^1n + \text{gamma radiation}$$
- Nuclear fusion of helium (^4_2H) nuclei produces beryllium (^8_4Be), carbon ($^{12}_6C$) and oxygen ($^{16}_8O$) atomic nuclei.
- Nuclear fusion of silicon ($^{28}_{14}Si$) nuclei produces calcium ($^{40}_{20}Ca$) and iron ($^{55}_{26}Fe$) atomic nuclei.

Thus a chain nuclear fusion reaction is set up for producing heavier atoms.

As the mass number of atoms increases, the required nucleus fusion energy also increases. The nucleus of the heavier atoms has very high binding energy. The prevailing conditions in the star do not provide enough activation energy to initiate nuclear reactions of fission or fusion of stable heavier atoms. Therefore, the further creation of atoms heavier than iron is not possible at this state of the star. The severe conditions required for the generation of atoms heavier than iron are met at the time of the supernova explosion of the star. The very severe conditions of temperature and pressure at the time of a supernova explosion facilitate the formation of heavier elements.

Artificial elements are produced by artificial nuclear reactions, for example by smashing atomic nuclei together in a sophisticated machine called an accelerator. Artificial elements are also produced by colliding fast-moving neutrons, with fissionable atomic nuclei such as uranium ($^{238}_{92}U$) in a cyclotron machine. Artificial nuclear reactions have many scientific, industrial and medical applications.

1.7   ELEMENTS DISTRIBUTION IN THE UNIVERSE

Most of the elements are produced in the star as a result of nuclear reactions. During the supernova explosion of the star, the elements spread in space and become seed elements for the creation of other cosmic bodies. All the members of the universe, galaxy, star, solar system, including earth, were created from the same raw material generated and transported from the stars and galaxies. They have a common origin and are expected to show common elemental composition along with variation from one cosmic body to another.

The study of variation in elemental composition among, earth, solar and cosmic bodies may be utilized to show the pathway of the processes during earth formation.

Samples for determining the elemental compositions are obtained from the following sources.

- Radiation coming to earth from heavenly (universe) bodies including the sun
- Sample of meteorites from space striking the earth
- Space research program

Radiation and collected samples are subjected to spectroscopic analyses. A number of spectroscopic techniques are available. The universe is a dynamic system. Therefore variation in number is expected from location to location and from time to time. The average relative data could be used for practical purposes. Element distribution in the universe and solar system is as follows:

- There is greater abundance of lighter elements (H, He) in the universe and solar system. These two elements outweigh other elements. Most of the hydrogen and helium lighter atoms are produced and concentrated at the surface of the star/sun.
- Li, Be, B and Ne are present in appreciable quantity in the universe but their traces are found in the solar system. C, O, N and Si elements are present in appreciable quantity in the universe and solar system.
- As the atomic number of the elements increase, their concentration decrease in the universe and solar system. Heavier elements are present in trace quantity. Heavier atoms are produced in the core of the star and during a supernova explosion.

The universe as a whole is dominated by hydrogen and helium, and this is the main difference between the universe, sun and earth elemental contents. There is a clear distinction in the distribution of elements in the universe, galactic bodies, sun and earth.

The sun radiation contains about 85% hydrogen and helium, produced by nuclear reaction at the surface of the sun. The remaining 15% radiation originates from the nuclear reaction inside the sun, producing various elements. Radiations from the solar system are electromagnetic waves that also contain energy particles like neutrons, protons, ions and dust particles, released by the sun. The earth receives radiation from the sun. The radiation provides energy necessary for sustaining life on earth. The earth's atmosphere and magnetic field shield the earth from harmful radiation.

Much smaller quantities of the two lighter stable gases, helium and hydrogen, are found in the earth compared to the sun and universe. The lighter gases may have escaped from the gravitational force of the earth. Or the earth was created from cloud dust particles, which did not have sufficient gases. Or these gases have been converted to more stable substances through chemical transformation. Not only are the light gases less abundant in earth, but also heavy stable gases such as krypton and xenon are absent in earth. The absence of these gases from earth suggests that all the gases escaped, overcoming the gravitational pull of the earth. The heavier gases of ammonia and carbon dioxide are in low concentrations in the universe and earth. Iron, silicon, magnesium and oxygen are the major elements found on earth (92%). The abundant metallic elements in the earth (core), along with the earth's high density, suggest that the earth was formed by the aggregation of dust particles, containing metallic elements, probably originating from a cosmic dust cloud and supernova explosion. In the initial stage of the earth (4.0 bya), it was very hot and the earth's matter was in a homogeneous state. The molten state of earth facilitated the matter to separate according to its gravity differential. The heaviest matter settled down in the core followed by less heavy matter in the mantle and the lighter in the crust. The earth turned from homogeneous to a heterogeneous layered structure. The gas diffused from the crust and began to form the atmosphere.

1.8   ORIGIN OF THE EARTH

The earth is a part of the solar system. The earth was created about 4.6 billion years ago. The earth is believed to have been formed from the same cosmic material that formed the stars and solar system. Probably during a supernova explosion, the emitted particles from the dying star constituted radiation dust or cosmic dust or nebula cloud. The nebula cloud contained radiation photons, gaseous, light elements, ions, molecules and dust incandescent materials that were forcefully ejected from a supernova star and spread out in the universe with high temperature and velocity. Nebula cloud turned into solar nebula under the influence of gravitational and centrifugal forces. How did the constituent materials of solar nebula aggregate to form solid earth? Possibly by one of the following methods:

- Direct condensation and aggregation of solar nebula incandescent materials forming earth
- Separation from solar nebula

With time and a drop of temperature the incandescent gases condensed to solid mass. Similarly, with a drop of temperature and gravitational attraction, the cosmic dust particles aggregated to form the solid mass of the earth. Solar nebula swirled around with great speed; some of its mass concentrated at the center and conglomerated to form the sun. The outer part of the swirling sun nebula formed smaller high-density bodies. The center part was the sun and the outer small bodies were 'seed planets'. The seed planets were orbiting the centered sun, and as the speed and temperature dropped more particles were attached and adsorbed at the surface of the seed planets that grew in proportion. A number of these seed planets grew to different sizes. The planets with thick and concentrated mass exhibited more gravitational force than the smaller ones. They attracted more mass from the cosmos and became bigger than the others. Among the bigger ones was 'earth'. The moon is the earth's satellite, is a dark colored rock, and its gravitational pull is one-sixth of the earth's.

1.9   EARTH STRUCTURE

The earth's three zones and their boundary lines are shown in Figure 1.1. Earth as a whole must include its atmosphere, hydrosphere, biosphere and solid shell. The earth's atmosphere, hydrosphere and biosphere will be discussed later (Chapter 8). Restricting the subject to the solid shell, the earth is divided into three zones on the basis of their distinction in geo-physico-chemical nature as follows:

- Crust
- Mantle (lower and upper)
- Core (lower and upper)

The three zones of the earth are separated by two distinct boundary lines. The first boundary line, between the earth crust and earth mantle, is known as the 'Mohorovicic discontinuity'. The second border that separates the earth mantle from the earth's inner core is known as the 'Wiechert Gutenberg discontinuity'. Furthermore, the earth's mantle and core are each subdivided by geological discontinuities known as transition zones into upper/lower mantle and outer/inner core.

The earth is a non-uniform spherical body. The non-uniformity is mostly exhibited at the surface of the earth and occasionally in the internal structure. All the information about the earth's internal structure is gathered from geophysical studies. The earth's non-uniformity is geological discontinuity. Each of the three zones of the earth is well separated by two distinct boundary lines. Furthermore, the earth's mantle and core are each subdivided by geological discontinuities known as transition zones into the upper/lower mantle and outer/inner core.

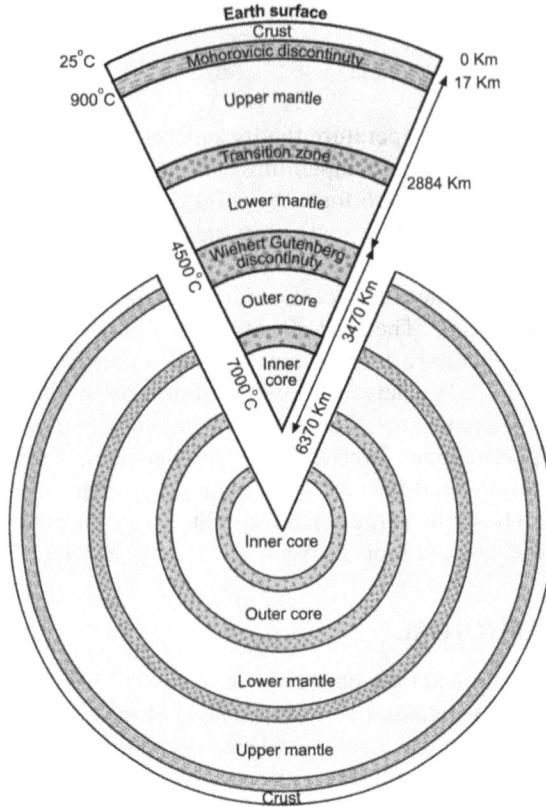

FIGURE 1.1   Structural zones of the earth.

1.9.1 EARTH CRUST

The earth's crust is defined as the distinct outermost solid shell of the earth. The crust is clearly distinguishable from the underlying bordering earth mantle on the basis of geology and chemical constituents. The earth's crust is important to mankind. The crust is directly linked to the atmosphere, hydrosphere and to all living species of the earth. The crust is like wrapping paper on the surface of the earth. It is the outermost zone of the earth's shell, and formed 1000 million years after the formation of the earth (4600 mya). The earth's crust is located above the thermodynamic zone of the earth's upper mantle. The earth's mantle is molten lava in plastic state along with some solids. Because of the mantle's thermodynamic state, the adjacent crust experiences various forms of stress and undergoes strain. That results in the breaking of the crust, into seven to eight different tectonic plates, faults and fissures.

The earth's crust is non-uniform and heterogeneous. The crust surface is markedly divided into two, land/terrestrial crust and underwater ocean bottom crust. The ocean bottom crust is hard and compact with a density of about 3.2 g/cm^3. The land crust is soft with a density of about 2.7 g/cm^3. Land is at a higher elevation so that the natural flow of water to the ocean is maintained. The thickness of land and undersea crusts varies from area to area. At the sea bed it is 10–12 km thick, and the land crust is 37 km deep. Below big mountains the crust is as deep as 60 km. On average the earth's crust is 17 km thick, which is 0.27% of the radius of the earth. The crust weight (0.42%), volume (0.74%) and thickness are minor compared to the mantle and core. The crust spreads from the surface of the earth to the Mohorovicic discontinuity.

1.9.1.1   Geology of the Crust

The earth's crust was formed by the upward flow of shallower molten lava material from the upper mantle zone of the earth. The upward flow of the lava is affected by volcano and tectonic processes. Subsequently, at the lower temperature of the crust, the lava undergoes solidification and is exposed to sedimentation and weathering at the surface conditions. At the sea bed the intruded mantle matter mostly remains unaltered. Another factor contributing to the formation of the earth's crust is that the less dense or lighter and incompatible solid matter in the molten lava separated from the liquid and due to buoyancy floats on the top of the molten lava. The solid floating matter passed through the void spaces, pores and cracks and faults to appear at the crust for solidification. The minerals in the mantle as well as in the crust may be of the same origin and chemically similar, but both types of mineral differ in the formed crust rock and in the mantle. Both types of minerals differ in mineral geology, seismic traveling velocity and crystal structural forms.

1.9.1.2   Chemical Constituents of the Crust

Most of the elements do not exist in a pure state. The majority of them are found as oxides. Oxygen and silicon are the major elements, constituting 74% of the crust. Other significant elements are magnesium, iron, calcium, aluminum, sodium and potassium, forming 23% of the crust. As mentioned above, no free elements exist; they are always in chemical bonding with other elements resulting in molecule and compound formation. The elements have greater affinity for oxygen and quickly form metallic as well as non-metallic oxides.

Oxides of silicon, aluminum and calcium are dominant (82%) in the crust. The next largest group (14.5%) of oxides is of iron, magnesium and sodium. The rest of the oxides form about 3.5% of the crust.

1.9.1.3   Crust Thickness and Weight Distribution/Iso-Static Balance

The earth's crust is formed by lighter minerals (density 2.7 g/cm^3) and floats in resting position on the inner hot magma, consisting of heavy-weight minerals of iron and nickel (density 3.1 g/cm^3). The thickness and weight of the earth's crust are not uniform and vary with location. The weight of the crust depends on its thickness at a particular location. Variation of the thickness and weight distribution of the earth's crust plays an important role in keeping the crust in position and in stable condition. The variation in thickness of the crust provides an 'iso-static (equal pressure) balancing line' between the crust and mantle that prevents the merger of the crust with the hot molten magma below.

Heavy-weight solid mountains do not fall down into the hot molten magma below due to the presence of thicker crust foundation below them. The thick crust acts as a foundation and keeps the mountain in a static position at its location. On the other hand, depressions and basins at the surface crust have a thin crustal layer below them. Smaller weight masses do not need a thick foundation.

1.9.2   EARTH MANTLE

The earth's mantle represents the bulk of the earth. It is 83% of the earth's volume and 68% of the earth's weight. The mantle is the layer between the crust and outer core. It envelops the earth's hot core in the same way the crust covers the mantle. The subdivision of the mantle is as follows:

- Upper mantle
- Transition zone
- Lower mantle

The earth's mantle starts from the lower end of the Mohorovicic discontinuity to the Wiechert Gutenberg discontinuity with a total thickness of 2884 km. The lower and upper mantles are separated by a transition zone.

1.9.2.1   Geology of the Mantle

The uppermost mantle and overlying crust are solid, rigid and rocky layers. Together they are known as the 'lithosphere'. The thickness of the heterogeneous lithosphere is about 200 km. The upper mantle just below the lithosphere forms rigid to plastic material.

The transition zone between the upper and lower mantles contains complex materials. The lower mantle is a homogenous molten zone with higher temperatures. The mantle consists of different layers of materials of different melting points at the existing pressure/temperature. The upper mantle mostly contains three types of silicate minerals, namely garnet, pyroxene and olivine. In the transition zone, all three silicate minerals undergo a phase transformation, forming different minerals and solid solutions from the same types of element molecules. In the lower mantle zone, due to high pressure and temperature, silicate mineral are split into simpler oxides of Si, Mg and Fe elements.

1.9.2.2   Chemical Constituents of Mantle

Chemical constituents of the mantle along with the crust are expressed in terms of oxides. The oxides of silicon, magnesium and iron are the major compounds (92%). Next are calcium, sodium and aluminum (6%). Other elements found in the mantle are less than 1.0% each.

1.9.3   EARTH CORE

The earth's core is the most inner part of the earth. It is divided into two zones. The outer zone extends inside from 2900 km to about 5100 km deeper, the inner zone from 5100 km to the center of the earth's shell at 6371 km. The total thickness of the core is 3470 km, which is 54% of the total radius of the earth (6371 km). The Wiechert Gutenberg discontinuity at the core–mantle boundary separates the two. The boundary/discontinuity spreads to a thickness of about 600 km. The core is 16% of the earth's volume and 31% of the earth's weight. The inner core mostly contains iron nickel alloy in solid state. Probably some iron oxide, sulfur and silicon are also present. The temperature of the inner core is about 7000°C and the pressure is about 3.3 million atmospheres. Iron is solid because of this high pressure, even though the temperature is higher than the melting point of iron. The inner core is believed to have been in solid state because of slow precipitation, solidification and settling of molten Fe–Ni alloy from the outer core. The outer core predominantly contains iron and nickel in molten states, above the solid inner core. The temperature of the outer core ranges from 5730 to 4000°C (average temperature 4500°C). The iron-nickel alloy exists in liquid state, because the pressure is not high enough to increase the melting point. The molten state of iron, nickel and their motion are believed to be the cause of the earth's magnetic field. Heat is transferred from the inner core to outer core and from the outer core to mantle and to the crust.

1.10   EARTH'S PHYSICAL STATE

The earth is the third planet from the sun, in between Venus and Mars.

1.10.1   DENSITY DISTRIBUTION

Earth is a heterogeneous solid, except the inner zone which is homogenous and in a molten (liquid) state. The variation of density is expected. The dependence of density on temperature, pressure and chemical structural forms is well known, especially in severe earth conditions. The average earth density is 5.5 g/cm³, whereas the crust, mantle and core zones have densities of 2.7, 4.5 and 10.7 g/cm³ respectively. It is interesting to see that the materials of both the inner and outer core are the same (Fe–Ni alloy). But in the outer core, molecules are in liquid form whereas in the inner core they are in solid state with much higher density than the outer core in liquid form.

The inner core is at a much higher temperature than the outer core, but the extremely high pressure keeps the molten materials in a solid state in the inner core.

1.10.2   TEMPERATURE VARIATION

Temperature and pressure both modify the minerals' crystal structural forms. The earth's mineral materials exist in solid as well as in liquid forms because of temperature and pressure effects. Temperature and pressure increase with depth but not uniformly. The temperature gradient in the crust is 10–50°C/km. The average temperature gradient of the earth is 30°C/km. In deeper zones the temperature gradient is much higher. The crust is made of heterogeneous solid minerals. The upper mantle is in a mixed solid/plastic state. This is because different layers of minerals have different melting points. The temperature ranges from 300 to 900°C at the boundary between the crust and mantle. The temperature is as high as 4000°C at the border of the lower mantle and outer core. The outer core contains mainly Fe–Ni alloy in a liquid state because of high temperatures. The temperature between the inner and outer core boundary is 5400°C. The inner core of the earth is a solid state of dominant Fe–Ni alloy. Deep inside the inner core the temperature is as high as 7000°C.

1.10.3 Pressure Variation

The pressure increases with depth. Pressure and temperature control the behavior and state of the materials of the earth. In the crust all minerals are in solid state. Mantle materials are in solid/molten mix states. The pressure effect is most pronounced in the core. The material of the outer and inner cores is the same (Fe–Ni alloy). The alloy is in solid state even at a very high temperature of 7000°C, many times above the melting point (1200°C m.p. of the alloy). The inner core is under extremely high pressure of about 333 million kPa (3.3 million atm). The extremely high pressure changes the crystal structure of the liquid minerals to a small compact mass of solid. The high pressure tends to compact the liquid crystal structure to form a high-density solid material.

The solid/liquid phase behavior at low and high temperatures/pressures is illustrated by the following figure:

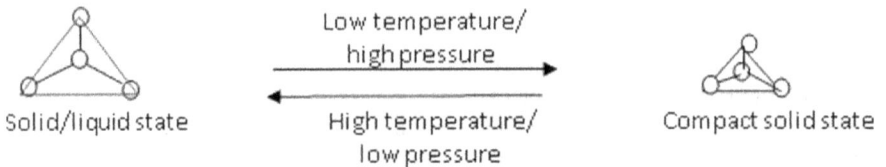

Solid/liquid state   Low temperature/high pressure →   ← High temperature/low pressure   Compact solid state

1.11 GEOLOGICAL TIME SCALE AND EVENTS

The geological time scale is an arrangement of events and strata according to the time they occurred. The time scale gives the history of the earth since its emergence 4.7 billion years ago. Particularly it focuses on the events that have occurred over the last 600 million years. The crust came into being about 3500 million years ago as a result of the cooling and solidification of hot matter. The crust underwent evolutionary stages and attained the present form about 600 million years ago with the formation of sedimentary rock. The geological events (geochronology units) had occurred over different durations of time, not on a uniform regular basis.

It is relevant and continuity demands that a brief account of the geological history of earth since the origin of the earth and its evolutionary processes be discussed, to the present time. Along with the earth's geological history it seems beneficial to include the creation and evolution of the universe during various eons and eras. The geological history of the earth is recorded in two units of time, 'chrono-geological time unit' and 'chrono-stratigraphic time unit'.

The chrono-geological time unit defines the time interval in which a lithological (rock type, origin, color, texture, grain size, mineral nature, composition, constituent distribution and rock structure) event occurred. The geological time unit is an arrangement of events according to the time they occurred. Evidence for a geological event is identified by utilizing different sources and procedures, for example, geological similarity and differentiation between rocks. Old rock layers are found at the lower portion and new layers at the top portion of the rock, according to the law of superimposition. Paleontology is a science of fossils, once-living organisms. Fossils are found in the strata of sedimentary rock. Fossils together with strata give useful evidence of an event, and geological conditions of the time. Each layer and fossil corresponds to a particular geological time. The measurement of time with the help of radiometric methods has greatly improved the accuracy.

The chrono-stratigraphic time unit defines the time interval for the formation of particular rock stratum describing the thickness and size of rock but not the type of rock. The time unit is the record of the formation of particular rock strata in a particular time interval.

Both units clearly define the event and rock formation intervals by using different terms as follows:

Chrono-Geological Time Unit	Chrono-Stratigraphic Time Unit
Eon	Eonothem
Era	Erathem
Period	System
Epoch	Series
Age	Stage
Phase	Zone

Geological time intervals are explained as follows:

- Eons are the largest geological time intervals. The four eons are the Phanerozoic, Proterozoic, Archean and Hadean. Together all four cover the geological history from origin of the earth (4600 mya) to the present time.
- Eras are the divisions of each eon. The Phanerozoic eon comprises the Cenozoic, Mesozoic and Paleozoic eras. The Proterozoic, Archean and Hadean eons all together are called Precambrian historical time.
- Periods are the subdivision of eras. For example, the Mesozoic era consists of the Cretaceous, Jurassic and Triassic periods, and Quaternary and Tertiary are the periods of Cenozoic era.
- Epochs are the further subdivision of periods. The Quaternary period's subdivision is the Holocene and Pleistocene. The periods may also be divided into upper, middle and lower epochs according to the event intervals.
- Ages are specific to each column (epoch) of Table 1.1 and reported as million years ago (mya).

TABLE 1.1
History of Universe and Earth in Geological Time (million years ago, mya)

Eon	Era	Period	Epoch	Age (mya)	Life (organic)	Mineral (inorganic)
Phanerozoic	(Cenozoic (new)	Quaternary (2.5 million years)	Holocene	0.117–present	Rise of modern civilization	Soil peat, clay, kerogen, nitrate. Glaciers ice formed in north and south poles.
			Pleistocene	2.5–0.117	Rise of mammals, birds and plants, *Homo habilis* and *Homo sapiens*. Pleolithic (old stone age) and Neolithic (new stone age). Metal stage	
		Tertiary (63 million years)	Pliocene	5.3–2.5	Rise of primitive mammals, birds and plants. Disappearance of dinosaurs.	Oil, gas, coal, oil shale phosphate, gold, silver.
			Miocene	23.0–5.3		
			Oligocene	33.9–23.0		
			Eocene	37.8–33.9		
			Paleocene	65.5–37.8		
	Mesozoic (middle)	Cretaceous (80 million years)	Upper	99.6–65.5	Dinosaurs, marine reptiles, flowering trees, plants.	Oil, gas, coal, lead zinc tungsten, uranium, radium.
			Lower	145.5–99.6		
		Jurassic (54 million years)	Upper	161.2–145.5	Flying, lizard, primitive birds, palm-like trees.	Fine-grained stone, gypsum, salt, coal, oil, gas. Super continent Pangaea broke apart.
			Middle	175.6–161.2		
			Lower	199.6–175.6		

(Continued)

TABLE 1.1 (CONTINUED)

History of Universe and Earth in Geological Time (million years ago, mya)

Eon	Era	Period	Epoch	Age (mya)	Life (organic)	Mineral (inorganic)
		Triassic (51.0 million years)	Upper	228.7–199.6	Giant dinosaurs, lizards, primitive mammals.	Small deposits of gas, oil and coal.
			Middle	245.9–228.7		
			Lower	251.0–245.9		
	Paleozoic (old)	Permian (48 million years)	Upper	260.0–251.0	Large amphibians, conifer plants, primitive seed plants, insects.	Potash, oil, gas coal, phosphate, gypsum.
			Middle	270.6–260.0		
			Lower	299.0–270.6		
		Carboniferous (61 million years)	Pennsylvanian	318.0–299.0	Mass carbon producer animals and plants, reptiles, sharks, insects, land flora, vascular and	Limestone, oil, gas coal, shale potash, tar sand.
			Mississippian	358.9–318.0		
		Devonian (57 million years)	Upper	385.0–358.9	Fern-like plants, trees, amphibians.	Oil, gas, glass, sand.
			Middle	397.0–385.0		
			Lower	416.2–397.0		
		Silurian (27 million years)	—	443.7–416.2	Air-breathing insects, aquatic animals, rise of fish.	Iron ore.
		Ordovician (45 million years)	—	488.3–443.7	Fish, land plants, corals	Lead, zinc, oil, gas.
		Cambrian (53 million years)	—	541.0–488.3	Algae, seaweed, marine invertebrate, shell insects.	Phosphate, marble, ocean, oxygen.

(Continued)

TABLE 1.1 (CONTINUED)
History of Universe and Earth in Geological Time (million years ago, mya)

Eon	Era	Period	Epoch	Age (mya)	Life (organic)	Mineral (inorganic)
Precambrian time		Proterozoic (eon)	Neoproterozoic Mesoproterozoic Paleoproterozoic	10,000–541.0 16,000–1000.0 2500.0–1600.0	Primitive aquatic plants, invertebrates and insects. Primordial form of life, metazoa, bacteria, algae, protozoa, biological evolution, photosynthesis, cyanobacteria.	Iron, copper nickel cobalt. Chemical evaluation.
		Archean (eon)	Neoarchean Mesoarchean Paleoarchean Eoarchean	2800.0–2500.0 3600.0–2800.0 4000.0–3600.0	Cyanobacteria, sulfate-resistant bacteria, stromatolite.	Inorganic materials, water on earth's surface.
		Hadean (eon)	—	5600.0–4000.0	Origin of earth.	Oldest mineral, 4570 million years old.
Universe evolution		Matter era	—	14,700.0–5600.0	Formation of galaxies, stars and solar system. Formation of atom, elements and molecules.	
		Radiation era and origin of universe	—	Lasted less than a few minutes (14,700 million years ago)	Neutrons – other fundamental particles – photons. Super dense neutrons (10^{79}) sphere exploded with velocity of light at high temperature.	

- The post-Cambrian era (Phanerozoic) corresponds to about 10% of the total history from the origin of the earth to the present time; 90% belong to Precambrian history. The oldest sedimentary rock is found to be 600 million years old. Sedimentary rock strata between 500 and 100 million years old are considered to be promising for oil/gas prospecting.

The history of the universe and earth in geological time units (mya) is shown in Figure 1.2 as well as in Table 1.1. The figure describes the history of the universe and geological time scale from the origin of the earth to date. The table is a further elaboration of the geological time scales. The longest geological time division is eon. The eon is further subdivided into era, period, epoch and age. The last two columns of the table describe the major event, corresponding to the time. Additional information on each eon, era, period and epoch along with the events is given as follows.

1.11.1 Phanerozoic Eon

The Phanerozoic eon consists of 3 eras (Cenozoic, Mesozoic and Paleozoic) and 11 periods. The periods are as follows.

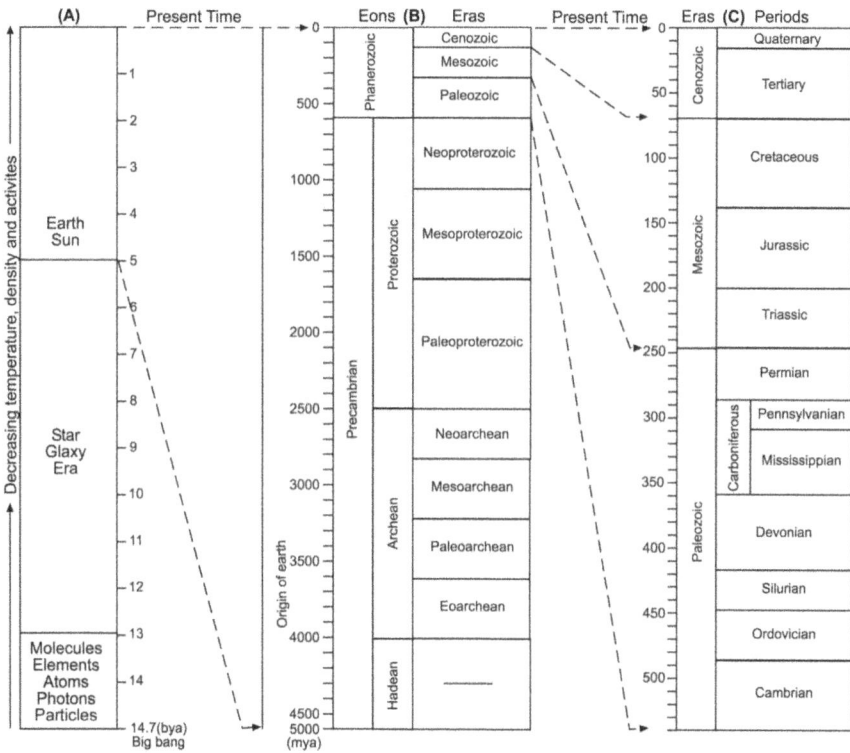

FIGURE 1.2 Universe and geological history. A = from origin of universe to present time. B = from origin of earth to present day. C = from Cambrian period to present time.

1.11.1.1 Quaternary Period

The Quaternary period is the fourth-order most recent period of the geological time. The period lasted from after the Tertiary (2.5 million years mya) to the present time. The periods are divided into two epochs, the Pleistocene (2.5–0.117 mya) and Holocene (0.117 mya to the present time). The earth's temperature had sufficiently cooled down. Glaciers and ice sheets were formed in north and south poles. The Holocene epoch is the beginning of modern civilization.

1.11.1.2 Tertiary Period

The Tertiary is a period of the Cenozoic era. It spanned from 65.5 to 2.5 million years ago. It consists of five epochs, namely the Paleocene (65.5–375.8 mya), Eocene (37.8–33.9 mya), Oligocene (39.9–23.0 mya), Miocene (23.0–5.3 mya) and Pliocene (5.3–2.5 mya). The period is marked by mammals' dominance on land and in the oceans. Mammals and birds evolved to different sizes and shapes. Reptiles were present but dinosaurs were absent.

1.11.1.3 Cretaceous Period

Cretaceous was the longest spanned period of 80 million years, from 145.5 to 65.5 mya, among the three periods of the Mesozoic era. It is divided into upper (99.6–65.5 mya) and lower (145.5–99.5 mya) epochs. Cretaceous refers to rock containing a major portion of limestone, along with sandstone and shale clay. In the period, the earth's surface, land and oceans were quite different than they are today. The sea level was high, and most of the land was submerged in shallow water. Flowering plants and pollinating insects evolved during the period. Flying creatures and dinosaurs appeared, but by the end of the period most of the big vertebrates and invertebrates vanished, due to climatic and biological factors.

1.11.1.4 Jurassic Period

The Jurassic period is the middle of the Mesozoic era, spanning from 199.5 to 145.5 mya. The period is divided into upper (161.2–145.5 mya), middle (175.6–161.2 mya) and lower (199.5–175.6 mya) epochs. The term Jurassic was derived from a sequence of rock in the Jura Mountains. The period was marked by booming forests, but plants were different than today and inhibited by big and small animals. Some call it the 'dinosaur's period'. Fish were abundant in oceans. Flying vertebrates and saury, long slender marine fishes multiplied manyfold. The super continent, Pangaea, broke apart in this period about 175.0 mya.

1.11.1.5 Triassic Period

The Triassic period of the Mesozoic era occurred between 251.0 and 199.5 mya. The period is divided into three, the upper (228.7–199.5 mya), middle (245.9–228.7 mya) and lower (251.0–245.9 mya) epochs. The term was first used to define a sequence of rock with marked threefold division without significant fossil content Land plants began to grow for the first time in this period. Alterations in climatic conditions and deviation in ocean currents took place. The period saw the rehabilitation of the surviving creatures from the mass extinction of life during the Permian period. New groups of animals and plants were born.

1.11.1.6   Permian Period

The Permian period is known for the mass extinction of life during the period. It is the early period of the Paleozoic era. It occurred from 299.0 to 251.0 mya and is divided into upper (260.0–251.0 mya), middle (270.5–260.0 mya) and lower (299.0–270.5 mya) epochs. A super continent named 'Pangaea' was formed. Much of its land was in the southern hemisphere and surrounded by ocean. The climate was extreme in the continent, from an ice-cold region to areas of intense heat. Rain fluctuated between arid and wet conditions. The once lush swamp forest of the Carboniferous period disappeared, replaced by emerging drought-resistant plants, conifers and seed ferns. Reptiles prevailed over amphibians. The sea was inhabited by a massive number of bony fish with thick scales and fan-shaped fins. The period marked the end of previous prehistoric animals. The period ended with massive loss of life. The period wiped out 70% of the land animals and 90% of sea life. A combination of extreme hot and cold climate, dust from volcano eruption shielding the sun, the release of underground methane by earthquake and massive material falling on earth caused climatic catastrophe for species. The next period saw the filling of the void by dinosaurs.

1.11.1.7   Carboniferous Period

Carboniferous refers to the prehistoric plants, converted to underground coal and found in various parts of the world. It is called the 'Coal Period'. The period was divided into two parts, namely the Pennsylvanian period from 318.0 to 299.0 mya and the Mississippian period from 358.2 to 318.0 mya. Massive forestation consumed carbon dioxide from the atmosphere and released excess oxygen; the atmospheric oxygen was increased to about 35%. Today the atmosphere contains 21% oxygen. Deadly poisonous insects grew in size and numbers. Similarly amphibians grew to different sizes and types.

1.11.1.8   Devonian Period

The name is derived from a stratum in a village of England, where the period was identified. The period lasted from 416.2 to 358.9 mya and is divided into upper (385.0–358.9 mya), middle (397.0–385.0 mya) and lower (416.2–397.0 mya) epochs. At that time the earth had three continents. America and Europe were one mass of land. South America, Africa, South Asia and Australia were together in the south with Antarctica. Similarly there was a continent was in northern Siberia.

The initial period had only small plants; the later period saw the development of bigger trees and forests. Shelled insects and fishes appeared in large numbers. Land life had some divergence.

1.11.1.9   Silurian Period

The Silurian period witnessed remarkable change in life and climate. The period spanned from 443.7 to 416.2 mya. Life started on land, with the appearance of insects and bushes. At the transition period between the Silurian and early Ordovician, mass extinction of marine species took place. The climate was warm but stable compared to later periods.

1.11.1.10   Ordovician Period

The Ordovician period is the second period of the Paleozoic era; it extended from 488.3 to 443.7 mya. Many new types of marine vertebrates evolved. Land was barren, but occasionally beginning to grow some plants. Land was available only in the south pole. The north was flooded with water.

1.11.1.11   Cambrian Period

The Cambrian is the late period of the Paleozoic era. It extended from 541.0 to 588.3 mya. The period witnessed massive but diversified emergence and evaluation of the water cycle, climate and life. Abundant oxygen generated by the photosynthesis process provided enough food material for both kinds of life, plant and animals. The living organisms diversified from primitive vegetation to modern animal phylum (a distinct line between plant and animal). All the life was concentrated in water. Land was devoid of life. The climate was warmer, and polar ice was rare or absent.

1.11.2   PRECAMBRIAN TIME

The Precambrian period encompasses the history from the origin of the earth, 5600 ma, to the beginning of the Cambrian period, 541 mya, before the Phanerozoic eon. The Precambrian time of earth history is spread over a long length of about 5000 million years. It covers about 90.0% of the history of the earth in terms of time. The period is so long that it is further divided into three eons.

1.11.2.1   Proterozoic Eon

Proterozoic time is from 2500.0 mya to the beginning of the Cambrian period, 541.0 mya. The eon is marked by the following characteristic events:

- Living organisms were small, simple and unicellular.
- Before the eon, the atmosphere was reducing containing hydrogen devoid of oxygen molecules. Only trace amounts of oxygen atoms were present. The atmosphere was non-reactive and stable.
- About 2400 mya mass oxygen was produced from the photosynthesis process by cyanobacteria. By 2000 mya oxygen reached a level of 1.0% in the atmosphere.
- By 800 mya, the oxygen content of the atmosphere was 21%. Ozone (O_3) molecules developed to shield the earth from harmful radiation from the sun.
- The first multi-cell organism was formed about 600 mya.
- With the availability of sufficient oxygen, chemical conversion became easy. First iron and similar metals oxidized to their stable oxide minerals.

After the consumption of oxygen by oxidizing material, the surplus oxygen produced ozone.

* The Proterozoic eon was the time when the modern tectonic plates developed in the earth.

1.11.2.2 Archean Eon

The Archean eon lasted from 2500 to 4000 mya. The main features of the eon are:

* The molten materials of the earth's crust solidified to the present level of solid rock.
* The earth was sufficiently cooled down to the level where condensation of water vapors from the atmosphere became possible. Oceans were formed.
* The earliest living organism was formed about 3400 mya, when the atmosphere was filled with hydrogen and helium gases and the oxygen was absent. The atmosphere was reducing, and the first living cell (micro bacteria), appeared 3400 mya.

1.11.2.3 Hadean Eon

The Hadean eon was from the beginning of the earth, 5600 mya, to 4000 mya. The earth totally came into being. It was hot, with massive radiation bombardment from the universe and sun.

The pre-Hadean period before 5600 mya saw the evolution of radiation, particles, matter and the universe as a whole after the Big Bang explosion 14,700 mya.

1.12 CONCLUSION

The universe was created about 14.7 billion years ago, as a result of the Big Bang explosion of a super dense spherical mass, containing neutron fundamental particles (10^{79}) of negligible volume with the speed of light and at high temperatures. From where the super mass came into being is debated. The neutron particles (10^{79}) in the form of radiation with high speed spread out in the vast universe. This radiation era was of very short duration followed by the generation of heavy fundamental particles, neutrons, protons and mesons. The interaction of heavy fundamental particles with radiation produced light fundamental particles, electrons, positrons and neutrinos.

Matter era started with the interactions of fundamental particles (electron, proton, neutron and radiation) among themselves to produce real matter atoms, ions, molecules and elements. Recent discoveries suggest that the universe was created and evolved through four natural basic forces of the universe, the gravitational force between particles with appreciable mass, electromagnetic forces between particles having electric/magnetic charges, strong nuclear forces between quarks (nuclear sub-particles) and weak nuclear forces between quarks and leptons (light sub-particles). The heavy fundamental particles are subdivisible into 'quarks', and light fundamental particles are sub-divided into sub-particles known as 'leptons'. The force of gravity between two bodies is carried by sub-particles known as 'gravitons', a neutral particle with virtually negligible mass. The electromagnetic radiation is carried by massless sub-particles 'photons'. The strong nuclear force is carried by sub-particles called 'gluons' and weak nuclear forces are conducted by sub-particles 'bosons'. Hydrogen and helium atoms dominated in the early stage along with radiation. Atoms converted into stable and neutral elements, molecules and compounds. Under the gravitational pull the conglomeration of fundamental particles, atoms, ions, elements and molecules contracted to form galaxies and stars. Stars are born and die due to an agglomeration of radiation dust (particles ions, molecules, gases and radiation) and supernova explosion. High-density black holes are the residue of dead stars, formed by the supernova explosion of the stars. Both heavy and light elements are formed in the cosmos by nuclear fission and fusion reactions. The solar system is one of the latest star systems and was probably formed due to the supernova explosion of a dying star and under the influence of gravitational and centrifugal forces. Most likely the earth was separated from the solar nebula about 5.6 billion years ago. The solar system including the earth consists of materials drawn from the universe (cosmos). The earth passed from a different level of evolution, from gases/solar nebula to the present solid stage. The earth consists of an inner & outer core, lower and upper mantle and crust. The earth's crust was exposed to the atmosphere, and soon converted into solid inorganic rocky materials. The core remained molten and the earth's mantle in a semi-solid plastic state. After the establishment of inorganic materials, the stage was set for the evolution of organic matter and biological life.

The geological time scale describes the earth's history, since its origin about 5.6 billion years ago (bya). It divides the history into the Phanerozoic and Precambrian eons. Each eon is divided into eras. The era is divided into period, epoch and age that describe the particular geological event that occurred during that time.

2 Petroleum Origin and Generation

2.1 INTRODUCTION

Petroleum is a naturally occurring complex mixture of hydrocarbon organic compounds and derivatives of hydrocarbon compounds containing sulfur, oxygen and nitrogen atoms, including small amounts of metallic organic compounds. Petroleum occurs both at the surface and subsurface of the earth at varied levels up to 10 km deep. Petroleum is found in all three states of matter, i.e. solid, liquid and gas. Polymeric asphaltic materials are solid as well as semisolid; crude oil is liquid and natural gas. The terms 'petroleum' and 'hydrocarbons' include all states of petroleum, gas, condensate, oil, semi-solid (bitumen) and solid (asphaltite).

All plants and animals are made of organic compounds except their bone, teeth and shell which are inorganic materials. It is most likely that the remains of dead plants and animals buried underground sedimentary rock converted into petroleum (asphalt, oil and gas) over geological time. Thus the origin of petroleum is related to the mass production of organic matter (plant and animals). How organic matter came into being is briefly described below.

2.2 INORGANIC TO ORGANIC COMPOUND

An organic compound is a chemical predominantly containing carbon and hydrogen atoms. The science of organic substances is related to living organisms, plants and animals. Carbon is the fourth group element in the periodic table. Its atomic number is 12 and it has 4 electrons in its outer orbit. The four outer shell electrons form an enormous number of compounds through a variety of chemical bonds with other elements, including carbon itself. It is said that one carbon element can form millions of compounds. The remaining 114 elements together form comparatively less, thousands of compounds. The concentration of carbon elements was very minute in the initial age of the earth. The majority of the earth's constituents (92%) were oxides of iron, silicon and magnesium. Organic carbon could have evolved through any of the following methods.

2.2.1 INORGANIC THEORY

The concept of the formation of organic carbon from inorganic substances has been derived from the production of acetylene gas (organic) from calcium carbide (inorganic).

$$2CaC(s) + 2H_2O(l) \rightarrow CH \equiv CH(g) + 2Ca(OH)_2(s)$$

The letters (s, l, g) represent solid, liquid and gas states. Acetylene gas is a highly reactive substance, and rapidly undergoes various additive, substitution, oxidation and polymerization reactions, producing a variety of organic chemicals. Similarly other inorganic chemicals, carbon dioxide, water and alkali metals could be used to produce sodium acetylide (Na–C=CH). Acetylide is convertible to organic acetylene gas. Additionally, a number of hydrocarbon compounds are commonly synthesized from two inorganic gases, namely carbon monoxide and hydrogen, in the presence of a catalyst at suitable temperature and pressure. The same concept of organic synthesis could be applied in the natural generation and accumulation of hydrocarbons in the subsurface rock through the interaction of inorganic chemicals. Did such conditions exist in the deep earth crust for the formation and preservation of organic chemicals? Having said that, is this inorganic phenomenon enough to correlate with the large quantities with varying chemical and structural properties of organic matter found the world over? The answer is most probably no. Alternate theories are given below.

2.2.2   COSMIC THEORY

Some believe that life first started in outer space. The argument says that atoms and molecules present in space synthesized certain molecules/chemicals, somewhere up in the universe. Methane, hydrogen and helium have been detected in Jupiter, Saturn and Neptune. The presence of methane (organic gas) and hydrogen in the cosmos suggests the formation of hydrocarbons in space. Through a chemical evolution, cosmic atoms and elements first formed simple molecules. With the passage of time bigger molecules evolved; methane, ammonia and carbon dioxide were formed that led to the production of amino acids and proteins (complex organic matter).

2.2.3   VOLCANIC THEORY

The volcano activities originate from subsurface magma and igneous rock. Several gases are found in the emission of volcanic eruptions. The volcanic gas components are methane, hydrogen, carbon dioxide, carbon monoxide, sulfur dioxide, nitrogen, hydrogen sulfide, water vapors and traces of bituminous material. Trace amounts of hydrogen chloride and hydrogen fluoride gases have also been detected. The quantity of methane (1–2%) and trace bituminous materials is so small compared to the huge quantities of organic matter found in sedimentary rocks. Volcano gases hardly have a significant quantity of hydrocarbons (organic matter). The necessary raw materials (carbon monoxide and hydrogen) for Fischer–Tropsch's synthesis for producing liquid hydrocarbons are present in volcanic gases, but their proportion and the conditions of the environment are not conducive for the reaction.

The mass and commercial production of petroleum from inorganic, cosmic and volcanic sources is impossible.

2.3   CHEMICAL ERA/BIOLOGICAL ERA

The biological era (organism) started much later after the matter era. When the inorganic matter was fully developed, the stage was set for the emergence of complex molecules, necessary for the synthesis of life. It is not established as yet that life exists anywhere else in the vast universe.

The chemical interaction of different elements produced simple and complex molecules. The period of formation of simple/complex molecules is known as the 'chemical era'. A living organism is composed of complex molecules. The generation of complex molecules paved the way for the 'biological era'. For example, amino acids (a biological chemical) were synthesized naturally from hydrogen, oxygen and nitrogen atoms. Amino acids are a component of a complex biological protein molecule. Along with amino acids, water, carbon dioxide and phosphate are also considered as biological chemicals.

2.3.1 EARLY LIFE

The earth's oldest life occurred about 3.5 billion years ago, with the emergence of prokaryote (incomplete life) organisms, namely cyanobacteria. Cyanobacteria are a single-cell, asexual organism. The cyanophyta bacteria family organisms obtained energy (food) from the photosynthesis process in a reducing atmosphere (absence of oxygen). The fossilized cyanophyta bacteria are known as stromatolite. The stromatolite constitutes a layered rock at the shallow marine depth. The mineral sediments and the fossilized bacteria, on being compressed, banded and cemented in geological time, changing to a stromatolite layered structure. Stromatolite is a well identified, dated and studied fossil.

2.3.2 BACTERIAL PHOTOSYNTHESIS

Photosynthesis is a natural process used by plants and some organisms to convert light energy from solar into chemical energy (food). Bacterial photosynthesis used light and carbon dioxide in hydrogen sulfide reducing conditions, to produce carbohydrates.

$$CO_2 + 2H_2S \xrightarrow{\text{bacteria/light}} \text{Organic compound } (CH_2O)_n + 2S + H_2O.$$

At that time the conditions were reducing, devoid of oxygen. Autotrophs bacteria performed photosynthesis by using hydrogen sulfide. The chemical combination of hydrogen sulfide and atmospheric carbon dioxide produced a carbohydrate organic compound (glucose). Sulfur was produced as a by-product, not oxygen. The oxygen is a product of green plant photosynthesis process.

2.3.3 GREEN PLANT PHOTOSYNTHESIS

Green plant photosynthesis uses the atmospheric carbon dioxide, water and sunlight in the presence of a green plant (chlorophyll) to produced carbohydrate (food) and oxygen. During the photosynthesis process in green plants, light is captured and is used to convert water and carbon dioxide into oxygen and energy-rich organic carbohydrate food. This was the beginning of the mass production of oxygen, in oxidizing instead of reducing atmosphere. Almost all plants, most algae, seaweed and 'certain bacteria' are capable of performing photosynthesis. Photosynthesis is a very complex process, but basically it is a chemical reaction involving the transfer of hydrogen from water to carbon dioxide to produce hydrocarbon organic matter in the form of glucose. The oxygen is freed from a water molecule. The chemical conversion reaction is as follows:

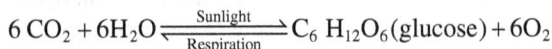

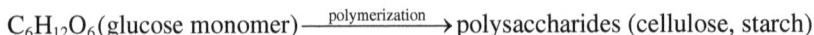

$$6\,CO_2 + 6H_2O \xrightleftharpoons[\text{Respiration}]{\text{Sunlight}} C_6\,H_{12}O_6(\text{glucose}) + 6O_2$$

$$C_6H_{12}O_6(\text{glucose monomer}) \xrightarrow{\text{polymerization}} \text{polysaccharides (cellulose, starch)}$$

Glucose is a monosaccharide or monomer, and undergoes polymerization to form polysaccharides. Polysaccharides consist of at least ten monosaccharide units interlinked by glucoside ($C_6H_{10}O_5-$) linkages. Important polysaccharides, occurring naturally, are starch and cellulose. Starch ($C_6H_{10}O_5)_n$ provides carbohydrates in our food. Starch is stored in the seeds, roots and fibers of the plant. Wood is about 50% cellulose, while cotton and wool are almost pure cellulose ($C_6\,H_{10}\,O_5)_n$. Straw, corn-cobs, bagasse and other agriculture wastes contain sizeable amounts of cellulose.

There are two types of primitive photo bacteria, one is known as 'autotroph' (self-nourishment) and the other heterotroph (different nourishment). Photo autotroph bacteria can make food (carbohydrate) from simple inorganic materials water and carbon dioxide in the presence of light through a photosynthesis process. Heterotroph bacteria have duel characteristics. They can act like autotrophs by conducting a photosynthesis process and also can utilize nearby fossilized organic material as food.

The photosynthesis process initiated the mass production of atmospheric oxygen and biomass (food) necessary for the survival and growth of flora and fauna.

2.4   BIOLOGICAL EVOLUTION/LIVING ORGANISMS

See also Chapter 8. Chemical evolution led to the formation of potential or seed biological inorganic molecules. Ammonia (NH_3), phosphoric acid (H_3PO_4), water (H_2O) hydrogen sulfide (H_2S), carbon dioxide (CO_2) and carbon monoxide (CO) are inorganic raw materials for the production of organic matter. At higher temperatures the organic matter cannot survive. The suitable conditions for the synthesis of organic matter and its preservation slowly evolved through changing environmental conditions during the earth's history. Under conducive moderate temperatures, inorganic molecules interacted with each other, first to create simple organic molecules, such as amino acids, glucose and carboxylic acid. These interactions were followed by more complex reactions resulting in the formation of proteins, carbohydrates and lipids. Amino acids, glucose and carboxylic acid are known as biological monomers that can undergo polymerization, forming proteins, carbohydrates and lipids which are known as biopolymers or bio-macromolecules. Once this happens and under suitable conditions, these macromolecules can convert or form biological living organisms, known as cells. The following hypotheses are proposed as possible explanations for the formation of living organisms.

2.4.1 PROTEINS HYPOTHESIS

Proteins are large biological macromolecules; they consist of more than one long chain of amino acids. Two amino acids (glycine and alanine) combine with each other to form glycylalanine dipeptide (protein) with the elimination of water. In the same way more than one amino acid molecule (from 5000 to millions) combines through peptide linkage to give a protein polymer. The proteins are also called polypeptides. Proteins are present in muscle, skin, hair and other tissue of the body except bony structures. According to their molecular shape and structure proteins are classified as fibrous and globular proteins; each class performs a specific role. They are required by all animals, to synthesize tissues, enzymes, hormones and blood compounds. Additionally they are used for improving and maintaining the existing tissues and are a source of energy for the body. They are the 'reaction center' or reactors for the conversion of light energy into chemical energy, through the photosynthesis process.

2.4.2 RIBONUCLEIC ACID HYPOTHESIS

Ribonucleic acid (RNA) is a polymeric macromolecule. It is made from a monomer known as a nucleotide. Each nucleotide has three components: a five-carbon sugar, a nitrogen base hydrocarbon group and a phosphate ion. If the sugar is deoxyribose in the ribonucleic acid, the polymer is known as deoxyribonucleic acid (DNA). RNA and DNA macromolecules are essential for life. They regulate genes and help to replicate and build cells, possibly through biological cell evolution and metabolism.

2.4.3 LIPID MEMBRANE HYPOTHESIS

A lipid membrane is a layer of molecules that surround a cell interior, separating it from external exposure. The membrane acts as a selective barrier that helps to speed up chemical reactions inside the cell. The selective barrier membrane consists of phospholipid, which is created by a lipid and a phosphate. The phospholipids create a hydrophilic head (phosphate) and hydrophobic tail (fatty acid). This is the major component of the membrane and organelle of the living cell.

Lipids are essential components of animals and plants, insoluble in water but soluble in organic solvents. Fat and oil are saponifiable lipids whereas steroids are non-saponifiable lipids. Isoprene and fatty acids units are the building blocks of lipids. The functions of lipids include the sharing of chemical energy, absorbing sunlight like pigments and responding and communicating between cells. A summary of the above biological hypothesis is stated below:

- Evolution of biological monomers
- Evolution of biological giant macromolecules (biopolymers)
- Evolution of living cells from biological active polymers

2.4.4 BACTERIAL CHEMOSYNTHESIS

In the deep ocean in the absence of sunlight, photosynthesis is not possible. Another process by anaerobic chemoautotroph (chemo self-nourishment) produces organic matter (food). In moderate thermal conditions, carbon dioxide and hydrogen sulfide gases react in the presence of a limited supply of dissolved oxygen and chemoauto-troph bacteria.

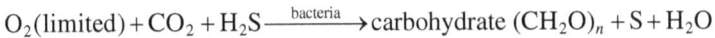

$$O_2(\text{limited}) + CO_2 + H_2S \xrightarrow{\text{bacteria}} \text{carbohydrate } (CH_2O)_n + S + H_2O$$

The synthesis process provides a large amount of carbohydrate food, in the darkness of the ocean bottom, for the development of undersea plant and animal organisms (fauna, flora).

2.5   DEVELOPMENT OF BIOSPHERE

Biosphere, biomass and organic matter are terms used for plants and animals includ-ing bacteria and algae found in the land and water of the earth. The type, nature and history of biomass (living organism), that later converted to petroleum in sedimen-tary rock, have profound effects on the quantity and quality of petroleum generated.

During the period from 10 million to 500 million years ago, there was a tremen-dous increase in biomass, especially algae, bacteria and small vegetable-like species, in comparison to big animals and terrestrial trees (big land trees). The contribution to petroleum generation by big trees and animals is rather limited. The main contrib-utors are algae, bacteria, small weeds, bushes and small insects. All these organisms normally grow and survive in calm marine environments and to a lesser extent on land and the high seas. The dead or alive aquatic organisms, floating on the water's surface or drifting in quiet marine water ultimately settle down at the bottom and become part of the sedimentary rocks. Vegetable or plant species drifting or floating in water are called 'phytoplankton' (wandering vegetable). The availability of green vegetation food (flora) led to the growth and abundance of fauna population bacteria, marine animals, insects, worms, mollusk, anthropoids and tiny invertebrate species.

The animal kingdom developed and evolved side by side with phytoplankton. These small marine animal species were called 'zooplankton'. Many groups or kinds of bacte-ria, algae, phytoplankton and zooplankton have been identified as the main contributors to the total organic matter. Algae are supposed to be the precursor of the whole plant kingdom. Algae first developed into sea weeds, sea bushes and higher plants.

Higher plants or vascular plants were the other sources which significantly con-tributed to the build up of organic matter. Vascular plants evolved first in the marine environment and then spread out on land, during the Devonian period. Terrestrial (land) animal appeared after the plants probably in the Permian or Triassic period.

The evolutionary step of flora (plants) formation is the foundation of the 'food chain'. This was followed by the evolution of fauna (animals). According to the law of the food cycle, zooplankton and animals mostly existed in region of phytoplankton and plants. Bacteria are found in greater numbers in or around a zone of organic mat-ter, so that they get their food and survive. It is a fact that the organisms, both plants and animals, thrived and developed side by side mostly in the marine environment.

The factors responsible for the development and concentration of organic matter in aquatic environments are given below.

2.5.1 MARINE

'Marine' refers to the portion of the sea/ocean adjacent to land where some human activity is carried out. The oceans store a vast amount of saline water. The ocean bottom is structured in three layers. Coastal shallow water consists of two descending bottom surfaces. The first one extends from zero meters (land surface) to 200 meters deep. It is called the continental shelf, which is flat or slightly inclined. The continental shelf is 7.5% of the total sea bottom. The second layer is called the continental slope from 200 to 4000 m deep (8%) and includes the continental rise (5%). The deep sea flat bottom starts from 4000 m deep to the lowest bottom of the sea. The deep sea bottom constitutes 77.5% of the total bottom along with 2% trenches (hole) of various sizes. The continental shelf and continental slope together are called the continental margin. This is the potential area for the growth of organic matter, particularly the continental slope. Calm, quiet, free from turbulent waves and currents, the aquatic pool is conducive for the evolution of organic matter, rather than the open sea with rapid water circulation and current.

2.5.2 TEMPERATURE

Temperature plays an important role in the generation and preservation of organic matter. Biological productivity in general depends on the temperature. The temperature of ocean depends on water circulation, turbidity and water column exposure to sunlight. The temperature at the water surface, which is exposed to sun and atmospheric conditions of wind and air, is around 25°C. The temperature gradually decreases with depth. At the lowest bottom of the sea, it may be between 0 and 5°C.

Various phytoplankton (flora) grow favorably at temperatures around 25°C, whereas calcareous and siliceous hard-shelled organisms are comfortable in the cold climate of polar regions. It is observed that large species of organisms are found in warm regions. A comparatively smaller number of species exist in the colder region. The solubility of oxygen in water depends on temperature and depth. Oxygen is an essential ingredient for living organisms including in water. At a high temperature of water, there will be low solubility, and low temperatures mean more oxygen content in the water. The oxygen content of water decreases with depth. However, apart from temperature and depth, other factors such as water circulation control the oxygen solubility.

2.5.3 SALINITY AND pH

Marine water is saline. It contains dissolved minerals and salts. The mineral salt composition of sea water is expressed as salinity. Salinity is expressed as grams of salt per 100 grams of water. It is 35% and is almost constant throughout the oceans. Salinity does not affect the total growth of organisms, but the variation of salinity affects the number of organic species. All organisms survive in and require slightly alkaline conditions. Nature has maintained the pH of sea as 8.1 ± 0.2, by the buffering action of carbonate/bicarbonate/silicate ions. Salinity decreases the solubility of dissolved gases in water. A less saline water dissolves more gases and more saline water contains less dissolved gases.

2.5.4 DISSOLVED GASES

The marine water surface is in constant contact with the atmosphere. At the interface between the water surface and atmosphere, the gases from the air enter the marine water and gases from the water escape to the atmosphere. So the gases are in exchange constantly with the water and atmosphere. The dissolved gases found in marine water are nitrogen, oxygen, carbon dioxide, hydrogen sulfide, ammonia, methane, hydrogen and some noble gases. Nitrogen is inert and as such not used in any biological/chemical process, except for a small quantity of nitrogen that is used by nitrogen-fixing organisms. The noble gases are also inert. Oxygen and carbon dioxide are important gases for metabolism/conversion processes in marine water, including the important photosynthesis process. Hydrogen sulfide, methane, hydrogen and ammonia are produced by the decomposition of organic matter at various stages at different depths. In deeper waters, where anaerobic (absence of oxygen) conditions prevail, hydrogen sulfide is produced by a sulfate bacteria reduction process. In the absence of free oxygen, anaerobic bacteria take the oxygen from sulfate ions to oxidize the organic matter, releasing hydrogen sulfide as a by-product.

2.5.5 WATER TURBIDITY

The turbidity of water is caused by suspended particles and colloidal particulate matter. Suspended particles are silicon dioxide, alumina, calcium carbonate, fragments of rock, biological organic debris, bone skeleton, shell pieces, fecal drops and pollen grains. The colloidal particulate matters are mostly drawn from silicate minerals, clay, talc, quartz and feldspar. Turbidity affects the temperature, sunlight reaching the water column and oxygen content. So as a result, biological processes including photosynthesis are also affected.

2.5.6 WATER CIRCULATION

The circulation of water may be beneficial or harmful for the growth of organic matter. Rapid water circulation adversely affects the growth of organic matter. On the other hand, water circulation may bring fresh nutrients for the organisms. Water currents, waves, ripples, whirls and tides are the main factors for water circulation and the mixing of marine water. Circulation plays an important role in climatic change, marine activities and maintaining the uniformity of water. On the other hand, water circulation and mixing act counter to the stagnant and calm water pool. As stated earlier the stagnant and quieter water helps the deposition and evolution of organic matter. Constant water circulation does not give sufficient time for biological transformation. It is evident from the fact that enormous organic matter is found in restricted and quieter gulf and lake water with a limited supply and circulation of water. Much smaller quantities of organic matter are found in open sea. Conditions prevailing in open sea are the constant circulation and mixing of water by wind, wave and current. However, this is not the whole story. Sometimes water circulation, particularly in stagnant water, may enhance the growth of the organic matter. The circulating water brings fresh nutrients and new seed organisms to the area.

2.5.7 WATER DENSITY

The density of water is temperature-dependent. Warm water is lighter than cold water. The sun keeps the surface water warm. Warm water moves from warm areas toward colder poles. Colder water, being heavy, sinks down to the bottom and moves along the surface of the sea bottom. As such, a current is generated, and mixing and circulation take place between the surface layer and deeper water.

2.5.8 NUTRIENTS

All organisms originated from unicellular algae, and all needed nutrients for their growth, survival and multiplication. These nutrients are minerals containing phosphorus, nitrogen and some other micronutrients. Due to more demand in thick phytoplankton regions, the quantity of nutrients is depleted. Simultaneously in the region oxygen is liberated and carbon dioxide consumed in the photosynthesis process. The law of the food chain comes into play. Remains of dead organisms, both plants and animals, sink down from the water column and are deposited at the bottom. The decomposition of dead matter takes place, consuming oxygen and liberating carbon dioxide, and producing minerals containing nitrogen and phosphorus. The minerals dissolve and return to the water pools to function again as nutrient fertilizers.

2.5.9 SUNLIGHT

All plants, including underwater plants, carry out a photosynthesis process under sunlight and respiration in the absence of sunlight. Photosynthesis produces oxygen and consumes carbon dioxide. The photosynthesis process takes place near the water surface, up to 200 meters deep in clear water. The depth is reduced by turbidity and high-salinity water. Light is unable to travel long distances into the sea. So the oxygen production area is limited to the photosynthesis zone. Marine animals are found in all levels of depth. The animal by respiration consumes oxygen and exhales carbon dioxide. Thus carbon dioxide is available throughout the water volume. In the absence of oxygen in deep sea, sulfate bacteria reduction processes provide oxygen energy required by organisms.

2.5.10 BIOLOGICAL SUBSTANCES

This includes all living plant and animal species. The organisms produce and introduce the bio substances as their execratory product, dead residue and decomposed parts into the region. The bio substances are important because they interact with each other and affect the life cycle of all marine organisms; they modify and affect the marine food chain supply, survival, death, production and consumption of organisms. Bio substances provide all raw materials that subsequently undergo transformation to petroleum oil and gas.

2.5.11 GEOGRAPHICAL ZONES

Different geological regions of the world differ in the climatic conditions, and as such all the above-mentioned factors also differ. Therefore, organisms also differ from region to region and carry the specific character of the region.

2.5.12 GEOLOGICAL TIME

All geological processes and events are time-dependent. Primitive organisms evolved during the Precambrian period about 3000 million years ago. During the Cambrian period about 600 million years ago, algae, bacteria, marine invertebrates and shelled water insects were well developed. The plant kingdom extended its domain towards land during the Ordovician period about 500 million years ago. By the end of the Carboniferous period about 270 million years ago plant life was well developed. The Cretaceous period saw the emergence of big animals. Primitive mammals appeared during the Cenozoic era (10–50 million years ago). Evolution is a continuous process.

2.6 LIVING ORGANISM CHARACTERISTICS AND DEAD ORGANIC MATTER

The main contributors to the organic matter in sedimentary rock are algae, bacteria and phytoplankton, zooplankton and to some extent land terrestrial plants. Phytoplankton and zooplankton are small plants and animal organisms of special characteristics. They are found in aquatic environments. Their weight is so small in comparison to their volume; they hardly or slowly sink in water. They are so light that they hardly retain their position in water, due to circulating water. So they drift or wander in water and are at the mercy of nature. The term plankton means drifter or wanderer. Phytoplankton and zooplankton together are called marine plankton.

Phytoplankton usually stays in sunlight in the water column. They carry out photosynthesis to create carbohydrates and form the food chain for zooplanktons. Planktons do not have roots. They take nutrients from the surrounding water for their survival and growth. Marine plants are fundamentally different from land plants; they are simple and of very small size. On the other hand, land plants are large trees, herbs, bushes and grasses. They have roots to take water and nutrients from soil. They conduct photosynthesis and are rigidly bounded to soil through roots. Each part of the plant has different characteristics, compositions and functions. Root, stem, bark, leaf, spores, flower, fruit and seed all have different characteristics and chemical composition.

The major elements found in both plants and animals are carbon, hydrogen and oxygen. In addition to these other elements exist: nitrogen, phosphorus, potassium, calcium, magnesium, zinc, sulfur, chlorine, boron, iron copper, manganese,

TABLE 2.1
Elemental Composition of Organic Matter (dry basis)

	Elemental Composition in Weight Percent					
	C	H	S	N	O	H/C
Carbohydrates	44	6	–	–	50	1.66/1
Proteins	53	7	1	17	22	1.59/1
Lipids	76	12	–	–	12	1.90/1
Lignin and cellulose	63	5	0.1	0.3	31.6	0.95/1
Petroleum	85	12	1	0.5	0.5	1.83/1

Average H/C atomic ratio of organic matter is around 1.58/1.

molybdenum and other trace elements. All these elements are found in various proportions in organic matter. Some are found in trace amounts. The elements are not found in organisms as such; they are present as chemical compounds. Such chemical compounds in organic matter (organisms) can be divided into four major groups, namely carbohydrate, protein, lipid, lignin and cellulose, apart from minerals and water. The element compositions of these groups are given in Table 2.1. For comparison, the element composition of petroleum is also included. Lipids closely follow petroleum. It may be noted that the major constituents of plants and animals are water and inorganic bone structure. Some plants and animals may contain up to 90% water. The bone and muscle of animals constitute the animal body structure. Minerals and proteins are responsible for the development of bone and muscle. Cellulose performs the same function for plants. Oil and fats or lipids are an important component of animals, acting as reserve food materials. Starch acts as reserve food material in plants. Variable amounts of mineral components are found in plants, whereas in animals the variation in minerals is less. Land-derived plants are more aromatic because of the presence of lignin. Therefore they have a smaller H/C ratio of 1.2 to 1.4. Proteins contain more hydrogen and lipids still more hydrogen, so their H/C ratio is higher, ranging from 1.7 to 1.9. Carbohydrates, cellulose, proteins and lipids are all macromolecules or biopolymers. The individual molecular weight of each constituent varies considerably. The average molecular weight may be quoted in the range of 2000–100,000. Land and aquatic organic matter have different features and the mode of assimilation and the fate of residue of dead organisms into sedimentary rock also differ. Land conditions are variable and sometimes very harsh, whereas the marine environment is steady and moderate; dead animals and plants undergo the same process of decay with time, but to different extents in land and water. Dead terrestrial plants may undergo complete oxidation/decomposition so that they are completely annihilated into carbon dioxide, water and inorganic ash. Or they may undergo little or moderate decomposition and be partially transported to the basin and sea side. After reaching the basin, the land-derived biomass along with aquatic dead biomass is incorporated into sedimentary rock. This was the first transformation of little or partially decomposed dead biomass into sedimentary rock. The biomass in the sedimentation process undergoes fossilization and possibly conversion to oil and gas in geological time under overburden pressure.

2.7   CARBON CYCLE

Carbon exists in many chemical forms on earth. It is abundantly found in nature in gaseous and solid states, both in organic and inorganic forms. Diamond and graphite are pure carbon solids and carbon dioxide is gas. Carbon is associated with all forms of organisms. Carbon is closely linked with other elements of life, hydrogen, oxygen, nitrogen and sulfur, etc. Living organisms are mostly composed of carbon compounds, and their survival is linked to the carbon-rich organic food materials. Carbon compounds are continuously being created, consumed, transferred and decomposed in the land and ocean environments. Balancing and following the fate of carbon during these transformations is known as the 'carbon cycle'. If one balances the creation and consumption of oxygen in land and ocean that is known as the 'oxygen cycle'. Similarly following the fate of nitrogen during transformation on land and ocean is 'nitrogen cycle'. Terms like 'food chain' or 'food cycle' are interrelated to the organic carbon cycle. Phytoplankton is a source of food and subsistence for zooplankton. Zooplankton in turn is consumed by larger carnivorous animals and fishes (food cycle).

The atmosphere and ocean are in a 'dynamic state' and partially in a 'static state'. So the carbon elements associated with them are in a 'dynamic or static state'. Carbonate minerals and fossil fuels are examples of 'static or storage carbon'. Organic carbon (plant, animal) is an example of 'dynamic carbon'. Static and dynamic carbons are interrelated through the carbon cycle. The decomposition and weathering of carbonate minerals release carbon dioxide gas. The inorganic carbon dioxide is consumed in the photosynthesis process, producing organic carbon in the form of carbohydrate, and takes part in the development of the plant and animal kingdoms. A small part of organic carbon is assimilated in sedimentary rock and converted to fossil fuel. A large part of the dead plant and animal residue is incorporated in sedimentary rock and over geological time transforms into carbonate mineral. The carbon cycle is the exchange of carbon compounds among the atmosphere, biosphere (life on earth), hydrosphere (ocean) and lithosphere (terrestrial carbon).

2.7.1   ATMOSPHERIC CARBON

The earth's atmosphere begins from surface sea level and extends up to a height of 500 km. It consists of a mixture of gases and water vapors. Carbon is present in the atmosphere as carbon dioxide gas and volatile hydrocarbons. Carbon dioxide in the atmosphere is found because of the respiration of plants and animals, decomposition of organic matter, burning of fossil fuels and weathering of carbonate minerals and sediments. Carbon dioxide and hydrocarbons are desirable to a certain extent in the atmosphere, so as to maintain a comfortable temperature on earth. Human activities and the mass utilization of fossil fuels in industries and transport have increased the emission of carbon dioxide and hydrocarbon manifolds, causing global warming and the greenhouse effect. The atmospheric carbon dioxide is consumed in the land and ocean by the plants in the sunlight region through the photosynthesis process. At the interface of the atmosphere and water surface, carbon dioxide dissolves into the water or releases from water into the atmosphere, depending upon the conditions of the environment. Carbon dioxide in the atmosphere is washed down by rain water, forming corrosive carbonic acid. The acidic water causes the erosion of rocks and other structures of the earth. The acidic water ultimately pollutes lakes, rivers and oceans. The greenhouse effect brings about more rain in certain regions and dry spells in other regions. The uneven rises in sea level have also been noticed in certain regions. The use of fossil fuels, coal and petroleum has disturbed the ecosystem, that is, the balance between the emission and consumption of carbon dioxide by natural processes.

2.7.2   TERRESTRIAL/LAND CARBON

Terrestrial or land carbon includes all plants and animals and their remains, as organic carbon. Soil carbon is identified as inorganic carbon (calcium carbonate). Plants consume carbon dioxide (inorganic carbon) from the air and convert it to carbohydrate (organic carbon). Animals consume plant carbon and release carbon dioxide through respiration or the decomposition of dead organisms. Combustion also releases carbon dioxide.

2.7.3   OCEAN CARBON

The turbulent motion and temperature variation of the ocean are responsible for the evolution (release) of carbon dioxide from the ocean to the atmosphere. On the other hand, wind and the basic (alkaline) character of ocean water facilitate the absorption of carbon dioxide into ocean water from the atmosphere. A portion of the gas is precipitated as carbonate in the water and the rest undergoes the photosynthesis process through the phytoplankton. The photosynthesis process creates more organic carbon as biomass (carbohydrate).

Carbon dioxide absorption by the ocean may balance extra carbon dioxide produced by human activities. However the dissolution of carbon dioxide is limited to a certain extent. One reason is that the ocean becomes less alkaline and may not absorb or may not be able to precipitate the already dissolved carbon dioxide. Perhaps that is the natural way of maintaining the required carbon dioxide and alkalinity in the ocean. It is apprehended that more carbon dioxide in the ocean may create the same type of effect as the carbon dioxide-created adverse greenhouse effect on land. Inorganic carbon is introduced from the dissolution of carbonate rock structure into the ocean. Carbon dioxide is produced by the decomposition and oxidation of dead organisms (plants and animals). The shell and bone of animal structures are converted into calcium carbonate and silica minerals.

2.7.4   GEOLOGICAL CARBON

Carbon associated with the lithosphere (crust and mantle) is known as geological carbon. The earth's geological structure is known as the geosphere. The carbon of the geosphere is identified as follows:

- Inorganic carbon associated with the earth's mantle dates back to the earth's origin.
- Drastically oxidized organic matter is reduced to almost pure carbon as graphite or anthracite, which is settled down deeper as metamorphic rock.
- Carbon from the geosphere is released during volcano eruption, as carbon dioxide gas.
- Most of the carbon in the geosphere is present as calcium carbonate formed from the shells and skeletons of marine cretaceous organisms.
- Part of the carbon in the geosphere is organic carbon known as kerogen, shale oil, coal and oil/gas. These are formed from the burial and sedimentation of dead organisms (organic matter) under high temperatures and pressure.

2.7.5   INDUSTRIAL CARBON

The natural carbon cycle is affected by the industrial revolution, especially by the increasing use of fossil fuels. The carbon from the geosphere (fossils/kerogen) is directly transferred to the atmosphere through burning and carbon dioxide emission. For the last three to four centuries human activities have adversely affected biodiversity and the ecosystem. Natural processes of maintaining the balance of carbon dioxide in environments are greatly disturbed. Additionally industrial activities are increasing the carbon (CO_2) in the atmosphere. With a growing population there is a need for more consumption. Deforestation for the settlement of increasing populations and for agricultural purposes has greatly reduced the natural carbon storage (plant). The consumption of carbon dioxide and the production of the food chain has been depleted due to the reduction of the green belt. The production of carbon dioxide is increasing with increasing population. Air and soil pollution are adding to soil corrosion and washing out the organic carbon content. This is increasing acidity and limiting agriculture and farm products. The decomposition of organic matter is accelerated by higher temperatures. There is more release of carbon dioxide to the atmosphere. Population control, alternate fuels and massive plantation are the answer to the above adversaries.

2.7.6   POTENTIAL CARBON FOR PETROLEUM

Carbon is accumulated in sedimentary rock in two forms, organic carbon of biological origin and as carbonate minerals. The former carbon is known as reduced carbon and the latter as oxidized carbon. After due consumption and production, only a small quantity of the total organic carbon survives to be accumulated and preserved in sedimentary rock as potential for conversion to petroleum.

2.8 PETROLEUM SOURCE ROCK

Petroleum source rock is an organic, rich, fine-grained sedimentary rock from which hydrocarbons are generated or are capable of being generated. The coarse-grained rock is unfit to act as source rock. In coarse-grained rock supply, the drain and loss of the sediments along with organic matter are too rapid for them to be preserved for conversion to petroleum over geological time. Source rock is capable of preservation, conversion and expelling the formed oil and gas, under geological conditions of temperature, pressure and time. The definitions of 'source rock', 'potential source rock' and 'effective source rock' are given below:

Petroleum source rock. 'It has the capability to form an accumulation of oil and gas. The source rock can generate and expel enough hydrocarbons'.

Potential source rock. 'It is the one that is too immature to generate petroleum in natural shallow setting. But the rock will form significant quantities of petroleum when heated in the laboratory or during deep burial with time'.

Effective petroleum sources rock. 'It is the one that has already formed and expelled petroleum to carrier/reservoir rock'.

Source rock is the first component of the underground petroleum system. The petroleum system consists of five components, i.e. source rock, carrier rock, trap rock, seal rock and reservoir rock, along with three processes, i.e. the generation, migration and accumulation of petroleum.

Four types of source rock are identified, on the basis of original organic matter, as follows:

Type I source rock. Type I source rock is formed from algae and weed remains, deposited in lakes and river estuarine basins. The organic matter is hydrogen rich, and it generates paraffinic (waxy) crude oil.

Type II source rock. The type II rock is formed from marine plankton and bacteria in marine basins. It generates both oil and gas.

Type III source rock. Type III rock is the result of the decomposition of terrestrial plants and their parts by bacteria and fungus, mostly in coastal and terrestrial basins. It creates gas and hydrogen-deficient residue, i.e. coal and shale oil.

Type IV source rock. Type IV source rock is the organic residues from types I, II and III. The rock is highly oxidized and devoid of functional groups. It is rich in carbon content and deficient in hydrogen. Under subsurface conditions of high temperatures and pressure, type IV source rock generates gas and coal.

2.9 FORMATION OF SOURCE ROCK

The formation of organic rich source sedimentary rock in a typical continental shelf is illustrated in Figure 2.1. The factors governing the formation of organic rich source rock are:

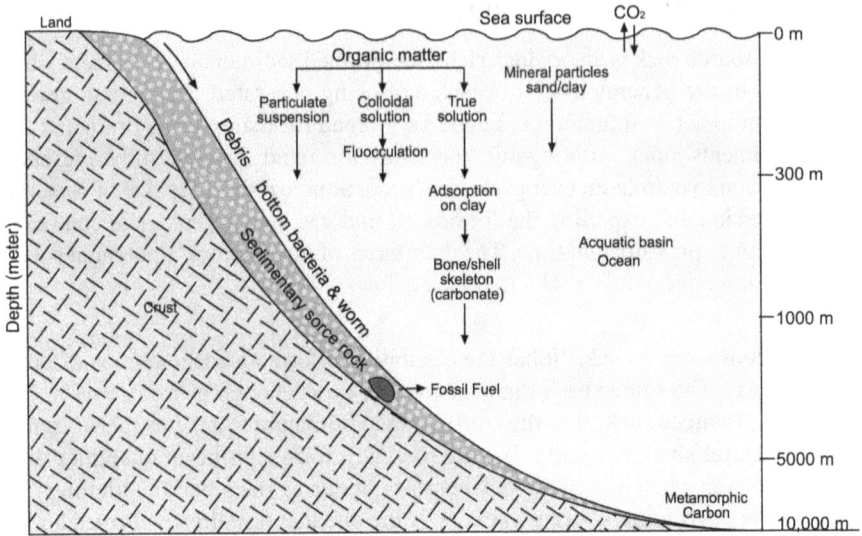

FIGURE 2.1   The formation of organic rich source rock in a typical continental shelf by the accumulation of (i) debris (inorganic and organic sediments) brought from land surface drained into the basin by wind, rain, flowing river water, (ii) continental shelf aquatic minerals and organisms and (iii) benthonic (marine bottom) bacteria and worms. The final transformation of organic matter to fossil fuel and dead carbon is also shown in the Figure. Source: Modified from Tissot B.P. and Welte D.H., *Petroleum Formation and Occurrence*, Springer Verlog, New York, 1984 (Figure no I.4.17).

- Land debris, rock fragments, sediments, dead animal/plant residue, spore, pollen and terrestrial organic matter brought from distant land surface and drained into the basin by wind/rain/flowing river water.
- Continental shelf aquatic organisms/minerals.
- Benthonic (marine bottom) bacteria and worms.

The ultimate transformation of a small portion of organic matter to fossil fuel and bulk organic matter residue to metamorphosed dead carbon and graphite is also shown in the Figure 2.1.

Petroleum sedimentary source rock is formed by a series of physico-chemical, biological and geological process, usually in an aquatic environment (basin). The organic matter that is incorporated in the source rock originates from two sources:

- Local organic matter in the petroleum basin, for example in marine or lacustrine basins.
- Organic matter brought to the basin from far and wide land surfaces through rain and river.

A large quantity of organic matter is produced in aquatic basins and marine conditions. Dead land animal/plant residue consisting of spores, pollen and terrestrial organic matter is brought from distant land surfaces and drained into the basin by wind, rain and flowing river water. The marine basin water may be turbulent or calm. Turbulent water does not support the deposition, accumulation and preservation of organic matter along with the sediments. Turbulent conditions facilitate the erosion and dispersion of organic matter and sediment deposit. Likewise, still and quiet water does not help to bring and accumulate the ingredients necessary for the formation of source rock. Both turbulent and calm conditions of water do not support the formation of organic rich source rock in the basin. In between, a basin that receives a controlled and restricted supply of water containing sediments and enough organic matter is suitable for petroleum source rock formation. The conditions facilitate the accumulation, deposition, preservation and formation of organic rich source rock. Oceanic organic matter mainly originates from phytoplankton and zooplankton. Both of these contribute over 70% of the aquatic living organisms. A considerable amount of land terrestrial organic matter is added to oceanic organic matter. Diversified oceanic and terrestrial organic matter and various types of sediments along with benthonic (sea bottom) organisms interact with each other in their own way. New species and organic matter are produced. Sediments get consolidated. All types of organic matter, oceanic or terrestrial, and sediments are trapped and preserved in source rock. The quantity and quality of the preserved organic matter depend on the many geological and environmental factors such as the thickness of strata, volume and area of source rock, water current and characteristics of organic matter.

2.9.1 ORGANIC MATTER IN AQUATIC ENVIRONMENT

Organic matter in water can exist in one of the following forms:

- Completely dissolved true solution
- Colloidal forms, colloidal solution/emulsion
- Suspension or dispersion of particulate matter

Solute particles are dissociated in ionic form or as individual single molecules in true solution. A suspension or dispersion contains an aggregate of a large number of molecules in a medium (water). The colloidal solution is intermediate between a true solution and suspension. There is still an aggregate of molecules in colloidal solution but the number is less than in the suspension.

Solute particle size distribution in three forms is as follows.

True solution = 0.0001–0.001 μm (1–10 A°)
Colloidal emulsion = 0.001–0.2 μm (10–2000 A°)
Suspension particles = >0.2 μm (>2000 A°)

The incorporation and subsequent settlement of organic matter in these three forms determine the characteristics of the bottom source rock. All three forms of particles in the water column take different paths and times to settle down into the bottom.

Considerable variation in the concentration of organic matter in water is observed. However the concentration of dissolved matter is much greater than that of the particulate suspension matter. On the basis of total organic carbon (TOC), it is reported that the dissolved organic matter is around 1.0 mg/liter, compared to the particulate matter of around 0.01 mg/liter in water. Heavier particulates can settle in less time. On the other hand, lighter particles take more time for settlement. Lighter organic particles may drift away to other places and be lost by water waves and other physical processes. Colloidal particles are prone to physico-chemical adsorption and aggregation. The particles get coagulated and flocculated to form heavier weights and settle down in the bottom rock. Most of the time water is saturated with suspended inorganic sediments, mostly of clay and carbonate minerals. The surfaces of clay particles are good adsorbents and have affinity for organic matter. The adsorbed clay particles become heavier with the buildup of adsorbed organic layers at the surface. The heavier particle settles down in the bottom rock.

The maximum concentration of organic matter is found near the water surface. The surface and just beneath the surface area are associated with more organic growth and biological activities. The species get maximum oxygen and a supply of fresh organic matter (food). On the other hand, free availability of oxygen enhances biodegradation and the destruction of organic matter at the surface. Physical disturbances near the surface can alter and interfere with organic matter systems, leading to rapid annihilation. The oxygen content of the water decreases with increasing depth. Below a depth of 200 meters, the availability of oxygen is reduced to nominal value and a reducing environment is beginning to form. The reducing environment is enhanced by the fresh availability of hydrogen sulfide and the depletion of oxygen. The reducing environment eliminates the severe condition of oxidative destruction, and instead facilitates the preservation and accumulation of organic matter for a longer period of time. A proper proportion and quality of the organic matter in association with shale/clay sedimentary rock, sometimes with carbonate minerals, are needed to form a commercial source rock. A source rock with more than 2.5% TOC is considered a good commercial prospect.

2.9.2 Quantity of Organic Matter

The quantity of organic matter in a source rock is variable. The content of organic matter is expressed as a percent of TOC. Inorganic carbon exists as carbon dioxide, carbonic acid and carbonate minerals. Considerable variation in the content of organic matter is found in source rock that may range from 0 to 25% (TOC). The variations depend on many factors as indicated below.

2.9.2.1 Type of Mineral/Rock

Different sedimentary rocks have different capacities for holding organic matter. A very small amount of TOC is found in sandstone rock of high porosity (large pore size). This is due to rapid oxidative destruction and the draining of organic matter from the void of the minerals. The majority of commercial source rock belongs to low porosity clay/shale sedimentary rock that may be associated with calcareous minerals.

2.9.2.2 Mineral Grain Size

Not only does the type of mineral affect the TOC content in sedimentary rock but also the size of the mineral. A higher content of TOC is found in fine-grained sediment than in the course-grained deposit. High TOC content in fine-grained rock is due to the greater holding and preserving nature of the sediment. Coarse-grained sediment provides enough space for transportation, seepage and chemical degradation. For example, sandstone has 1.0% TOC with a larger grain size of 100–200 µm and clay having a smaller grain size of 5.0 µm, holds 15% TOC. Silt stone sediment with a greater grain size contains less TOC whereas the same silt stone sediments with a smaller grain size contain more TOC.

2.9.2.3 Geochemical Oxidation/Reduction Conditions

Geochemical oxidation/reduction reactivity plays a role in source rock. Oxic (presence of oxygen) conditions enhance the destruction of organic matter. A geochemical anoxic reducing environment (absence of oxygen) favors the preservation of TOC. The rate of degradation of organic matter is reduced in conditions of deficient/absent oxygen. High contents of TOC are found in stagnant, silted basins in anoxic conditions. Quantitatively oxidation/reduction activity is defined by the relative amounts of ferric (Fe+++) and ferrous (Fe++) ions in the environment. A high Fe+++/Fe++ ratio indicates high oxidation, and a low ratio refers to a reducing environment. A geochemical basin with a Fe+++/F++ ratio of 10 corresponds to 0.25% TOC. As the ratio progressively drops, the TOC increases accordingly.

2.9.2.4 Turbulent Zone

Low TOC quantities are found in high turbulent zones of coastal areas of open sea. There may be enough productivity of the organic matter but, due to water turbulence, the organic matter is transported to some other place. More oxygen content is observed in turbulent water than in quiet and calm water. Therefore, the possibility of oxidative degradation is enhanced in strong turbulent current water. Quiet and stagnant water favors high TOC, and turbulent water washes away the organic matter.

2.9.2.5 Color of the Mineral

Generally a light color rock contains less TOC than a dark/black rock. The pure limestone is white and correspondingly its TOC is very low or negligible. Brown limestone is comparable to the TOC (3.0%) content of grey shale, 4.0. Different colors of shale, red, green, grey, black, indicate different amounts of TOC. The color of the mineral is a rough indication of TOC content in the rock. The use of color can be misleading. A mineral itself can be of various shades and colors, for example iron pyrite and manganese dioxide minerals are black.

2.10 TRANSFORMATION OF ORGANIC MATTER IN SOURCE ROCK

Transformation of organic matter to oil/gas in source rock depends upon the following:

- Quantity and quality of organic matter.
- Prevailing geophysical conditions in the source rock.

The quality of generated oil and gas depends on the quality of the organic matter trapped in a source rock under constant geological conditions. Organic matter is classified into two categories.

- Organic material derived from marine plankton.
- Organic matter derived from terrestrial cellular plants.

The former is known as 'saprogenic' and the latter as 'humic' materials. Both are characterized on the basis of their H/C ratio. Saprogenic material (higher H/C ratio = 1.5) supports crude oil formation. On the other hand, humic material (lower H/C = 0.9) facilitates hydrogen-deficient and carbon-rich asphaltic material and coal formation.

Source rock is formed by the deposition and accumulation of organic matter that survived atmospheric, aquatic and biological decomposition and destruction. The deposition and accumulation of organic matter in the source rock are a continuous process. With geological time, millions of years, the thickness, overburden pressure and source rock temperature are progressively increased. Under these conditions, various geological, physical and chemical transformations of organic matter took place. These transformations are catalyzed by the following factors.

- A strong reducing environment in stagnant, quiet water helps to preserve and improve the quality of organic matter with time. The oxidation environment destroys the organic matter.
- The organic matter accumulated in the source rock is the biopolymers, carbohydrates, proteins, lipids, lignin and cellulose. The lipids, fat, wax and their homologues behave differently in geo-physico-biological conditions than the other carbohydrates and proteins. Lipids are the precursor of geochemical, and they will be discussed later, after the organic matter containing carbohydrates and proteins.
- The biopolymers undergo many alterations. They are consumed by living organisms in the environment. The biopolymer gets adsorbed and associated with the sediment. The biopolymers are degraded to form simple monomers. Monomers are the original molecules from which the biopolymers were formed earlier. Some of the monomers initiate further condensation, polymerization reactions to form high molecular weight geopolymer. Geopolymer and un-degraded biopolymer are the precursor of kerogen.

The sequence of transformations from organic biopolymer to geopolymer to humin substance to kerogen and finally to oil/gas, along with carbon residue generation, are attributed to the following four distinct geological processes:

- Diagenesis process, kerogen formation
- Catagenesis process, oil/gas generation

- Metagenesis process, gas and carbon formation
- Metamorphic process, graphite formation, sediment modification.

The four processes are continuous, one after the other. There are no sharp boundaries among them. Overlap exists in each case.

2.11   DIAGENESIS STAGE, ORGANIC MATTER TO KEROGEN

The diagenesis process is the first stage of conversion of the accumulated organic matter and sediment to sedimentary rock. The initial changes in organic matter from the water surface to the bottom water–sediment interface and down to 150 meters deep are known as diagenesis processes. In general terms, the diagenesis process is defined as 'the sum of all geo-bio-physico-chemical processes that occurred during sediments (organic and inorganic) deposition period followed by initial compaction and consolidation'. The formations of sedimentary source rock and diagenesis process are acting simultaneously and are interrelated.

Environmental conditions during the diagenesis process are low temperatures (20–60°C) and low overburden pressure. The process does not alter the form of minerals, but the consolidation of sediments takes place from one form to another. The whole environment consists of organic matter, mineral sediments (clay, mud) and aerobic/anaerobic bacteria. These components are continuously coming in and incorporated into the sediment, thus increasing the thickness of the fine-grained source rock. The conditions prevailing create an unstable, dynamic and heterogeneous system. The system tries to attain equilibrium through various geo-bio-chemical processes. Physical forces can affect the abrasion and compaction of all the components involved. The environmental conditions lead to controlled decomposition and decaying of organic matter along with the consolidation of all the sediments (minerals and organic matter). The following discussion is concerned with the fate of organic matter consisting of carbohydrates and proteins only. Three different kinds of transformation take place during the diagenesis process.

- Aerobic (oxic) bio-chemical decomposition
- Anaerobic (anoxic) bio-chemical fermentation
- Transformation due to geological factors

The above processes occur almost one after another. One process activity almost ceases before the next process sets in. However some overlapping exists.

2.11.1   Aerobic/Oxic Bio-Chemical Decomposition

Aerobic (oxic) biochemical oxidation and decomposition occurs throughout the water column, the bottom water–sediment interface and to about one meter deep into the sediment. Below this limit, no oxygen is available. Aerobic (oxic) conditions are said to exist when the oxygen content of the water is more than 1 ml/liter of water. Oxygen content less than 0.1 ml/liter of water is considered as devoid of oxygen, that is, anaerobic (anoxic) conditions. In anoxic conditions, the oxidative metabolism of organic matter is taken up by anaerobic bacteria. Oxic, sub-oxic and anoxic conditions in water and their corresponding TOC may be stated as follows:

Aerobic/oxic conditions = oxygen content of the water is more than 1.0 ml/liter = 0.05–1.0% TOC
Sub-aerobic/sub-oxic conditions = oxygen content of the water around 0.5 ml/liter = 1.0–3.0% TOC
Anaerobic/anoxic conditions = oxygen content of the water/sediment is less than 0.1 ml/liter = >3.0% TOC

2.11.1.1 Aerobic Biological Fermentation/Metabolism Process

The fermentation process is a biological process. It is the chemical breakdown of organic compounds by bacteria, yeast and other organisms. The reaction is exothermic, involving effervescence gases. A related term is the metabolism process. It describes all the biochemical reactions needed for maintaining the life of all cell and organisms. Anabolism and catabolism are two kinds of metabolism process. Anabolism synthesizes all compounds needed by a living organism. The catabolism process works by breaking the larger organic compounds into smaller ones to meet the energy requirements of the cell. The environment of aquatic bottom sediment is conducive to fermentation and metabolism processes. The environment is alkaline and muddy with availability of oxygen; full aerobic (oxic) conditions prevail. The conditions are ideal for bacterial activities as well as slower and sustained oxidation and degradation of organic matter. The macromolecules, carbohydrate and protein, are hydrolyzed and decompose into smaller molecules, forming glucose, amino acid, alcohol monomers and gases. The monomers become a source of energy for the living organisms, bacteria, fungus, algae, worms, etc., in the bottom water–sediment zone. The metabolism reactions are further catalyzed by enzymes. The enzymes are biological catalysts and produced by living organisms. Chemically enzymes are proteins. The environmental conditions in the zone resemble the rotten swamp and pond that produce marsh gases (biogases), carbon dioxide, hydrogen sulfide, hydrogen, nitrogen and carbon monoxide, oxides of nitrogen, methane and water. Through hydrolysis and decomposition reactions, the organic matter loses the reactive and unstable parts (moiety) of the macromolecules (biopolymers). The functional groups are reactive and unstable components of biopolymers. They are the first to be removed during the metabolism process. Functional groups contain hetero atoms, namely nitrogen, oxygen and sulfur atoms. At the end of aerobic decomposition, the organic matter is stripped of its most reactive hetero atoms (S, N, O). By losing the reactive groups, the remnant organic matter becomes more stable, condensed and compact. It is said that the organic matter has been transformed from biopolymer to geopolymer.

2.11.2 ANAEROBIC BIOLOGICAL FERMENTATION

Anaerobic (anoxic) fermentation is biological decomposition in the presence of anaerobic bacteria, and in the absence of free oxygen. With increasing depth to a few meters, the oxygen content of the sediment gradually decreases to a minimum and finally reaches to the zero mark. Simultaneously, aerobic bacteria also decrease with depth. Beyond a depth of 1 meter in fine-grained sediment rock, the aerobic bacteria decreases and is replaced by anaerobes. At a depth of 2 meters, only anaerobes exist. With further depth, anaerobic bacteria also decreases rapidly. The environment is devoid of oxygen and living organisms.

Earlier transformation processes of fermentation and decomposition of organic matter in the presence of aerobic/oxic conditions continue but at a much slower rate in the presence of anaerobic/anoxic conditions. The severity and extent of chemical changes are much less compared to oxic oxidation. The severe oxic oxidation leads to the complete destruction of organic matter in the atmosphere. The oxic (aerobic) environment acts harshly and tries to destroy the organic matter with the evolution of carbon dioxide gas, water vapors and the formation of ash. In aerobic subaquatic sediment, the oxidation is slow, whereas in the anaerobic subsurface, the oxidation is much slower and sustained, so the organic matter is preserved. The anaerobic process is slow but acts steadily and smoothly on organic matter. It helps to preserve the organic matter, with small alteration of chemical structure.

Carbohydrates and proteins further decompose, producing remaining unstable monomers, glucose, amino acid, alcohol and gases, through the anaerobic fermentation process (sulfate bacterial reduction). It is said that anaerobic bacteria respire on sulfate instead of oxygen, but it is not true. Actually anaerobic microorganisms use the oxygen of the sulfate ions to oxidize organic matter, producing sulfide ions.

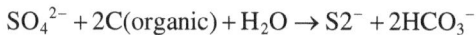

$$SO_4^{2-} + 2C(organic) + H_2O \rightarrow S2^- + 2HCO_3^-$$

The reactive sulfide (S^{2-}) ions immediately convert into stable molecules of hydrogen sulfide gas. In the presence of iron, iron sulfide is produced. Some elemental sulfur can also be found, and that may incorporate into the remaining organic matter. Hydrogen sulfide gas is extremely poisonous, but anaerobic bacteria manage to survive in the zone. Some sulfide ions diffuse through sediment into the water column and get oxidized back to sulfate ions by dissolved oxygen. Therefore, a balance is established between the consumption and supply of sulfate ions.

The rate of decomposition of organic matter by anaerobic bacteria continues to decrease with sediment depth. At a certain depth it is completely stopped. This is because of the following reasons.

- The organic matter lacks unstable parts and is uncreative.
- Lack of living organisms (bacteria) and food/nutrients in the environment.
- Primary organic matter is more condensed and insoluble.
- Incorporation of poisonous material in the condensed matter, such as sulfur and certain metals.

So far the transformation of organic matter is brought about by aerobic and anaerobic oxidation and fermentation processes. Both processes are combinations of multiple reactions, dehydration, hydrolysis, decomposition, cracking, condensation and the removal of functional groups and hetero atoms (S, O and N) under mild conditions. At the end of the diagenesis process, the degraded geopolymer (remaining organic matter) is sparingly soluble and stripped off reactive functional groups. It is a darker, compact, heterogeneous complex solid. The organic solid is given a new name, 'humic compounds', a first stage for kerogen formation in subsequent processes. Kerogen is a precursor of oil and gas generation. The total duration of bio-chemical diagenesis transformations of organic matter in the presence of aerobes and anaerobes is very short on a geological time scale.

Humic compounds can be categorized into three constituents on the basis of their solubility.

- Fulvic acid. It is an acid (low pH)-soluble portion of the humic compound. Fulvic acid is formed at an early stage of diagenesis and may support nutrients for organisms in soil.
- Humic acid. It is insoluble in acid but soluble in alkali (high pH).
- Humin substance. It is insoluble at all pH values, and survives to produce kerogen.

Fulvic acid and humic acid are not acids in the normal sense. Both are neutral solid material and become part of the soil. At the end of the anaerobic process the organic matter is fully converted to humin substance. The humin substance is transformed to kerogen due to geological factors.

2.11.3 Transformation Due to Geological Factors

After the formation of the more resistant, compact humin substance, it requires severe geological conditions to undergo transformation to kerogen. The earlier mentioned oxic and anoxic processes are also facilitated by the geological factors given below.

2.11.3.1 Water Circulation

Controlled water circulation supply is needed for the availability of oxygen in the bottom area of the basin. Oxygen supply is provided to a deeper basin by the circulation of dissolved oxygen in water. In all aquatic environments flora and fauna are formed and survive due to photosynthesis, which is possible to a certain depth. Below this limiting depth sunlight is not available and there is a shortage of oxygen, because of consumption of oxygen by respiration and the biological decomposition of organic matter. The geological design and conditions of the basin provide oxygen to greater depths. The circulation of 'oxygen dissolved water' from the surface to bottom and 'oxygen stripped water' from the bottom to the surface provides oxygen to the deeper bottom. The natural evaporation of surface water makes it more salty and denser. The denser water along with dissolved oxygen settle down and provides oxygen below in the bottom zone. The bottom water free from oxygen rises to the surface to complete the water circulation cycle.

2.11.3.2 Change in Sea Level and Transgression

Geological factors, such as tectonic movement, sea floor spreading as a result of magma movement and formation of deltas, contribute to the rise of sea level and water transgression on land. The sea transgression on land helps to build organic rich source rock and create suitable basins.

2.11.3.3 Overburden Conditions and Fossilization

The transformations are now greatly affected by the overburden of thicker sediment comparatively at high temperatures and overburden pressure. Fulvic, being less stable, is first to vanish from the humin compound. The destruction of fulvic acid is followed by that of humic acid, which takes a longer period of time for its annihilation. Fulvic acid and humic acid both are supposed to be converted into a stable humin substance with the elimination of functional groups.

The transformation of 'humin substance' to kerogen occurs at a greater depth of sediments over a longer geological time of millions of years. The humin substance is compact total organic mass. The substance is now comparatively more stable, non-reactive and insoluble. It survives and is consolidated into the rock. The compact mass of humin substance is said to be fossilized into the sedimentary rock. The rock continues to build up by incoming sediments, both mineral and organic.

The fossilized humin substance has undergone compaction, consolidation and insolubilization, according to the following sequence of reactions:

- Decarboxylation, denitrogenation and desulfurization reactions remove the oxygen, nitrogen and sulfur atoms with the release of gases. With these reactions the ratio of hydrocarbon to non-hydrocarbon content of the organic mass increases.
- Polymerization of hydrocarbon through free radicals generated by the cracking of hydrocarbons.
- Poly condensation of aromatic, naphthenic and heterocyclic rings.
- Cyclization of straight chain hydrocarbons.
- Cracking of some liable (week) compounds.
- Decomposition and dehydrogenation and dehydration of molecules.
- Solidification and insolubilization.

During the above-mentioned reaction conditions, some active non-hydrocarbon elements, such as iron, sulfur and metals, are incorporated into the fossilized organic matter (kerogen). Polycondensation and polymerization reactions lead to insolubilization. Higher molecular weight substances tend to become insoluble. The light-colored organic mass (geopolymer) is converted to form dark-colored immature kerogen. During the period small quantity of lipids are also incorporated into the kerogen. Later in the catagenesis process, the lipid molecules are released on cracking of the kerogen. At the end of the diagenesis period the organic matter is fully humin and tends to form kerogen. There is an increase of carbon content and a decrease of H/C, O/C, S/C and N/C atomic ratios in newly formed immature kerogen. At this stage hydrocarbon generation is very low; only biogenic methane and other bio gases are produced. Near the end of the diagenesis period, small quantities of lipids are also incorporated into the kerogen. Later in the catagenesis process, the lipid molecules are released by the thermal cracking of the kerogen.

The source rock is termed as young and is unfit for the mass production of petroleum hydrocarbons, so it is called 'immature rock'. The total alteration of the organic matter, at the end of diagenesis process, is summarized below:

- Organic matter → biopolymer → geopolymer → humic compound.
- Humic compound → fulvic acid + humic acid + humin substance.
- Humin substance → kerogen (immature).

Organic matter as humin substance in recent sediment is the first stage of kerogen which is not mature to generate oil and gas. The 'recent and immature kerogen' is extractable through a suitable solvent, whereas old mature kerogen rigidly lithified in rock is hardly extractable by any solvent.

2.12 CATAGENESIS PROCESS, OIL AND GAS GENERATION

The catagenesis process is the main oil and gas generation stage. The catagenesis process is a thermal transformation (temperature and overburden pressure) of immature kerogen to mature kerogen and finally to oil and gas generation. There is continuous accumulation and deposition of organic matter and sediments into the sedimentary rock, resulting in greater thickness of the sediment from millimeters to several km over geological time. Overburden pressure and temperature increase substantially to about 1500 bars and from 60 to 160°C, respectively. Under these severe conditions, the rock porosity, permeability and pore water content are greatly reduced. Underground tectonic movement is also responsible for an increase of temperature and pressure. Stress on the rock increases, and the prevailing system loses equilibrium and becomes unstable. To adjust the equilibrium position and balance the situation, a natural force, that is, the breaking of the compact mass of kerogen into liquid and gas, comes into play. The generation of liquid and gases creates more volume to counteract the stress force on the rock.

The depth and temperature of sedimentary rock at which most of the oil and gas are generated and expelled from the source rock is known as the 'oil window'. The oil window is the subsurface oil/gas producing zone in terms of depth and temperature, in a vertical direction. In most cases, the temperatures for oil windows are 60–160°C at a depth of 1.5–5.5 km. The volume of the source rock that generates and expels oil and gas is known as the 'generative depression', or 'hydrocarbon kitchen' or 'cooking kettle' of the rock. The conversions that take place in the oil window are as below:

$$\text{Kerogen} \rightarrow \text{bitumen} \rightarrow \text{oil and gas}$$

Strongly lithified kerogen (mature kerogen) contains more than 20% insoluble organic solid mass.

It is believed that the bitumen is first produced by the thermal cracking of mature kerogen. Bitumen is a dark-colored non-volatile, semi-viscous solid organic mass and is soluble in organic solvent. In the second stage of conversion, bitumen is thermally cracked to oil and gas. The thermal cracking of the kerogen takes place according to the following reaction scheme.

- Paraffin hydrocarbons in the kerogen are first to be cracked/removed to form methane, ethane propane and butane (C_1–C_4) hydrocarbons gases along with liquid (pentane plus) hydrocarbons. Straight and branched chain paraffin hydrocarbons do not exist as free molecules in the kerogen matrix. These hydrocarbons occur as side chains attached to aromatic or naphthenic rings. They are broken (cracking) under thermal stress.
- The multiple-ring structure of kerogen yields oil on cracking, containing one- to four-ringed molecules of aromatics, naphthenic and heterocyclic hydrocarbons.
- Several multi rings (5–12) of aromatics, naphthene and heterocyclic hydrocarbons are formed as asphalt bitumen on the thermal cracking of kerogen. This fraction of hydrocarbon oil contains most of the hetero atoms (oxygen, sulfur and nitrogen).
- The residual kerogen is almost free from functional groups and side chain paraffin hydrocarbons. Residual kerogen is a black solid carbon residue, devoid of any appreciable amount of hydrogen. Liable (unstable) part of the kerogen has been removed. Now the remnant kerogen is most stable and unreactive, unfit for generation of bituminous material (oil/gas). This is the completion of the catagenesis process. Under severe conditions of temperature and pressure, in the next metagenesis stage, only methane is produced.

Thermal cracking/alteration in catagenesis stage is purely a chemical reaction, under geological conditions of moderate temperature, pressure and depth of the sedimentary rock.

2.13   METAGENESIS STAGE, GAS AND CARBON FORMATION

Metagenesis is the last process associated with the study of the fate of organic matter. The metagenesis of organic kerogen occurs at a greater depth of about 7500 m and at temperatures between 200 and 250°C. With the conditions of severe temperature and pressure only dry gas (methane) is generated, and the kerogen is converted to carbon residue with very low H/C ratio of 0.4. Low H/C ratio indicates that the kerogen residue is highly cyclic and aromatized. At the end of the metagenesis process, hardly any organic matter exists. Kerogen is almost converted to pure inorganic carbon. Carbon becomes the part of mineral inorganic sediment.

2.14   METAMORPHISM, SEDIMENT MODIFICATION

Metamorphism is a geological process under greater depth and severe conditions of temperature and overburden pressure. The process is not related to petroleum; it is purely of geological interest. The metamorphism process is related to the action of heat and pressure on the minerals at greater depths (>8000 m), high overburden pressure (>10,000 psi, 700 bar) and higher temperatures (>250°C). At this depth, the inner crust is pushed close to the heated earth upper mantle. The dynamic magma from the upper mantle transfers heat and motion to the crust. The thrust at the overlying crust rock increases the severity of temperature and pressure conditions. The structure of the rock mineral undergoes drastic modification under metamorphic conditions of high temperature, high pressure and geological time. The minerals are compacted under high pressure. The void space between the rock grains fill with water (pore water), are squeezed under pressure and expel almost all the pore water. The breaking of old and the making of new minerals along with the dissolution or recrystallization of minerals take place. The destruction of old rock structures and the formation of new structures of metamorphosed rock appear. For examples, clay mineral (kaolinite) loses its pore and interlayer water and assumes a greater crystalline form as muscovite mineral. Hydrated iron oxide mineral turns into an anhydrous hematite mineral. Carbon residue progressively changes into anthracite mineral. The anthracite finally metamorphoses to graphitic carbon. Similarly other minerals evolve into new shapes and structures. Metamorphism is the thermal alteration/ modification of inorganic mineral structures at very high temperatures, pressure, greater depth and over a long geological time. Metamorphism is the evolution of sediments to metamorphic and igneous rock with depth and time.

The transformation path ways of organic matter in the diagenesis, catagenesis, metagenesis and metamorphism processes are explained in Figure 2.2. The initial transformation of organic matter mostly drawn from aquatic organisms, bacteria and algae, consisting of carbohydrate and protein biopolymers, takes place in the diagenesis stage also known as the bio gas zone or the immature kerogen zone at a depth up to 1,500 meters at 20–60°C. Biopolymers are progressively transformed to fulvic acid, humic acid, humin compounds, geopolymer, humin substance and possibly immature kerogen under various geo-bio-chemical decomposition and fermentation processes, generating bio gases. It is said that no liquid hydrocarbon is produced during this stage, or near the end of the diagenesis stage small quantities of oil with wet gases are produced. The catagenesis stage, also known as the oil/gas zone or oil/gas window or mature kerogen zone, works under a depth from 1500 to 3500 meters at 60–150°C. Oil and gas are generated by the thermal cracking of mature kerogen. The metagenesis stage (3500–5500 meters deep, 150–250°C) generates dry gas, mostly methane, by the severe thermal cracking of residue kerogen. At greater depths (5500–7500 meters) and temperatures (250°C plus), residue kerogen is transformed to graphite (carbon) and anthracite in the metamorphism stage, also known as the carbon zone or post-mature kerogen zone.

2.15 TRANSFORMATION OF LIPID GEOCHEMICAL

Lipids are natural occurring constituents of animals and plants. Lipids represent a large group of substances known as geochemical. The substances fat, oil, wax, isoprenoids, steroids, oil-soluble vitamins A, D, F and K, phospholipids and porphyrins are important lipid compounds. The first three mentioned, oils, fats and waxes, are simple lipids, whereas the others are complex lipids.

Oil, fat and wax lipids are saponifiable (neutralization) where as other lipids, such as steroids and isoprenoids, are unsaponifiable. Lipids are different in their chemical behavior compared to carbohydrates and proteins. Carbohydrates and proteins are hydrophilic (water-loving), the lipids are hydrophobic (water-hating) like petroleum oil.

Lipids resemble petroleum oil in many respects. Like petroleum oil, lipids are combustible and are used in illumination. The lipids are used as lubricant. Lipids are used as bio-fuel. They are used as a replacement for diesel fuel. Plant lipids are usually liquid and are unsaturated hydrocarbons. The animal fats are solids and saturated hydrocarbons. A saturated paraffin hydrocarbon of molecular formula $C_{21}H_{44}$, is solid at room temperature, having a melting point of 41°C, whereas an unsaturated hydrocarbon of similar molecular formula $C_{21}H_{42}$ is liquid (MP 3°C). Marine organisms use and produce unsaturated hydrocarbons (olefins). The land organic life uses and produces solid fat saturated hydrocarbons (paraffin). Both kinds of lipids, marine and land, contain variable amounts of paraffin/olefin hydrocarbons. The complex lipids are more stable and more insoluble than the simple lipids. They exist as particulate organic compounds (POC) in water. The lipid particulates settle down much earlier into the bottom sediments, without undergoing appreciable change or decomposition by aerobic and anaerobic aquatic environment conditions. Small amounts of hydrocarbon produced during the diagenesis process are attributed to lipids and not to carbohydrates and proteins. However a major portion of lipids remain unaffected during the diagenesis stage. The transformation of lipids is not complete, as that of carbohydrates and proteins in the catagenesis stage. A limited chemical transformation of lipids occurs during the catagenesis process, generating hydrocarbons n-paraffin, isoparaffin compounds from medium C_{15}, to high molecular weight C_{45}. Derivatives of steroids and isoprenoids are found in the generated oil. The remaining (residual) lipid contains small quantities of heterocyclic rings containing sulfur, oxygen, nitrogen and metal elements. Residual lipid is a resinous semi-solid resembling asphaltene/resin. The residual lipids submerge into and become a part of kerogen. The contribution of lipid organic matter in the generation of oil is very small compared to carbohydrate and protein organic matter.

One of the significant aspects of lipids is that they provide the history of the origin of crude oil. Lipid compounds are resistant to alteration and are comparatively chemically stable. On the other hand, carbohydrates and proteins undergo tremendous changes during the diagenesis and catagenesis processes. The complex carbohydrate/protein appears in crude oil with much simpler and drastically altered forms. The lipids retain the characteristics of their original organism in the generated petroleum. For example, porphyrins are formed in the green leaves and steroids present in bacteria, appearing in crude oil without much chemical change. Geochemicals are known as bio-markers used for the determination of the origin of crude oil.

Refer to Figure 2.2 to see the path way of lipid transformation. Most of the lipid compounds remain unaffected and incorporated into the kerogen without significant chemical change. Only small amounts of the compounds are altered and incorporated into the kerogen along with carbohydrates and proteins.

temperature & depth	Stage	zone	status	reaction type	Bio polymer			geoproduct
					acquatic organism bacteria, algae / carbohydrate, protein	animal, plant lipids / fat, oil, wax	terrestrial plant / cellulose, lignin	
20 - 60 °C (diagenesis, biogas zone, immature kerogen)				biological metabolism decomposition	fulvic acid humic acid → • glucose • aminoacid			→ biogases
				polycondensation	humin substance	geochemicals, fossils		
1500m				solidification insolubilization	geopolymer		stable kerogen	→ coal, dry gas
60 - 150 °C (catagenesis, oil and gas zone, mature kerogen)				thermal cracking	kerogen			→ oil and gas
3500m								
150 - 250 °C (metagenesis, gas zone, post mature kerogen)				severe thermal cracking	kerogen residue			→ dry gas
5500m								
250 °C plus (metamorphism, carbon zone, post mature kerogen)				compaction crystallization	carbon residue		anthracite	→ graphite
7500m								

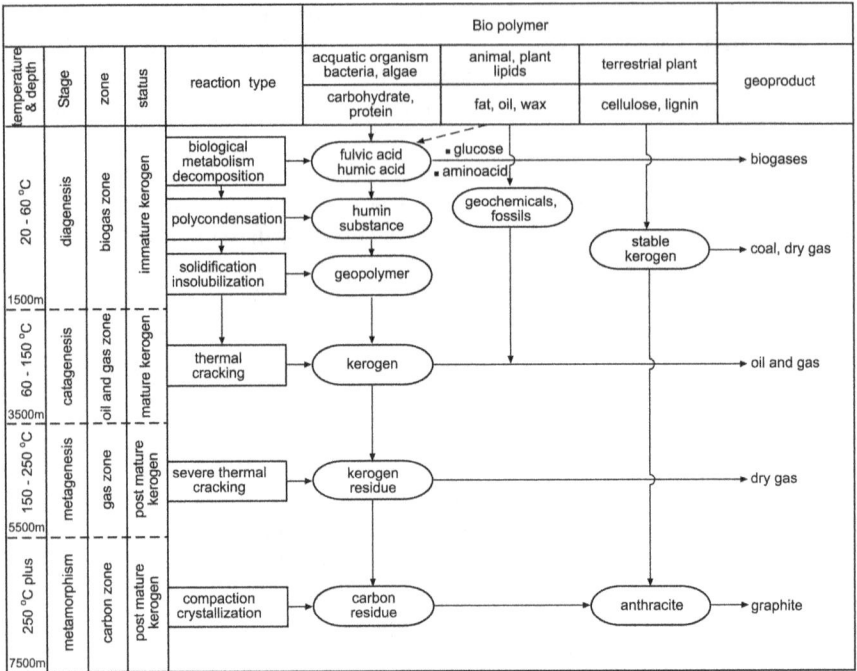

FIGURE 2.2 Transformation path way of organic matter into geo-products during diagenesis, catagenesis, metagenesis and metamorphic geological processes. Source: Modified from Tissot B.P. and Welte D.H., *Petroleum Formation and Occurrence*, Springer Verlog, New York, 1984 (Figures no II.1.1, II.1.2, II.2.10, II.3.1).

2.16 KEROGEN

Kerogen is one of the most complex macromolecules, a highly condensed mass of impermeable black solid organic substances. Kerogen invariably exists in combination with varying proportions of minerals in sedimentary source rock. The physical characteristics and chemical composition of kerogen differ considerably from one source rock to another. Kerogen is characterized on the basis of its insolubility. Contrary to the usual belief that an organic compound dissolves in organic solvents, the kerogen organic part does not dissolve in any organic solvent. The solvent molecules cannot penetrate into the kerogen matrix to dissociate the kerogen molecules in soluble form. In general, as the molecular weight of an organic mass increases, the solubility in organic solvent decreases. Kerogen is a high molecular weight organic compound, so it is not soluble in any solvent. It can be defined as follows:

Kerogen is a complex, amorphous macromolecular organic compound, consisting of poly-condensed naphthene, aromatic and heterocyclic rings all are bonded together by hetero atoms O, S, N and metal. The paraffin and iso-paraffin hydrocarbons are attached to the rings as side chains; they do not exist as independent molecules in kerogen.

All the major organic groups, paraffin, olefin, aromatics, naphthene, heterocyclic and organo-metallic compounds, are bonded together in a single macromolecule of kerogen.

2.16.1 Kerogen Type

The considerable variation of properties and chemical composition occurs in kerogen because of its varied original organic matter and subsequent several transformation processes during the diagenesis, catagenesis and metagenesis stages in different geological environments. Kerogen generates petroleum in source rock over geological time. Petroleum contains a greater hydrogen percentage than kerogen. Hydrogen-rich organic matter is likely to produce kerogen with a comparatively higher percentage of hydrogen. The hydrogen-rich kerogen will produce more oil and gas.

Kerogen is closely related to coal. Coal consists of three types of distinct microscopic organic units called macerals along with inorganic materials. 'Maceral' particles are the building block of coal similar to 'mineral' particles in rock. A maceral group of coal which is also found in kerogen consists of three constituents: liptinite, vitrinite and inertinite macerals. All three maceral groups have sub-groups as well. On the basis of hydrogen content, atomic H/C and O/C ratios and maceral group, kerogen is classified into four generic types as shown in Table 2.2.

2.16.1.1 Type 1 Kerogen

Type I is saprogenic kerogen. This type of kerogen is formed mostly in freshwater basins (lacustrine) containing mostly algae-derived organic matter. The kerogen has high H/C (>1.5) and low O/C (0.15–0.10) ratios. The produced crude oil is paraffinic and waxy in nature with few cyclic compounds. The oil is volatile and, on evaporation (500°C), leaves negligible residue. It contains a high percentage of lipid-derived straight chain paraffin hydrocarbons.

2.16.1.2 Type II Kerogen

Type II is a mixed kerogen of both saprogenic and humic organic matter. Type II kerogen is made from planktonic organic matter of marine origin as well as a small quantity of terrestrial origin. The spores, pollen and cuticles of leaves and herbaceous plants from land also contribute to kerogen formation. Both oil and gas are generated. On evaporation (500°C) of the oil, considerable asphaltic residue is obtained. The H/C is 1.3–1.1, and the C/O ratio is 0.20–0.15.

2.16.1.3 Type III Kerogen

Fibrous terrestrial plants, containing cellulose and lignin, are the main raw materials for type III kerogen. It has a low H/C (<1.0) ratio and a high O/C (0.3) ratio. The kerogen contains mostly poly-aromatic rings along with a heterocyclic nucleus. Among hetero atoms, the most abundant is oxygen in the form of ketone (=C=O), carboxylic (–COOH) and ether (–O–) functional groups. Type III kerogen is a source of gas and little or no oil.

TABLE 2.2
Generic Properties of Kerogen

	Type I	Type II	Type III	Type IV
1. Origin	Saprogenic Lacustrine (lake).	Planktonic Marine.	Humic Coal-like.	Highly oxidized. Like anthracite coal.
2. Source	Produced from fresh water algae, also from wax, resins, lipids and protein organic matter.	Phytoplankton, zooplankton, bacteria, spore, pollen, resins, cuticle, carbohydrates, proteins, lipids.	Land plant, vegetable, cellulose, lignin, terpene and phenolic compound.	Obtained from residual kerogen type III at greater burial more than 5 km depth.
3. Atomic ratio	H/C = 2.0–1.5 O/C = 0.15–0.10	H/C = 1.3–1.1 O/C = 0.20–0.15	H/C = 0.95–0.90 O/C = 0.30–0.25	H/C = <0.5 O/C = >0.30
4. Environment conditions	Anoxic anaerobic, sub aerial, sub surface.	Anoxic anaerobic sub aerial, sub surface.	Oxic aerobic sub aquatic.	High overburden pressure and temperature.
5. Chemical structure	Molecules consist of bigger paraffinic chains with few naphthenic, aromatics and heterocyclic rings.	Molecules consist of short paraffinic chains with poly condensed (more rings than type I) system.	Highly poly condensed cyclic and heterocyclic hydrocarbon, with very short chain.	Highly cyclic, condensed structure. No paraffinic side chain.
6. Linkage	Oxygen atoms are bonded as ester (–C–O–O–R)).	Ketone (=C=O) and ester (–C–O–O–R) are oxygen containing group. Sulfur in heterocyclic ring as sulfide (–S–).	Carboxylic (–COOH), ether (–O–) and ketone (=C=O) groups. Sulfur in heterocyclic ring.	Oxygen and sulfur exist in heterocyclic rings.
7. Pyrolysis at 500°C	Produced up to 80% paraffin dominated oil usually waxy, high pour point and low sulfur with few naphthenic and aromatics rings.	Yields 50–60% mixed crude oil, little gas. The crude oil is naphthenic with high sulfur, low pour point mixed oil.	Generates mostly gases and little oil.	Decomposes to yield only carbon. Little gas but no oil but dead carbon.

(Continued)

TABLE 2.2 (CONTINUED)
Generic Properties of Kerogen

	Type I	Type II	Type III	Type IV
8. Maceral group	Liptinite.	Liptinite.	Vitrinite.	Inertinite.
9. Elemental composition				
Carbon	67–70%	75–85%	90–95%	>95%
Hydrogen	10–12%	8–5%	4–2%	<2%
Oxygen	0–5%	2–1%	<0.5%	<0.03%
Nitrogen	<2%	<1%	<0.4%	<0.02%
Sulfur	Variable	Variable	Variable	Nil

2.16.1.4   Type IV Kerogen

This is the highly oxidized form of kerogen I, II and III. There is no potential for oil and gas. It may be called residual kerogen, having very low H/C (<0.5) and high O/C (>0.3) ratios. The hydrogen content of residual kerogen is reduced to the extent that it is not combustible; therefore, it is termed as 'dead carbon'. It is formed at the end of metagenesis.

2.16.2   KEROGEN TRANSFORMATION

The maximum transformation of kerogen occurs during the catagenesis stage. Maximum oil and gas are generated along with the residual kerogen. The oil and gas generation zone is the 'oil window'. The 'oil window' is defined as the range of the subsurface temperature and depth during the catagenesis process in which maximum thermal cracking (chemical reaction) of kerogen occurs and bulk oil and gas are generated. The main oil window temperature is 60–150°C at a depth of 1500–5500 meters and a time scale of around 100 million years. The oil window temperature range is not fixed. It varies with the type of kerogen and geological factors of the sub-environment. Normally the crust temperature varies 30°C/km depth. However, it can change to 50°C/km, due to basement rock uplift from volcanos and tectonics. The rate of chemical reaction increases exponentially (more and more rapid) with the increase in temperature. The range of maximum temperature during catagenesis is from 150 to 200°C.

2.16.3   KEROGEN COMPOSITION

Kerogen is a highly condensed organic solid mass. Its generic components are hardly required in the practical world. The composition of kerogen is broadly defined on the basis of solubility using acid and alkali solutions as solvents. Selective acid and alkali treatment of minerals is carried out to separate the complex kerogen mass into its simpler components. Hydrofluoric acid dissolves all the inorganic minerals content, but not the organic matter of the kerogen. Mineral-free kerogen is obtained by treating with hydrofluoric acid. Mineral-free kerogen is successively treated with sodium hydroxide and hydrochloric solution of appropriate strength to get the humin substance, humic acid and fulvic acid. These components are much simpler than kerogen. Simpler molecules are easily analyzed and elucidated for their chemical structure. Humin substance, humic acid and fulvic acid compounds are progressively formed during the early stage of the diagenesis of carbohydrates and proteins. The carbohydrates and proteins are soluble in water, and on decomposition yield simpler molecules, carbon dioxide, water, glucose and amino acid. The remaining part of carbohydrates and proteins progressively undergo poly-condensation and polymerization to, first, fulvic acid, then humic acid and humin substance (geopolymer), immature and mature kerogen. The path way of chemical transformation during the diagenesis, catagenesis and metagenesis stages can be followed by studying the relative hydrogen, carbon and oxygen elemental composition and H/C and O/C atomic ratios of the kerogen. From elemental composition and atomic ratios, the structural information of different types of kerogen can also be obtained. As the carbon content increases or H/C ratio decreases, the kerogen tends to become more condensed and cyclic. Higher hydrogen kerogen type I generates paraffinic and waxy crude oil. The carbon and hydrogen content of type II corresponds to mixed naphthenic/aromatic crude oil. The carbon and hydrogen contents of kerogen of types III and IV are close to those of coal or dead carbon. Kerogen cannot be assigned a specific chemical formula. Long paraffinic chains in type I kerogen are attached to the sides of cyclic hydrocarbons. Shorter paraffinic chains are seen in type II which has more cyclic hydrocarbons than type I. Paraffinic hydrocarbons decrease and cyclic hydrocarbons increase from type I to type VI through types II and III.

The relative quantities of elements and different hydrocarbon types and their distribution in the macromolecule determine the characteristics and type of kerogen. Type I kerogen contains one to three aromatic nuclei linked with long paraffin side chains to form one layer. Type II kerogen contains two to eight nuclei including heterocyclic hydrocarbons with medium-range paraffin side chains. Both types I and II are oil-generating kerogen. Type III is gas-generating kerogen and contains more than ten fused aromatic/naphthenic rings including heterocyclic nuclei with very short paraffin chains. Type IV kerogen is devoid of paraffin side chains; it is an almost condensed ring structure. Types III and IV kerogen are found in molecular multilayered systems. Each layer of the kerogen is linked or bridged either by alkyl groups or by functional groups mostly containing hetero atoms as follows:

- Alkyl linkage ($-CH_2-$). Both normal and iso-paraffin chains are attached to nuclei of two layers.
- Oxygen functional groups, hydroxyl ($-OH$), ketone ($=C=O$), carboxylic ($-C-O-O-H$), ester ($-C-O-O-R$), methoxy ($-OCH_3$), ether ($-O-$), sulfide ($-S-$) and disulfide ($-S-S-$), etc.

The progressive decrease of atomic ratios is correlated with the chemical structures of respective types of kerogen. The initial H/C ratio of kerogen type I is 1.5 in the diagenesis stage. The H/C ratio progressively drops to 1.3, 0.96 and 0.47 during catagenesis, metagenesis and metamorphism, converting type I and II kerogen to type III and IV kerogen.

2.17 KEROGEN VERSUS BITUMEN

The formation of kerogen is a progressive process from the surface to a depth of 5500 meters; there is no sharp start or end. Similarly, different gases and crude oil (bitumen) are gradually generated from diagenesis to catagenesis and end up in the metagenesis stage. The generations of oil/gas attain maximum intensity in the catagenesis oil window. Although bitumen and kerogen are both organic masses, they differ widely. Kerogen is a high molecular weight solid mass having more than 1000 carbon atoms (C_{1000}) per molecule, whereas bitumen is a low to moderate molecular weight hydrocarbon having carbon atoms from C_1 to C_{100}. The bitumen is a colloidal system of solid asphaltene dispersed in liquid (oil, hydrocarbon up to C_{45}) in the presence of a resin (dispersing agent). The resins are the dispersing or solubilizing agent of asphaltene in oil medium. Asphaltene is a black amorphous solid consisting of fused hydrocarbon rings. The resins consist of aromatic and heterocyclic hydrocarbons along with a small paraffin side chain. Broadly the constituents of bitumen are as follows:

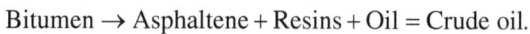

$$\text{Bitumen} \rightarrow \text{Asphaltene} + \text{Resins} + \text{Oil} = \text{Crude oil.}$$

Crude oil contains dissolved gases as well.

2.17.1 KEROGEN TYPE VERSUS GENERATED OIL QUALITY

Hydrocarbons in crude oil and as well as in kerogen are classified into four hydrocarbon groups:

- n-Paraffin hydrocarbons (n-P)
- Iso-paraffin hydrocarbons (iso-P)
- Naphthene hydrocarbons (N)
- Aromatics heterocyclic hydrocarbons (A)

These groups appear in crude oil in much simpler form than in kerogen. The four groups are interconnected by bonding between atoms and linkages involving hetero atoms in a kerogen macromolecule. The kerogen is a complex three-dimensional network macromolecule of four hydrocarbon groups. Obviously the characteristics of kerogen are related to the relative distributions and proportions of the four groups in the kerogen macromolecule.

The four hydrocarbon groups are related to the type of kerogen.

- Type I kerogen. The paraffinic hydrocarbons chains are dominant with a smaller number of cyclic hydrocarbons in type I kerogen. The same trend is reflected in the generated oil. The oil is of low molecular weight and low boiling point and waxy in nature.

- Type II kerogen. Generates mixed crude oil containing naphthene, aromatics and paraffin hydrocarbons. The kerogen molecule contains less paraffinic hydrocarbons and more cyclic compounds, which is reflected in the generated oil (crude oil) composition. On distillation the kerogen yields oil as well as asphaltic residue.
- Type III kerogen. Generates gases, mostly methane and some ethane. The little oil that is generated contains mostly aromatic components.

The quality and quantity of the generated crude oil (bitumen) depends on the relative distribution of the four hydrocarbon groups in kerogen.

2.17.2   DIAGENESIS/CATAGENESIS/METAGENESIS VERSUS GENERATED OIL

Typical quantitative distribution of generated oil/gas in diagenesis, catagenesis and metagenesis stages is shown in Figure 2.3. To quantify the conversion in all three stage is difficult. Conversion of kerogen differs considerably from one source rock to another. The quantity of the generated hydrocarbon is variable in all three stages, diagenesis, catagenesis and metagenesis. Additionally, the quantity depends on the quantity and quality of organic mass (kerogen) available during these conversion processes. The catagenesis process covers the full spectrum of oil/gas and kerogen residue. The diagenesis stage produces mainly C_1 hydrocarbon and small quantities of C_2 hydrocarbon gases. Carbon residue (dead carbon) and dry gas are mainly generated in the metagenesis stage. The transformation of kerogen and generation of different geo-products in three stages are summarized below:

Stage	Transformation	Geo-Products
Diagenesis	Organic matter (biopolymer)	Fulvic acid
	↓	Humic acid
	Insolubilization	Humin substance
	↓	Geo-polymer
	Immature kerogen	Bio and ethane gases
Catagenesis	↓	-
	Mature kerogen	Oil
	↓	Wet gas (ethane to butane)
	Thermal cracking	Dry gas (methane)
	↓	-
	Residual kerogen	-
	↓	-
Metagenesis	Severe thermal cracking	Dry gas
	↓	-
	Carbonization	-
	↓	-
	Dead carbon	-

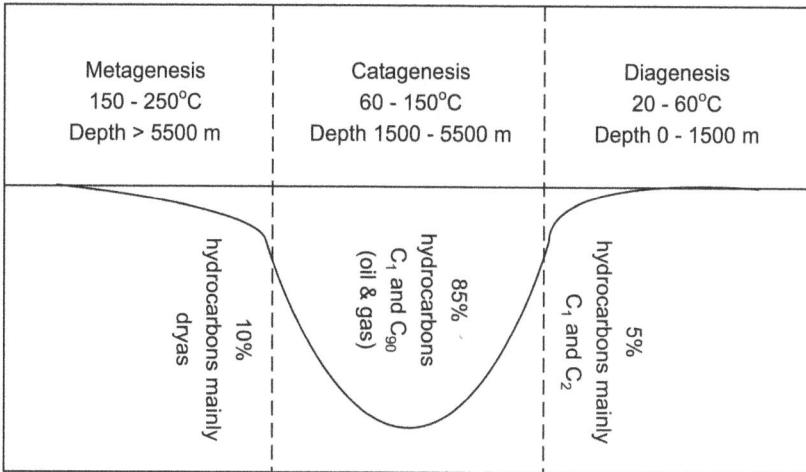

Metagenesis	Catagenesis	Diagenesis
150 - 250°C	60 - 150°C	20 - 60°C
Depth > 5500 m	Depth 1500 - 5500 m	Depth 0 - 1500 m

10% hydrocarbons mainly dryas

85% hydrocarbons C_1 and C_{90} (oil & gas)

5% hydrocarbons mainly C_1 and C_2

FIGURE 2.3 Typical quantitative distribution of generated oil/gas during diagenesis, catagenesis and metagenesis stages.

2.18 THERMAL CRACKING OF KEROGEN

The overall transformation of complex kerogen and geochemical, to simpler asphaltene, resin, oil and gases, through the thermal cracking process is shown in Figure 2.4. The cracking is the main conversion process of kerogen to oil and gas during catagenesis and metagenesis. Thermal cracking is a chemical conversion of macromolecules of kerogen into smaller molecules in the presence of heat and overburden pressure. The process is equally applied to the geochemicals that have been incorporated into the kerogen. The geochemicals are more resistant to change, so the conversion of geochemical components is far less than that of the carbohydrate and protein

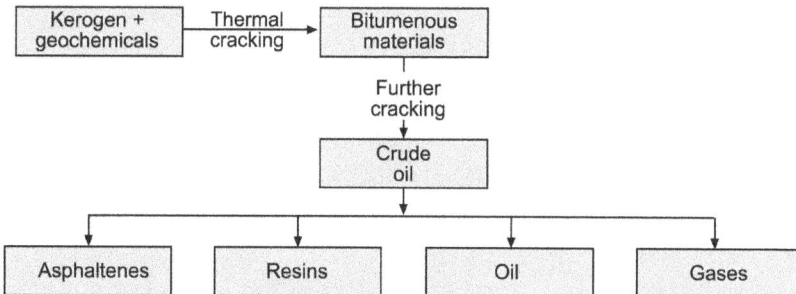

FIGURE 2.4 Kerogen and geochemical thermal cracking, conversion to bituminous material and crude oil (asphaltene, resin, oil, gas).

components of the kerogen. The majority (85%) of the hydrocarbons are generated in the catagenesis stage (within the oil window) and 10% of the hydrocarbons are generated by the cracking process under severe subsurface thermal and pressure conditions during metagenesis. Thus 95% of hydrocarbons are generated by the cracking process. It is appropriate to see some details of thermal cracking.

- Normal paraffin cracks easily. Long chains crack still more easily. The long chain cracks in the middle and forms two smaller molecules.
- Iso-paraffin cracks similarly to n-paraffin.
- Isolated naphthenic rings or aromatic rings are difficult to crack. They do not respond to cracking.
- Side n-paraffin and iso-paraffin chains in naphthene crack easily.
- Side n-paraffin and Iso-paraffin chains in aromatics crack easily to give a lower molecular weight hydrocarbon.
- The above cracking reactions involve the scission of chemical bonds, yielding fragments of lower order molecules.
- Simultaneously, cracking is accompanied by the reactions that give rise to higher molecular weight material. The higher molecules become part of the residue kerogen. Under thermal conditions free radicals are produced. Free radical and unsaturated hydrocarbons polymerize with other molecules instantaneously, producing bigger molecules. Condensations of two or more aromatics or naphthene rings produce a bigger molecule (condensed rings).

Free radical polymerization and condensation produce more stable higher molecular substances, which become part of the residual kerogen. The cracking process can be summarized as giving rise to three types of hydrocarbons (gas, oil and residue). Hetero atoms (O, N, S) in the macromolecule are progressively reduced in the forms of non-hydrocarbon gases such as hydrogen sulfide, carbon dioxide and water vapors.

Under conditions of high temperature and pressure, prevailing in catagenesis, the macromolecules in the kerogen become unstable. The heterogeneous macromolecules of the kerogen try to attain stability by splitting into smaller, simpler and more stable fragment molecules. In the same condition of temperature and pressure, smaller and simpler molecules are more stable than the macromolecules. The net result of fragmentation is in the generation of various types of lower to moderate molecular weight hydrocarbons. All the generated hydrocarbons may be called bitumen or whole crude. Bitumen is best understood by classifying it, on the basis of its physical constituents, into three types as follows.

2.18.1 ASPHALTENE

Asphaltenes are soluble in polar solvents such as carbon disulfide and insoluble in non-polar solvents such as n-pentane. Asphaltenes are amorphous black solid materials.

2.18.2 Resin

Resins are viscous liquids with yellow to brown color. They are a mixture of naphthene and aromatic rings containing hetero atoms (S, N and O). Resins are a homogeneous mixture of low to moderate molecular weight substances.

2.18.3 Oil

Oils are white to brown liquids at atmospheric conditions. Oil consists of low to moderate molecular weight saturated paraffin and naphthene hydrocarbons. Low amounts of aromatics and heterocyclic rings render the oil as colored liquid. The components of oil are gasoline, kerosene, diesel fuel and lubricating oil.

Asphaltene colloidal particles are dispersed into the oil medium. The resins act as the solubilizing agent (emulsifier). All three constituents form a homogenous solution of crude oil as bitumen including dissolved gases.

2.18.4 Gases from Cracking

The gases generated during the cracking of kerogen can be either paraffin hydrocarbons or non-hydrocarbons. All gases are dissolved in oil. Hydrogen sulfide is produced during the thermal cracking of kerogen. The more sulfur in the source rock (calcareous), the more hydrogen sulfide is produced. The reaction of free sulfur with hydrocarbons also produces hydrogen sulfide. Nitrogen gas is produced through the denitrogenation of kerogen at higher temperatures and pressure as in the metagenesis stage. Carbon dioxide is produced by the decomposition and cracking of hydrocarbons containing carboxyl, hydroxyls and carbonyl groups. Hydrocarbons containing ether (–O–) and ester (–O–O–) functional groups are inert.

2.19 GAS BEHAVIOR AND ORIGIN

There are a number of underground sources from where the gases originate, including the cracking of humic and saprogenic kerogen.

Gases behave differently than liquids and solids. Gas can expand to any limit to fill the space available. Likewise, it can be compressed to a small volume by the application of pressure. For example, a gas can expand to several times its original volume if brought to the surface from an underground, higher-overburden deposit. High pressure creates a smaller volume of gas. A gas can diffuse rapidly around, due to weaker cohesive forces among gas molecules. It is more truer for methane because of its small size. Methane is the lightest of all hydrocarbon gases and inorganic gases except helium and hydrogen gases. Methane is the major component of natural gas. Methane's high hydrogen to carbon atomic ratio (H/C = 4) makes the gas light & buoyant. Methane gas goes upward rapidly and diffuse in all directions. Therefore, gas reservoir formation with proper seal/cap rock is limited. A rock that may act as a proper seal for oil cannot do so for gas. Extremely low porosity and low permeability are needed to function as a gas seal/trap rock. The origin of gases can be traced back to many sources. A gas can be organic or inorganic. The interaction of certain inorganic minerals and elements generate hydrocarbons gases as well as non-hydrocarbon gases. The properties of gas are different from the properties of liquid; both are treated as fluids.

Natural gas is a simple mixture of a few (about ten) organic and inorganic gases. On the other hand, petroleum oil is a complex mixture of several hundred hydrocarbons compounds. The generation of gas has been dealt with earlier in the chapter along with oil.

The total generation of hydrocarbon gases from organic matter is due to the following processes:

- Biological fermentation of organic matter
- Thermal cracking of kerogen
- Thermal cracking of oil/condensate
- Thermal cracking of residual kerogen

The progressive transformations of organic matter to gases take place during formation of geopolymer, kerogen and carbon residue, in diagenesis, catagenesis and metagenesis stages. This takes place over a period of geological time with the increasing burial of organic matter into sediment. The main processes that generate various gases are as follows.

2.19.1 Diagenesis Gases

Diagenesis is the initial stage of transformation of biopolymer organic matter into simple gas molecules. Most of the transformations during diagenesis are the biodegradation and fermentation of the biopolymer carbohydrate and protein with evolution of several biogenic gases. The process continues till the end of the diagenesis zone. Biogenic gases are methane, carbon dioxide, carbon monoxide, nitrogen, hydrogen, hydrogen sulfide and ammonia. The rotten organic mass below the stagnant water column or mud is deprived of oxygen. Here the mass undergoes anaerobic (anoxic) fermentation (biological decomposition). The anaerobic fermentation process is defined as the breakdown of biodegradable rotten organic mass by microorganisms. The sequence of biodegradable process is as follows:

- Bacterial hydrolysis of the insoluble organic polymers, carbohydrates and proteins, into more soluble sugar and amino acid.
- Sugar and amino acid are consumed as food by other microorganisms.
- Acidogenic bacteria convert the remaining sugar and amino acid into carbon dioxide, hydrogen, ammonia gases, alcohol and organic acid (acetic acid).
- Methanogenic bacteria convert the organic acid into methane and carbon dioxide gases. A small quantity of ethane is also formed.
- During diagenesis, only biogenic methane hydrocarbon gas is formed. However, below a depth of 1000 meters and with an increase of temperature,

some hetero atoms bonds are progressively broken. Toward the end of diagenesis (1500 m) a small quantity of light hydrocarbon gases methane (C_1) and ethane (C_2) are produced through mild cracking of kerogen.

2.19.2 CATAGENESIS GASES

Different gases are produced during the catagenesis stage by the thermal cracking of kerogen.

2.19.2.1 Methane Gas (C_1)

Methane gas is generated in all stages by the thermal cracking of kerogen/bitumen. At the more severe conditions of temperature and pressure in the final stage of catagenesis, the carbon–carbon bonds of the remaining (residual) kerogen and bitumen are broken, creating low to moderate molecular weight oil and hydrocarbon gas, mostly methane. The final generation of methane occurs in the cracking of residual kerogen during catagenesis.

2.19.2.2 Wet Gases (C_2–C_4)

C_2–C_4 hydrocarbon are known as wet gases. At a depth greater than 1500 meters, higher temperatures and pressure disturb the thermodynamic equilibrium. At this depth oxygen and bacteria are absent. Thermodynamic equilibrium is established through the thermal cracking of macromolecules of kerogen into smaller molecules. Hydrocarbon gases are generated in the catagenesis stage throughout the oil window. The primary cracking of kerogen (types I and II) produces oil along with C_2–C_4 hydrocarbon gases.

2.19.2.3 Condensate Pentane Plus

Condensate hydrocarbons are bigger molecules (C_5 plus) than the gaseous (C_1–C_4) hydrocarbons. Condensate liquid occurs in association with gases (C_1–C_4) in a reservoir. Under pressure the liquid is in a dissolved state in gas. At the surface, condensate liquid and gases are separated. Condensate hydrocarbons are formed at greater depths and temperatures. Condensate is a product in between oil and gas on a molecular weight basis. Condensate/gases are produced by the cracking of oil within a reservoir, not through the cracking of kerogen. Conversion of vapor into liquid is condensation process. Retrograde condensation occur due to reduction of pressure in reservoir condition.

2.19.3 METAGENESIS GAS

Metagenesis gas is produced from deeper rock (5500–10,000 m). On land it is easy to draw gas, but off-shore production of metagenesis gas is an expensive exercise from deeper well (> 5000 meter). A smaller rig, erected on barge/platform, is used to cut down the cost.

Residual kerogen types I and II were stripped of their major portion of paraffin side chain hydrocarbons, during the earlier catagenesis process. Now both kerogen types (I and II) are similar to humic kerogen type III, more condensed rings with very short alkyl chains. Kerogen is unable to generate oil and wet gases (C_2–C_4) in significant quantities. Under higher temperatures (150–250°C) and higher pressure during metagenesis, the small side chains are broken, evolving large amounts of dry methane gas. The generation of methane, along with little oil, is the last dehydrogenation process of kerogen organic matter in metagenesis.

2.19.4   GAS IN SEDIMENTARY ROCK

No other gas is found in the sedimentary rock than those mentioned here; only their behavior is reiterated. Rapid gas diffusion is the most distinguished property. Small molecule size light gases diffuse much faster than the large molecular weight gases. Hydrogen, helium and methane diffuse more than other organic and inorganic gases. Gases are much more sensitive to underground geothermal variation in sedimentary rock than oil. Gas accumulation may be destroyed by geothermal slight variation. On the other hand migration and gas reservoir formation is much faster than the oil. All organic and inorganic gases form a homogeneous mixture. But in mixture they retain their characteristics; the most prominent is their partial pressure. Inorganic gases are more reactive than the hydrocarbon gases except noble inert gases. Inorganic gases interact and are distributed differently than the hydrocarbon gases in sedimentary rock.

2.19.5   GEO THERMAL AND VOLCANIC GASES

Subsurface geothermal, magma movement, volcanic activities and intrusion of igneous rock into sedimentary rock are reported to generate many gases. Several gases are found in the emission of volcanic eruptions. The volcanic gas components are methane, hydrogen, carbon dioxide, carbon monoxide, sulfur dioxide, nitrogen, hydrogen sulfide, water vapors and traces of bituminous material. Trace amounts of hydrogen chloride and hydrogen fluoride gases have also been detected.

2.19.6   PHOTOSYNTHESIS GASES

Green plant photosynthesis uses the atmospheric carbon dioxide, water and sunlight in the presence of green plant (chlorophyll) to produced carbohydrate (food) and oxygen. The carbohydrate is partially used as food by living organisms and the remaining carbohydrate is used for the development of cellulose by phytoplankton that respires at night to produce carbon dioxide gas.

2.19.7   NON-HYDROCARBON GASES FROM KEROGEN/OIL

The major reactions involving oil/gas generation are the dehydrogenation, desulfurization, denitrogenation, deoxygenation and decarboxylation of kerogen; all contribute to producing non-hydrocarbon gases such as carbon dioxide, hydrogen sulfide, nitrogen, ammonia and hydrogen gases. In addition to the hydrocarbon gases, there are a number of other factors that contribute to the mixing of non-hydrocarbons gases in the oil and gas pool of the reservoir. Examples of non-hydrocarbon gases are given below.

2.19.7.1 Hydrogen Sulfide and Sulfur

The quantities of hydrogen sulfide and sulfur derivative of hydrocarbons in a crude oil depend on the amount of sulfur in the original organic matter. This is one of the sources of the sulfur compound in crude oil. Other reasons are types of sedimentary rock. High sulfur minerals are associated with carbonate/evaporite sedimentary rocks. A high quantity of hydrogen sulfide gas is generated from the catagenesis middle stage to the end of the period by thermal cracking of sulfur-containing organic kerogen. The bacterial sulfate reduction process generates hydrogen sulfide gas during diagenesis. Underground chemical reactions involving minerals containing sulfide radical produce H_2S. Hydrogen sulfide gas is soluble in water. The gas can be reduced to elemental sulfur (S) or converted to iron sulfide (FeS). Hydrogen sulfide reacts with sub-surface oil to form organo-sulfur compounds.

Sulfur is the only element found in crude oil in elemental form. The elemental sulfur is generated by the oxidation of hydrogen sulfide in deeper rock. The element sulfur reacts with hydrocarbons in the presence of hydrogen sulfide to produced sulfur derivatives of hydrocarbons such as thiophene, thiol and organic sulfide. Elemental sulfur is found in the reservoir as molten sulfur. In the reservoir the temperatures exceed the melting point (115°C) of the sulfur. Crude oil containing a high quantity (2–5 %) of sulfur is asphaltic and aromatic in nature. It is thick and black in color.

2.19.7.2 Nitrogen Gas

Generally, nitrogen gas is associated with crude oil. It comes into oil from many sources. Nitrogen is found in sediment pore water. It originates from atmosphere. At the water–atmosphere interface, the nitrogen gets into water. From this water, nitrogen enters into sediments and crude oil. Another source of nitrogen is the oxidation of dissolved ammonia to nitrogen gas. During metagenesis, nitrogen is produced from the thermal cracking of residual kerogen.

2.19.7.3 Carbon Dioxide Gas

Carbon dioxide is found as dissolved gas in crude oil. The concentration of the gas varies widely (1–95%) in different crude oil from different fields. The wide variation is attributed to the following factors:

- Solubility
- Reactivity
- Production source

Carbon dioxide gas is readily soluble in water and forms carbonic acid. Carbonic acid can dissolve some minerals, especially carbonate minerals, readily. The gas can react with many minerals. There are various sources of generation of the gas. The generations by thermal degradation/biological decomposition of organic matter are the main sources of production. The higher the temperature, the higher the reaction rate. A reaction between carbonate mineral and clay around 160°C produces carbon dioxide in the subsurface. Geothermal and volcano eruption produces many gases including carbon dioxide. It exists in the reservoir as dissolved gas in oil.

2.19.7.4   Hydrogen Gas

Hydrogen is produced through the cracking of oil and kerogen where high thermal and pressure conditions exist. Dissolved hydrogen in oil/gas and water, from nil to 35%, has been reported, under subsurface conditions. Hydrogen is the lightest element and can diffuse more easily than the other lighter gases, helium and methane.

2.19.7.5   Helium and Argon

In only a few cases have helium and argon been found in crude oil. Both gases are generated by the disintegration of radioactive elements present in sedimentary rock. Helium soon disappears due to its fast diffusion rate. The presence of helium in oil/gas indicates current radioactivity in the near vicinity.

2.20   CONCLUSION

Plants and animals contain similar types of elements as petroleum. It is most likely that the conversion of the remains of plants and animals organic matter into oil and gas in sedimentary source rock has occurred in geological time (100–500 million years). A potential source rock consists of the remains of phytoplankton and zoo-planktons of mostly marine origin. The source rock formed by the deposition of dead big terrestrial plants produces mainly coal and gases. The contribution to fossil fuel by the remains of big animals is not worth mentioning. The quality and quantity of generated oil/gas depend on the quality/quantity and types of organic matter deposited and preserved in sedimentary source rock. A fine-grained mineral, for example clay, holds much more organic matter than any coarse-grained minerals of sedimentary rock. The transformation of organic matter to oil and gas takes place through a series of subsurface biological, physico-chemical and geological processes, under varied sub-environmental conditions. The initial transformation of organic matter, carbohydrates, proteins, lipids, called biopolymers, takes place at shallow depths, 0–1500 meters, under oxic/aerobic and anoxic/anaerobic conditions of diagenesis. Biopolymer turns into geopolymer, a more condensed and compact organic matter. Towards the end of diagenesis, geopolymer is converted to immature kerogen under conditions of higher overburden pressure and temperature. Organic matter is now fully fossilized and becomes part of the sedimentary source rock. Organic contents of source rock vary widely. A source rock containing more than 3% TOC is considered as potential source rock. Under higher temperatures (60–150°C), depth (1500–5500 meter) and overburden pressure (1500 bar), the kerogen becomes unstable. To attain stability, thermal cracking of kerogen takes place. First, the most liable parts of the kerogen organic matter are 'broken', forming bitumen during catagenesis stage. Further cracking leads to the generation of oil and gas. The unaltered portion of the kerogen, called residual kerogen, is devoid of most of its functional groups and hydrogen content. Residual kerogen has lost its ability to produce substantial quantities of oil/gas. Under conditions of higher temperatures (150–250°C) and higher pressure, during metagenesis, the residual kerogen is further cracked to yield dry gas and residual dead carbon. Lipid components of organic matter, for example, oil, fats, waxes, isoprenoids, steroids, etc., are slightly altered during diagenesis and catagenesis. The lipids appear in oil almost unaltered and are used as bio-markers for the determination of the origin of oil. Both hydrocarbon and non-hydrocarbon gases are generated under different subsurface conditions and appear at the surface along with petroleum. Major non-hydrocarbon gases associated with petroleum are carbon dioxide, nitrogen, hydrogen sulfide and hydrogen along with other minor quantities of inert gases.

The mature kerogen in the sedimentary rock generates sufficient oil/gas which is forced out of the rock by geological driving forces. This leads to subsequent oil/gas migration and accumulation in wider pore rock.

3 Petroleum Migration and Accumulation

3.1 INTRODUCTION

Petroleum is generated in sedimentary source rock and is found in reservoir rock. Source and reservoir rocks both are sedimentary rocks, but their characteristics are entirely different from each other. Source rock is fine-grained whereas reservoir rock is coarse-grained. The source rock contains low amounts of organic matter whereas a large pool of commercial quantities of oil/gas is found in reservoir rock. The lowest limit of TOC in source rock is quoted as 2.5% for practical purposes and for oil/gas prospecting. The solid organic kerogen mass is totally absent in the oil/gas pool of reservoir rock. The precursor of petroleum is solid kerogen/organic matter that has been left behind in the source rock by the migrating fluids. The absence of solid organic matter in the oil/gas reservoir provides necessary evidence that the oil/gas have been formed somewhere else other than their present accumulation (reservoir rock). Moreover solid rock and oil/gas fluid are two distinct states of matter. The former is rigid, immobile, solid lithified sedimentary rock, whereas the latter is mobile and compressible fluid. Fluid tends to move from an area of high energy levels (high pressure) to one of low energy levels and yields to expansion and contraction under mild variations of temperature or pressure. The subsurface ground is in a dynamic state; therefore, it is quite clear that oil/gas fluids have moved to a safer equilibrium zone.

Petroleum is generated in source rock, but commercial quantities are found in reservoir rock. There must have been a migration of oil/gas. The fluid is first generated in a source rock and then expelled from it, living behind solid kerogen mass. The expelled fluid takes a path via carrier rock to the reservoir rock which has proper trapping and seal rock with negligible permeability. The trap/seal/cap rock required appropriate permeability (low) to stop and prevent the movement of the fluid from the trapped reservoir.

All sedimentary rocks were formed through a series of geological processes of folding, faulting and deposition of sediments. The sedimentary source, carrier, reservoir and trap/seal rocks possess particular porosity/permeability. Petroleum migration and accumulation is affected by the characteristics of subsurface rock mainly porosity and permeability.

Oil/gas migrates from lower-depth source rock to a greater-height reservoir rock in the subsurface. The oil/gas takes a slanting upward direction which has both vertical and horizontal components. Normally the migration of fluid from source rock is very slow process; it takes geological time to form a commercial reservoir rock. It has been observed in a few instances that source rock may also function as reservoir rock.

Cracks or fractures or some other openings facilitate the flow of fluid. In such a situation source rock may be used for the production of petroleum.

Porosity and permeability are important characteristics of sedimentary rock. Generation in source rock, migration through carrier rock and accumulation in reservoir rock of the petroleum are controlled by the porosity/permeability characteristics of sedimentary rock.

3.2 POROSITY OF ROCK

The porosity and permeability of the sedimentary rock play an important role in the migration, accumulation and production of petroleum. Porosity signifies the fluid-holding capacity of reservoir rock, and permeability is related to the fluid transmission capacity of carrier rocks. Both porosity and permeability are related to each other and independent as well. Porosity represents the small void or space between the mineral grains. The grains are small mineral particles that make up all rock. A small detrital (broken) rock having appreciable porosity contains numbers of mineral (sandstone) particles/grains. The grains may be touching each other or separated to create void space (pore). The void space creates porosity. The void space is called interstitial space or pores. Clay/sand/silt or any other cementing materials partially or completely fill the void space between the grains. The cementing materials adhere to the pore wall, thus reducing or eliminating the pore space and porosity. The coating and adhering of salt with the pore wall is known as the 'cementation' process. Limestone, quartz and halite are common cementation agents. Cementation and pressure expulsion of water lead to the solidification of sediments into a hard, rigid rock. The pores of reservoir rock are occupied by fluid (oil/gas/water). In addition to the pores, there is another type of void space in the rock, known as 'cavity'. A cavity is an open space in the rock, created by geological disturbance and physical stress, such as joint fracture, solution dissolution and detrital (broken) effect. The size of cavity opening is very large compared to the pore volume. The capacity of a petroleum accumulation (reservoir) depends on the total volume of the pores and cavities, which is governed by many factors.

- The rock porosity depends on the size, shape, distribution and arrangement of grains.
- Irregular grains produce more void space. Grains with smooth surfaces come closer to each other and reduce the void.
- Less contact among grains creates more void space. Perfect spherical grains create more void space.
- More edges and corners in grains produce less void space compared to grains with fewer edges.
- Interconnected pores and cavities of higher diameter create high porosity.
- Small pores and cavities create less void space that holds the fluid inside and does not permit it to pass, due to capillary action. The rock is said to be of low permeability and bad as a reservoir.
- Introduction of filler minerals and cementing materials between the grains reduces the void space.

- A rock matrix with many types of minerals creates less pore volume than the pure or few types of minerals containing rock. Smaller size grains fill the space between larger grains.

Quantitatively porosity is defined as the fraction or percentage (%) of the volume of the pores/cavities, compared to the total volume of the rock.

$$\text{Porosity (fraction) } Ø_T = V_p \diagup V_T$$

$$\text{Total porosity(\%) } Ø_T = V_p / V_T \times 100$$

V_p = Total void space (pore + cavity)
V_T = Total volume of the rock

The volume ratio V_p/V_T is dimensionless. Pores and cavities of the rock may either be interconnected or may exist separately (discrete), not connected to each other. The discrete pores/cavities do not contain any fluid, they are empty. The empty space is called 'dead pore'. Dead pores do not contribute to porosity or permeability. Therefore effective porosity ($Ø_E$) is defined as follows.

$$Ø_E = \text{Total porosity } Ø_T - \text{dead porosity } (Ø_D)$$

- Macro width pore = >50 μm.
- Meso width pore = 2–50 μm.
- Micro width pore = 0.8–2 μm.
- Ultra micro width pore = <0.8 μm.

A cavity is a wide pore and may be more or less than 1 cm in size. Porosity is related to the density of rock, overburden pressure, burial depth and consolidation. Generally, a higher density (compaction) results in less porosity. The porosity of the rock is measured by a number of techniques. The simplest method involves the examination of the 'well cuttings' brought to the surface during drilling. The cuttings are examined through a microscope. A 'rock core' cylindrical sample, measuring 2.5–3.0 cm in diameter and 2.5–8.0 cm long, is used to measure the porosity by an instrument called a 'porosimeter'. Other methods for porosity determination are neutron, density and sonic-velocity wire line logging methods. Typical porosity values encountered in oil/ migration are as follows.

0–5% porosity = insignificant
5–10% porosity = poor
10–15% porosity = fair
15–20% porosity = good
20–30% porosity = excellent

Porosity continues to change with burial depth, overburden pressure and time. It is not static. Knowledge of porosity is applied in deciding the drilling site in addition to other geological data.

- Porosity is created by the gradual deposition of the heterogeneous sediments and organic matter.
- Dissolutions of components of minerals in the sedimentary rock by water.
- Fissure and fracture due to overburden pressure.
- Fissure and fracture due to geological process.
- The consolidation and compaction of the sediments with time and continued pressure build up reduce pore volume and bring about new arrangements of sediments, breakage of rock particles and expulsion of oil/gas and water from rock pore. The rock gets more compressed and dense, and pore volume, porosity and permeability are reduced.
- The porosity reduction is observed in every type of rock with burial depth, but the extent of porosity reduction is different with different types of rock. For example sandstone rock exhibits less porosity reduction and the porosity of shale rock is reduced drastically with increasing depth.
- Porosity alteration undergoes three distinct phases during burial. In the first phase, from the surface to a comparatively shallow depth (0–1000 meters), the porosity decreases rapidly; the porosity decreases slowly in the second phase between a depth of 1000 to 5000 meters. The third phase is the compact zone below 5000 meters deep, where porosity is reduced to a minimum and there exists little void space or there is extremely little space among the fine grains of the rock, indicating rock compaction is achieved.
- The same porosity treatment is applied to density increment with depth. The first phase witnesses a rapid increase in the density of the rock. Density increases slowly in the second phase. The density assumes a constant value (almost 3.0) in third phase below 5000 meters, indicating compaction is achieved.
- Gas expands and compresses instantaneously with temperature and pressure. The gas needs less volume under pressure to accumulate a large volume of gas compared to oil. The gas has the ability to diffuse rapidly. Gas can be produced from deeper tight rock with extremely low porosity and permeability by slight fracturing.
- Porosity is used to decide on a drilling site. It is reported that the minimum porosity requirement for drilling in a sandstone formation is about 10%. Limestone containing fissures and fractures needs lower porosity (5%) for drilling.

3.3   PERMEABILITY OF ROCK

Permeability plays an important role in the subsurface migration pathway and production of petroleum from reservoirs. Permeability controls the fluid flow through rock. Permeability is the ability of rock to resist or transmit the fluid. Many factors contribute to the permeability. One of the main factors is related to the effective porosity and not the total porosity. Large effective porosity means more permeability. Rock fractures also add to porosity and permeability. Other common geological factors that affect permeability are rock pore size and structure, pore saturation with fluid, composition, type, properties and phase of the fluid, mineral features and fluid density. Heavier fluid finds it difficult to flow. One-phase fluid flow is expected to be simpler than two- or multi-phase flow. One-phase flow occurs when the reservoir is filled with one fluid; accordingly permeability is simple. When the rock is filled with two or more than two fluids the transmission of the fluid through rock is hindered and permeability is reduced. Geo-physical parameters like the temperature and pressure of the subsurface also affect the permeability of rock. All these factors control the permeability and act interdependently.

Quantitatively, permeability is defined by Darcy's law:

One Darcy unit (d) would transmit one cubic centimeter (cm³) of fluid per second having viscosity of one centipoise (cP) through one square cm cross sectional area (cm²) of the rock under a pressure one atmosphere per cm of thickness (atm /cm) along the direction of fluid flow.

Mathematically the law is expressed as follows:

$$1 \text{ darcy (d)} = \{(cm^3/sec) \times (cP)\} / \{(cm^2) \times (atmos/cm)\}$$

One Darcy means high permeability. Most rocks exhibit much less permeability. Therefore permeability is expressed in mili-darcy (md).

$$\text{One darcy unit (d)} = 1000 \text{ milidarcy (md)}$$

$$\text{One milidarcy (md)} = 0.001 \text{ (d)}$$

The classification of rock on the basis of permeability is given below.

- Permeability less than 10 md renders the rock impervious. An example is trap rock.
- Permeability between 10 and 10,000 md: semi-pervious carrier and reservoir rock.
- Permeability more than 10,000 md: pervious, mostly fractured rock.

The permeability of compact clay sedimentary rock is small (less than 10 md). It is least permeable and impervious rock. The permeability of semi-pervious sandstone sedimentary rock is between 10 and 10,000 md. Gravel (round pieces of stone), of more than 10,000 md, is the most permeable and pervious rock. Due to wide differences in permeability, fluids travel different distances in different rock. Fluid in clay rock may travel a distance of 1 meter in 30 years, whereas a sandstone rock (semi-permeable) may facilitate the fluid to migrate a distance of 100 meters in 30 years. The permeability of sandstone rock is 100 times more than that of clay rock.

3.4 POROSITY VERSUS PERMEABILITY

Generally, it is said the higher the porosity higher the permeability. The permeability increases linearly with porosity. But it is not always true. Permeability is controlled by many factors other than porosity as well. The type of porosity (effective porosity) is more important than the total porosity. Grain size and pore shape are the other main factors. The pores in the rock are either interconnected or exist separately. The pores are interconnected through small channels between the grains. The small channels are known as 'pore throats'. The throats have very small diameters. Narrow throat channels offer hindrance to the fluid flow. Narrow pore throats are more common in rocks containing small sized particles. Some factors are more effective for a particular rock than others.

The porosity of shale rock may be high but its permeability is very low. The characteristics of shale are determined mostly by the sizes of the constituents, not by the nature of minerals. The mineral size goes on changing along with permeability because of fluid flow, hydrostatic pressure inside the pore, overburden litho-logical pressure and environmental condition.

Rock fracture plays a significant role in porosity and permeability. A fractured rock can be a reservoir rock. Fine-grained fractured shale rock acts as reservoir rock although the permeability is low. Limestone is of comparatively low porosity but fairly good permeability, because of un-consolidation and less compactness of the minerals and presence of fractures. Similarly fractured basement metamorphic and igneous rock acts as reservoir rock. Without fracture it will not be reservoir. Actually igneous and metamorphic rock that has uplifted and intruded into sedimentary rock functions as reservoir rock. The basement structures acting as reservoirs rock are rare. Porous course-grained sandstones rock is most common and possesses suitable permeability. The rock acts as reservoir rock even without a fracture. The rock has large pore throat and variable porosity between 10 and 30%. It forms good oil/gas permeable reservoir rock.

3.5   PETROLEUM SYSTEM

The petroleum system resembles a chemical factory as illustrated in Figure 3.1.

The whole underground system from generation in source rock to accumulation in trapped reservoir rock via carrier rock is an elaborate 'petroleum system'. The system consists of four components and three processes as follows.

The system components are:

- Sedimentary source rock
- Sedimentary carrier rock
- Sedimentary reservoir rock
- Sedimentary trap/seal/cap rock

The system processes are:

- Generation of hydrocarbon (already described in Chapter 2)
- Migration (from source rock to reservoir rock through carrier bed rock)
- Accumulation (oil/gas in reservoir rock suitably trapped by seal/cap rock)

Petroleum system

	Generation	Migration	Accumulation
Geological elements and processes	↑	↑	↑
	Source rock	Carrier bed	Reservior, trap, seal, cap rocks

In put
Feed stock → Raw material

Chemical factory

	Reaction kettle	Transport	Tank farm
Chemical elements and processes	↓	↓	↓
	Production	Supply	Storage

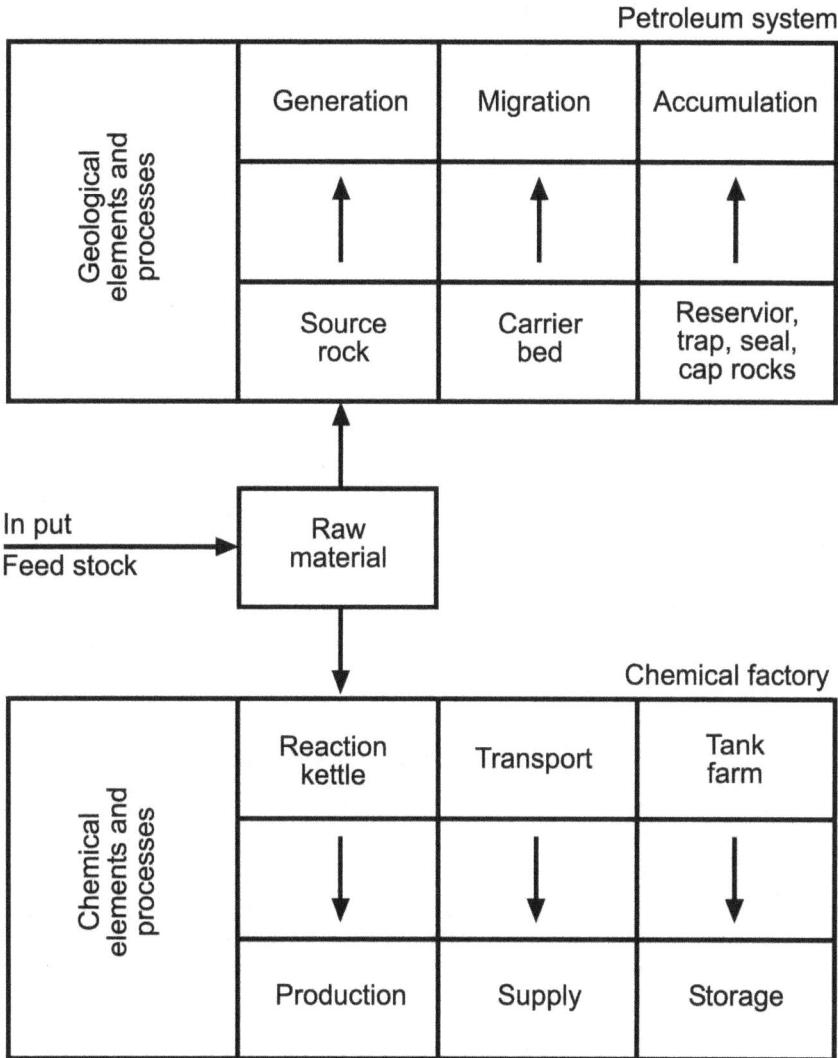

FIGURE 3.1   Chemical factory versus the petroleum system.

The above-mentioned seven elements are known as the underground 'petroleum system'. The factory is the source rock that manufactures the product (oil and gas), utilizing organic matter as feed stock. The oil/gas product is transported via carrier rock to bulk storage (reservoir rock) for commercial exploitation (drilling, production and sale). Petroleum generation has been described in Chapter 2, in detail. The petroleum field has been included as the eighth element of the petroleum system, in the coming discussion. The description of the petroleum system is as follows.

3.6   SEDIMENTARY ROCK

Sedimentary rock is composed of inorganic sediments and organic matter. Both sediments and organic matter are of varied kinds and sizes. Oil/gas on earth is found in selected sedimentary rock.

Sedimentary rock is classified into three categories according to the content of the dominant mineral as follows:

- Sandstone sedimentary rock
- Limestone (carbonate) sedimentary rock
- Shale sedimentary rock

Another type of sedimentary rock is composed of the comparatively unstable sulfate-halite-evaporite-anhydrite minerals group. Although calcareous and siliceous sedimentary rock is classified as carbonate rock, it has special characteristics owing to its biological origin.

Sedimentary rock is not a one-component (mineral) system but is a conglomerate of various types of sediments. There is no clear demarcation in the three categories of sedimentary rocks. The dominant sediment is associated with other types of sediment in variable proportions and sizes. For example limestone sediments are associated with shale sediments. Similarly, limestone minerals are mixed with sulfate and evaporite sediments.

3.6.1   RELATIVE DISTRIBUTION OF THREE MAJOR MINERALS

The relative distribution of the three major categories of sedimentary rocks on the earth's crust is given as follows:

- Shale sedimentary rock = 41% (forms good source rock but not reservoir rock)
- Sandstone sedimentary rock = 37% (forms good reservoir rock)
- Carbonate sedimentary rock = 21% (forms large reservoir rock)
- Other rocks = 1%

It may be noted that shale rock is the most abundant but it does not act as reservoir rock for the accumulation of conventional petroleum. Although the sandstone is more abundant than the carbonate reservoir rock, it produces less oil/gas as stated below:

- Oil/gas produced from carbonate rock = 62%
- Oil/gas produced from sandstone rock = 36%
- Other sources = 2%

The distribution of important minerals in the three well-known sedimentary rocks is tabulated in Table 3.1.

A description of the three common sedimentary rocks and calcareous and siliceous sedimentary rock is given to see which one is suitable as petroleum source rock.

TABLE 3.1

Distribution of Minerals in Three Sedimentary Rocks

	Minerals	Sedimentary Rock	Sandstone Rock	Limestone Rock
1	Clay/mud/shale	64%	5%	5%
2	Sand/quartz	22%	65%	20%
3	Feldspar	8%	20%	15%
4	Limestone	3%	5%	55%
5	Other	3%	5%	5%

3.7 PETROLIFEROUS SEDIMENTARY SOURCE ROCK

Petroliferous sedimentary rock (yielding petroleum) is associated with the generation, migration and accumulation of petroleum.

Source rock is the first element of the underground petroleum system. The petroleum sedimentary source rock contains oil/gas-prone organic (saprogenic) or gas-prone (humic) organic matter. A sedimentary rock containing enough organic matter is known as source rock. Shale rock is a conglomerate of big and small sediments. The sizes of sediments range from as big as a pebble (4–64 mm) to as small as clay particles (0.004 mm). The small sediments help to retain organic matter for longer periods of time in a changing geological environment. Sediments and organic matter are deposited and preserved. The source rock and its constituents (organic and inorganic) establish themselves in the changing subsurface thermodynamic conditions by bringing about necessary chemical changes in the organic matter, that is the conversion of organic matter into oil/gas under geological conditions and time.

3.7.1 SANDSTONE SEDIMENTARY SOURCE ROCK

Quartz/sand are the dominant sediments followed by feldspar mineral in sandstone sedimentary rock. Clay and limestone minerals are associated in minor quantity. Sandstone sediment is composed of sand grains (0.06–2 mm). It is not a conglomerate of big and small sized sediments as the shale rock. The grains are naturally cemented together. The rock has appreciable porosity and permeability. These types of sediments rarely accumulate enough organic material for the generation of commercial quantities of oil/gas. Sandstone sediments have very limited scope as source rock; small amounts of paraffin hydrocarbons have been generated by this kind of rock. Sandstone minerals form course-grained rock and are suitable for oil/gas accumulation formation.

3.7.2 LIMESTONE (CARBONATE) SEDIMENTARY SOURCE ROCK

Limestone rock is not suitable as source rock. Lime is the oxide and hydroxide of calcium whereas limestone is the carbonate of calcium. Calcite ($CaCO_3$) and dolomite {$CaMg(CO_3)_2$} are the principle and dominant minerals in this type of rock, it is mixed with quartz minerals. Chalk is another form of calcite; it is a soft white porous mineral. There is great diversity of grain sizes and shapes compared to sandstone. The carbonate matrix does not facilitate the accumulation of organic matter and formation of source rock; the organic matter soon drains out of the rock, before the thermal maturation is established. However due to variation of grain size from very fine to large size and mixing with sulfate and evaporite sediments, some organic matter is retained in the limestone sedimentary rock for the generation of oil/gas. Carbonate mineral forms course-grained rock and is suitable for the formation of oil/gas reservoir.

3.7.3 SHALE SEDIMENTARY SOURCE ROCK

Shale rock is the main type of source rock. About 75% of the earth's source rock belongs to shale rock. The rock is also known as argillaceous rock. The shale rock is more effective as source rock than any other mineral rock (limestone or sandstone). The basic and dominant mineral of the source rock is shale (clay). Shale is very small fine-grained clastic sedimentary rock. The typical mineral composition of the shale rock is 64% clay mud, 22% sand/quartz, 8% feldspar, 3% carbonate minerals and variable quantities of organic matter and other. Mud, mudstone, shale and alumina are different forms of clay. Mud is water-saturated shale (clay). With the elapse of geological time and increased burial, the shale loses water. Pores shrink, and the compactness of the shale increases gradually. In addition to these minerals, there are others that, when mixed with shale rock, produce good source rock. For example, shale sediment associated with a sufficient quantity of limestone sediments makes a promising source rock.

Low porosity and fine-grained minerals form good source rock. Shale sedimentary rock has such characteristics so it forms good source rock. Clay has the needed adsorption capacity to hold the organic matter till thermal maturation. Shale rocks are normally located in deeper parts of the sedimentary rock. It provides sufficient overburden pressure and thermal conditions to the organic matter. All of these are favorable factors to preserve the organic matter over long geological time for conversion to oil/gas.

The color of the shale is an indicator of potential source rock. The darker the shale, the higher the organic matter. Also, with age, the color changes from brown to light black and finally complete black. Blue and green mudstones have also been found.

3.7.4 CALCAREOUS AND SILICEOUS SEDIMENTARY SOURCE ROCK

Calcareous and siliceous sedimentary rock are also known as carbonate rock and coral rock. Coral is hard stony material formed by the deposition of the secretion and remains of the skeletons of calcareous and siliceous marine fauna organisms in an anoxic/anaerobic subsurface environment. The skeletons of the fauna organisms are made of calcareous (calcium carbonate) siliceous (quartz/silica) materials. Rock composed of primarily calcareous and siliceous minerals may be categorized as the fourth type of rock. This type of rock often mixes with iron and manganese sulfate, phosphate and sulfide minerals. Coral deposits are found in various colors at the bottom of the sea with appreciable porosity/permeability. They do not form source rock, but excellent reservoir rock is known from these types of sediments.

3.8   PETROLIFEROUS SEDIMENTARY RESERVOIR ROCK

Reservoir rock is an underground natural accumulation (pool) of hydrocarbons. It is sedimentary rock that can yield a commercial quality of oil/gas during production. Any other oil/gas accumulation in the sedimentary rock that cannot produce commercial amounts of oil/gas is not classified as reservoir rock. In a few cases, fractured, weathered, basement igneous and metamorphic rocks may form reservoir rock. The generated oil/gas from the source rock migrates through the pores of the permeable carrier rock. Carrier rock has similar characteristics as reservoir rock, but has no seal and trap rocks.

Sedimentary rock consists of three fundamental minerals: sandstone, limestone and shale. The three types of minerals may form a petroleum reservoir rock or may not. There are some basic requirements to fulfill, so that a sedimentary rock may act as a reservoir rock. The reservoir rock is a coarse-grained, porous or fractured formation, where the oil/gas pool is trapped by low-permeability seal rock.

Sandstone and limestone are the common minerals for forming a reservoir rock among the three types of sedimentary rocks. In connection with reservoir formation, the three basic rock-forming minerals are evaluated differently:

- Sandstone is characterized by the size of the grain.
- Limestone is characterized by chemical composition.
- Shale is characterized by fissure and fracture.

Often the limestone rock is mixed with sandstone and shale, whereas sandstone and shale rocks contain very minute amounts of limestone. Major reservoir rock minerals are sandstone or limestone sediments. Both types of rocks show a lot of variation in properties such as geometry of the grain (texture) and variety of depositional environments. The modes of formation and deposition of the two mineral types of the reservoir rocks are different. The sandstone/limestone rock with small permeability is known as 'tight rock'. The majority of the earth's reservoir rocks are composed of sandstone and limestone minerals, constituting about 90% of the total. But not all sandstone and limestone rocks are reservoir. That is elaborated below.

3.8.1   SANDSTONE SEDIMENTARY RESERVOIR ROCK

Sandstone consists of sand particles bonded together by cementing mineral. Sand is a crushed form of quartz, which is produced due to the weathering of rock. Quartz is a polymeric form of silica (SiO_2). Sand particles are hard, stable and insoluble without any specific crystalline structure and without any specific geometrical shape. Sand is a clastic particle that can be of any size between 0.06 and

2 mm in diameter. The thickness, size and area of sand reservoir rock are variable. Obviously factors like the porosity–permeability relationship, sand particle size and geometry, time and depth of sediment burials determine the quality and characteristics of the reservoir. Sandstone reservoirs are formed by the accumulation of sand particles in different environments. The sandstone sediment is formed by the erosion of rocks, and brought by blowing wind, tidal wave and water current. Therefore, environments of sand rock may be aeolian area (wind), fluvial land (river), sea coastal and deltaic regions.

3.8.1.1   Aeolian (Wind) Sandstone Reservoir Rock

Aeolian sandstone reservoirs are found in wind zone areas of desert and coastal belts. Aeolian sandstone rock formed by wind contains the maximum amount of sand particles compared to other types of sandstone reservoir rock. The wind can carry sand particles from one place to another far off or to a very short distance. The wind can hold up only small particles of sand less than 0.06 mm in size. When the wind carrying sand particles slows down or stops, the sand particles are released and deposited on land, forming sand dunes. The sand dune is a hillock of well-sorted fine-grained sand particles. The geometry and size of dune are controlled by the speed and direction of the blowing wind which depend on the weather conditions. Different layers (cross bed) with specific geometrical shapes of sand are formed in the dune, due to variation of wind velocity. The layers are flat (horizontal) at the bottom and slanting towards the top forming the crest (peak). The crest of dune is either parallel to the wind direction (longitudinal dune) or perpendicular to the wind direction (transverse dune) or a combination rounded shape.

3.8.1.2   Fluvial (River) Sandstone Reservoir

Fluvial sand reservoirs are formed by the transportation, deposition and erosion of sandstone sediments along river banks and bottoms. Land and island between flowing water streams are favorable locations for reservoirs. A river does not flow in a straight line: it changes its course with time. Water channels are formed when the main river carrying sediments is slowed down. The loaded sediments in the flowing river water settle down, forming deposits of sandstone. The deposit appears as an island or as a shallow part of the river or as mounds forming a sandstone reservoir. Subsequently the clay/shale minerals are deposited on the mound that forms a seal rock at the top of sandstone reservoir rock.

3.8.1.3   Coastal Sandstone Reservoir

Marine coastal areas provide a good example of sandstone deposits. Sea waves advancing toward the banks brings heavy course sediments with it and leaves the heavy course-grained sand sediment on the beach. The advancing wave brings a lot of fresh sandstone. The receding water wave takes back the lightest and fine sediment of silt and clay. The clay and silt size sediments are taken by the receding wave and deposited into deep water. Repeated to-and-fro motion of waves produce oil/gas sandstone reservoir in the coastal area. Deposition of clay/silt on top of the reservoir forms trap rock.

3.8.1.4 Deltaic Sandstone Reservoir

Deltaic regions are large deposits of sandstone (reservoir rock) and shale (source rock) sediments. At the delta both the deposition and erosion of sediments occur due to three distinct sedimentary phenomena, sea wave, sea tide and river flow. The delta is formed at the discharge point of a faster flowing river into comparatively slower or stagnant sea water. A flowing river brings a lot of sediments (organic and inorganic) from high land in the delta region. At the delta the river slows down and offloads and deposits the sediments. The delta being flat land and a low-lying area, the slow-moving river is divided into several channels called 'distributaries', creating marsh land between them (distributaries). Low sea tide facilitates the deposition and appearance of land. During high tide or turbulent waves, the marshy land is submerged with water and erodes the sediments. The shape of a delta depends on the relative abundance of deposition by river and erosion by sea wave/tide. When more sediments are deposited than eroded, mounds of sediments are formed often extending into the sea. When more sediments are eroded, new water distributaries and marsh lands are formed. The deltaic area is in a dynamic state. The sediment deposition by river and erosion by advancing sea waves continue to affect the delta and change the shape of region. A delta is a promising area for the formation of petroleum reservoir rock.

On the other hand, if the area is rich in shale sediments, it is suitable for the formation of source rock. Rivers bring with them enormous amounts of organic matter and discharge them in the nearby area of ocean and land. The land and sea bottom become a bloom for flora and fauna and organic matter. The presence of organic matter ultimately leads to the formation of petroleum source rock.

3.8.2 CARBONATE SEDIMENTARY RESERVOIR ROCK

A sedimentary rock having more than 50% carbonate mineral is known as carbonate sedimentary rock. The carbonate reservoir is formed by different methods having a wide range of porosity and permeability:

- Clastic carbonate particles. Broken rock particles called clastic particles. The particles are derived from 'clastic sedimentary rock' or 'detrital sedimentary rock'. Clastic or detritus are broken stone pieces of all kinds. Erosion, abrasion and weathering are responsible for making rock clastic particles and transporting them to far or near places for deposition and formation of carbonate reservoirs.
- Chemical precipitated carbonate. The rock is formed by chemical precipitation or available as naturally occurring mineral. Precipitated calcium carbonate mineral is formed by the interaction of carbonic acid with calcium soluble salt in shallow warm water.
- Organism carbonate. Calcareous (containing calcium) and siliceous (containing silicon) materials of biological origin composed of calcite ($CaCO_3$) and chert/silica (SiO_2) minerals contribute to the formation of organic carbonate sedimentary rock. Calcium carbonate and silica are the constituents of calcareous and siliceous shelled marine organisms (zooplankton). The

remains of the dead organisms are deposited in the aquatic bottom, forming sedimentary overburden. With the passage of geological time, it is converted to a calcite and chert sedimentary rock.

- Although the number of limestone rocks is much less than sandstone rock in the world, limestone rock produces more oil (62%) compared to sandstone rock (36%). The world's largest petroleum field is of limestone origin.
- The porosity and permeability of carbonate is excellent. Carbonate minerals form good reservoir rock for oil and gas.

The following different types of carbonate reservoir rock are identified.

3.8.2.1   Coral Reef Reservoir

A coral reef is a ridge of rock formed by the deposition of coral. The coral is hard stony material formed by the deposition of the secretion and remains of skeletons of calcareous and siliceous marine fauna organisms. Coral is found in various colors at the bottom of the sea with enough porosity and permeability to hold oil/gas. If the coral reef (rock) is properly capped by overhead shale rock it forms a good coral reef reservoir. Generally coral reefs are of particular geometry and form three types of reservoirs as follows:

- The first type are **barrier (hindrance) coral reef reservoirs** running parallel to the coast line. The reef separates the basin into two channels of shallow water.
- The second type, **atoll (ring) reef reservoir**, is a ring-shaped island formed by coral in shallow sea water. The periphery and center of the atoll reef are occupied by shallow sea water.
- The third type, **pinnacle (peak) reef reservoir**, is a tall cylindrical mountain of calcareous and siliceous minerals (coral) surrounded by aquatic basin. Normally a pinnacle reef is a small reservoir but greatly productive for petroleum.

3.8.2.2   Carbonate Flat Platform Reservoir

A platform is the deposit of limestone in a large flat area (topographic relief) submerged under shallow water, surrounded by a deeper marine basin. The warm waters in tropical locations are saturated with calcium carbonate salt. As water flows during high tide over the platform, calcium carbonate precipitates out and forms deposit.

3.8.2.3   Carbonate/Coral Rimmed Platform Reservoir

A rimmed carbonate reservoir is a flat bottom platform with a rim shaped deposits of calcareous and siliceous minerals at a shallow depth of marine water. The sediments are deposited along the edge of the platform reservoir. The reservoir is formed by the skeleton/shell sediments of marine origin (calcareous and siliceous remains of fauna organisms) along with precipitated calcium carbonate sediments.

3.8.2.4   Karsts Limestone Reservoir

Highly soluble limestone is known as 'karsts limestone'. Limestone is highly soluble in warm water in the presence of carbonic acid (H_2CO_3). Dissolved carbon dioxide gas in water forms carbonic acid. The dissolved karsts limestone is precipitated out under proper conditions. Karsts limestone, after precipitation and settling, forms reservoir rock of excellent porosity and permeability.

3.8.2.5   Chalk Reservoir

Chalk is a soft mineral, white in color, composed of calcium carbonate and derived from the microfossil shells of dead foraminifera (zooplankton) fauna and cocco-lithophores (phytoplankton) flora; both organisms grow in tropical seas. The dead organisms settle down and form a chalk deposit.

3.8.2.6   Dolomite Reservoir

Dolomite is a double carbonate mineral of calcium and magnesium. It is also called magnesium limestone [$CaMg(CO_3)_2$]. Dolomite is formed, by circulating water, rich in magnesium ions, by the exchange of calcium ions with magnesium ions in limestone rock. Limestone and dolomite are similar in color, shape and crystalline structure. Dolomite is a little bit harder than limestone. Its porosity is greater, and it is more stable than limestone under pressure and compaction. Therefore, the dolomite reservoir exhibits better porosity and permeability than limestone under varied conditions of overburden pressure.

3.9   SEDIMENTARY CARRIER ROCK/CARRIER BED

Petroleum carrier rock or carrier bed is course-grained, porous and permeable rock composed of sandstone and limestone minerals. The minerals and characteristics of carrier rock are the same as those of reservoir rock. Carrier rock is of wider porosity and greater permeability than the source rock but of the same values as reservoir rock. The hydrocarbon expelled from the source rock finds a wider pore and higher permeability in carrier rock; therefore, the fluid gets a pathway to migrate to a suitable accumulation (reservoir). The migrating path of oil and gas is through carrier rock to the accumulation.

   The flow of oil and gas from the source rock is mostly in the upward inclined direction. In this way the fluid covered long distances from the origin source rock. In addition to the normal permeability of the rock, a fluid carrier bed may also be fractured, fissured, jointed or faulted rock. The carrier bed is actually a reservoir rock but has not encountered a trap rock. The reservoir rock is intercepted by trap rock; the migration of fluid is terminated and a petroleum pool is formed. Fluid migrates underground hundreds of kilometers before it gets trapped and forms accumulation.

3.10   SEDIMENTARY TRAP ROCK

The trap rock is defined as 'The impervious and impermeable sedimentary rock which acts as seal and stops the migration of fluid in carrier bed and forms an oil and gas accumulation'. After expulsion from the source rock, the oil/gas migrates in a slanting upward direction in the carrier bed rock. If the migrating fluid does not encounter any hindrance in the form of trap rock, it continues to flow and appears at the earth surface as seep.

The pore, cavity and crack of sedimentary rock are filled with gas, oil and water. The fluid is in a dynamic state under the effects of pressure and gravity. The fluid follows a migration pathway in the carrier rock till it encounters a trapping rock. Whenever the flowing fluid encounters any hindrance in the form of impermeable rock, the fluid within the rock void space is trapped and forms a petroleum pool (reservoir). The trapped fluid, because of differences in gravity, separates into gas, oil and water. Gas is the lightest fluid and occupies the upper part of the trap, so the trap is called the gas cap. The water is the heaviest and settles down in the bottom of the trapped rock. The oil takes the space between gas and water and forms the oil pool.

A petroleum trap is a sedimentary rock mainly composed of fine-grained shale minerals with low porosity and permeability, overlaying the reservoir rock. Some traps are also formed by evaporite/halite minerals. A trap rock is usually the highest point in the reservoir rock. The fluid is prevented from migrating further upward. To complete the oil/gas accumulation (reservoir) a trap rock must overlie the reservoir rock and stop the flow. The trap should be near to oil/gas accumulation or surrounding the reservoir or along the petroleum migrating path way. The trap facilitates accumulation and the formation of a reservoir. It helps to retain and prevent the escape of oil/gas from the reservoir, over geological time.

No trap rock can be perfectly impermeable and impervious. Any cap rock would allow the fluid to seep up to the surface. The accumulation may not be completely free from leaks. If the rate of leakage is equal to the incoming (migrating) fluid, the accumulation will maintain equilibrium; otherwise it will not.

Trap rocks are of two type named according to their trapping functions.

- 'Cap rock' is a trap rock of low-permeability found at the top of the reservoir rock to prevent the upward leakage of oil/gas.
- 'Seal rock' is a trap rock of low-permeability and functions to seal or block fluid in carrier bed and reservoir rock.

Traps are formed either by tectonic activity (fault, fold) or by sedimentary depositional processes. The former is known as a 'structural trap' and the latter is the 'stratigraphic trap'. Sometimes the combined effect of tectonic and depositional patterns causes the formation of a trap; such a trap is known as a 'combination trap'. Different and various sizes and shapes of trapped accumulation exist, as given below.

3.10.1 Structural Trap Rocks

Most oil/gas fields of the world are formed by structural traps. Structural trap rock is formed when tectonic processes like faults, folds and bending of sedimentary rock occur. Movement of the plate produces compaction, contraction and expansion that bring about structural changes in the rock. The change in the rock may lead to the formation of trap rock. The movement of the plate sometime may cause a fracture and the leakage of fluid instead of forming a trap. The following are important types of structural traps.

3.10.1.1 Anticline Trap Rock

The majority of the petroleum traps belong to the anticline dome structure. Anticline traps are formed by compression stress due to tectonic and geological forces acting on the strata. When this force acts on sedimentary rock, folding occurs and the reservoir rock is deformed and folded upward, to make an arch or an anticline or a dome-shaped trap. If the rock layers bend downward, a syncline is formed. The syncline structure do not form petroleum trap rock.

3.10.1.2 Anticline Trapping and Description of Reservoirs

All the terms related to petroleum trapping and accumulation in an anticline reservoir are explained in Figure 3.2. The reservoir fluids are distributed in the accumulation according to their density. Gases are the lightest and so occupy the top portion. Water is the heaviest, so it settles down into the bottom. The oil is found in the middle of the accumulation. The quantity of trapped oil/gas in the reservoir depends on the reservoir capacity. The capacity of the reservoir is defined by two terms, 'closure height' and 'spill point'. The closure height is the maximum height from the spill point, located at the bottom, to the top crest point of the dome. The top of the anticline reservoir is the dome crest point. The 'spill plane' is the maximum circumference or rim of the dome base.

Drilling status has also been explained, on structure (oil/gas accumulation),of structure (away from the oil/gas accumulation) and dry hole (not hitting the reservoir).

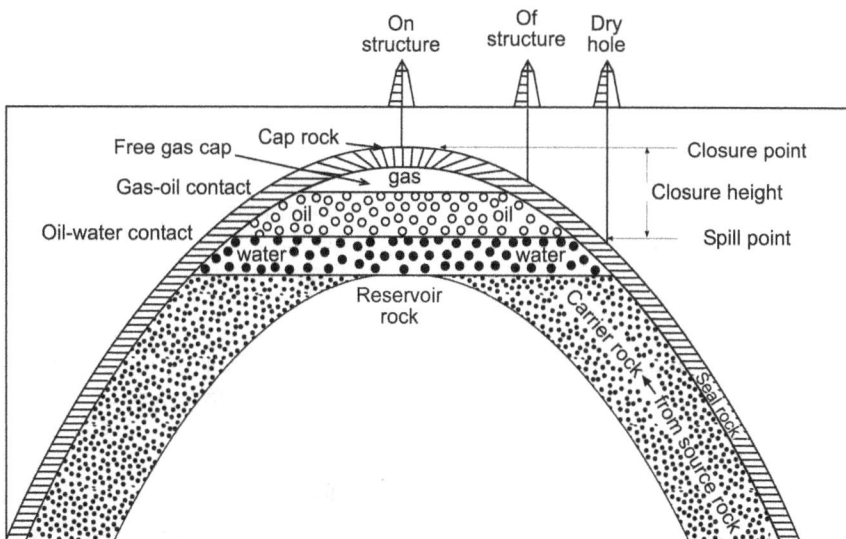

FIGURE 3.2 Detail of petroleum trapping and accumulation in an anticline reservoir. Source: Modified from Norman J. Hyne Ph.D., *Non-Technical Guide to Petroleum Geology, Exploration, Drilling and Production, Penn Well Book* (e-book), Oklahoma, USA, 1995 (Figure 11-5, page 155).

3.10.1.3   Symmetrical and Asymmetrical Anticline Structure

An anticline rock structure may be isolated or in combination with similar structures. Two anticline rock structures may be placed horizontally (side by side) or vertically one on top of another with respect to the earth's surface. The two structures may be also be symmetrical or asymmetrical with respect to the earth's surface. A straight vertical line passing through the peak axis of the dome crest to the base of dome divides the dome in equal parts and produces a symmetrical anticline structure with respect to the earth. When the vertical line passing through the crest of the dome divides the structure into two unequal parts, it produces an asymmetrical anticline trap with respect to the earth's surface.

3.10.1.4   Linear Fault Line Trap

A fault is a fracture or discontinuity in rock brought about by the movements of rock due to geological stress. A fault line on a rock surface is geological traceable line and it can be seen. The fault line is a perpendicular or oblique straight line (linear). It may also have a concave or convex shape. A fault may range from a few millimeters to thousands of km. Faults are formed as the earth's crust deforms due to stress. Stress is caused by plate tectonics. When the stress is greater than the strength of the rock, the rock breaks and a fault is formed. A fault line is produced when a crack develops in the rock due to the displacement of rock layers relative to each other. The fault line becomes the boundary between the two plates, such as the subduction zone, or between two blocks of strata.

When a reservoir rock moves to make an anticline or dome trap as a result of tectonics, the dome structure may be split into two parts along the fault line. One part of the reservoir rock slides up and another is displaced downward. When a permeable rock is displaced along the fault line it may face either an impermeable shale reservoir rock or permeable sandstone rock. The impermeable rock acts as a barrier (sealing rock) to oil flow across the fault line. The barrier is termed a 'sealing fault'. Thus the faulted structure leads to the formation of a petroleum trap along the fault line by the movement of rock. A fault line or the sealing fault divides the dome reservoir into two separate pools. Each pool works independently, provided both the reservoir face the impermeable sealing rock. If the reservoir faces the permeable sandstone rock the oil/gas leak out and no accumulation is formed.

3.10.1.5   Concave Growth Fault Trap

A concave growth fault trap is another example of the division of an anticline reservoir into two parts along the fault line. The curved fault line is concave, whereas the normal fault line is a perpendicular or oblique straight line (linear). The relative movement of the two rock parts is due to uneven overburden pressure. The left and right flanks of the anticline reservoir experience different sedimentation and overburden pressure. More sediment is being deposited and overburden pressure increases on one side of the rock compared to the other. The loose sediments are further compacted and move down from their original elevated position relative to the other flank, to form a concave fault line. The upper part of the fault line is known as a 'concave growth fault' and the downward part is a 'down-to-the-basin fault'. The fault grows further, bending inward, forming a concave structure as the pressure differential increases on both flanks of the dome due to more sedimentation and depositions.

3.10.2 STRATIGRAPHIC TRAP ROCK

Stratigraphic trap rock is formed when impermeable strata seal the permeable reservoir. Stratigraphic trap rock bears the characteristics of stratigraphic sedimentary rock. Stratigraphic traps are mostly formed in reef, beaches, river channels and coastal deposits. Common types of stratigraphic trap rocks are given below.

3.10.2.1 Primary Unconformity Stratigraphic Trap

Unconformity is the absence of a regular pattern in the rock structure. Unconformity is a break in the otherwise continuous structure of the geological record. A primary unconformity trap is formed due to the non-uniform and irregular deposition of sediments. It is the erosion of the underground surface of certain strata. Unconformity is caused by a period of sediment erosion and the absence of further sediment accumulation followed by new sediment deposition. Limestone or sandstone reservoirs are encircled by shale/clay trapping strata.

3.10.2.2 Secondary Unconformity Stratigraphic Trap

The secondary stratigraphic trap is formed by tectonics, where displacement or changes in the layers of sedimentary rock take place, but folding and faulting of the rock do not take place.

3.10.2.3 Discontinuity Stratigraphic Trap

Unconformity is a geological discontinuity between layers of strata. Layers are separated by a transition zone that is accompanied by a change in physical or chemical characteristics. The discontinuity may form an oil/gas trap rock or may lead to the leakage of fluids, depending on the permeable/impermeable structure and on the geometry of the trap rock. Discontinuous stratigraphic trap rock is created by the uplifting and erosion of the upper surface of reservoir rock which is followed by a layer of shale/clay sediments on the surface which act as a seal for oil/gas.

3.10.2.4 Angular Unconformity Trap

An angular unconformity is a region of geological unconformity between the underground eroded tilted rock layers and the overlaying sedimentary layers. It forms if the underlying rock is of sandstone or limestone minerals containing oil/gas and the overlaying sedimentary rock is a shale or salt rock that acts as a seal (cap rock) of the reservoir below. The oil and gas migrate from the source rock in an upward direction to the unconformity region and are trapped by the overlaying cap rock.

3.10.3 COMBINATION TRAPS

The combination traps are the third type of traps. They have both structural and stratigraphic features. Combination traps resulting from various factors are given as follows.

3.10.3.1 Salt Dome Trap

Salt dome rock structures form various petroleum trapping systems as illustrated in Figure 3.3.

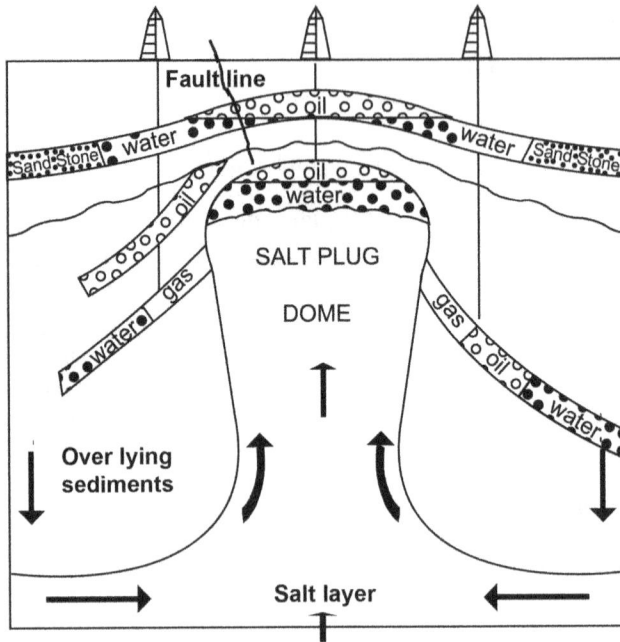

FIGURE 3.3 Multifunctional trapping in a salt dome. Source: Modified from Hobson G.D., *Modern Petroleum Technology*, Applied Science Publishers, London, 1975 (Figure 5, page 5).

A salt dome is a structurally formed dome. The structural salt dome is formed when a compact salt (evaporite) rock intrudes vertically upward into surrounding strata by geological forces, forming a diapir salt dome. Evaporite minerals are impermeable and lead to the formation of petroleum stratigraphic trap. Salt rock is an underground solid mass of halite and anhydrite minerals. A salt dome is an example of diapir rock, which is a geological intrusion in which a rock containing more mobile, ductile and deformable material is forced into brittle overlaying rock. Subsurface salt rock has the tendency to change and intrude (rise) into overlaying sediments through weak areas of un-consolidated sediments. The necessary force for the intrusion of salt rock originates from the following three sources:

- The downward pressure exerted by the weight of overlaying sediments.
- Lateral pressure of tectonic movements.
- Up welling pressure of mobile hot magma below.

The deep buried salt minerals are melted by high temperatures due to their close proximately with the earth's hot mantle. The salt rock melts and expands upward slowly to different heights within the sedimentary rock. Buoyancy helps the halite rock to move up, because the halite minerals are lighter than the minerals of sedimentary rock. Rising salt columns, usually of a dome shape, can pierce, deform and shear the original sedimentary layer. The salt dome continues to grow because of incoming compressed salt from the below salt layer till a balance is established in the operating pressures. The upward movement of salt mass into the sedimentary rock forms a plug (blocking)-shaped structure or plug dome. During upward movement the plug surface loses some of its salt minerals because of their solubility in water. The anhydrite minerals remain at the surface of the dome because of their insolubility in water. Anhydrites minerals have low porosity and permeability, and so act as a cap trap over the salt dome. However, anhydrite rock is prone to fracture. The salt dome can move upward into the sedimentary rock maybe up to 5 km, and sometimes break the entire sedimentary rock to appear at the surface. Since the salt dome is mostly water soluble, it is washed away by rain water at the surface.

A salt dome trap is a combination trap and can act in multiple ways. An oil/gas trapped accumulation may exist at the top of the salt dome or at a close shallow reservoir rock near the top of the dome. Oil/gas reservoir may be at the sides of the dome against the trap of impermeable surface of the salt plug. The forces and stress created by the uplifting salt dome may cause the above sedimentary layer to be deformed or faulted. Thus a fault trap is created above the dome. Thus various types of trap rocks exist at one place.

3.10.3.2   Bald Head Trap

The top portion of the anticline reservoir rock which may be nearer to the surface of the earth may be subjected to erosion. It is possible that most of the top portion of the rock is eliminated by erosion so it is said to be bald head rock. High tide from a lake or sea may bring fresh shale/clay to cover the bald head of the rock to make a seal and function as an anticline trap rock for oil/gas.

3.10.3.3   Hydrodynamic Trap

Hydrodynamic trap rock is formed by strong flowing water in the pores of the oil-bearing rock. The hydrodynamic water force disturbs the equilibrium of the horizontal oil–water interface in the reservoir. There is contamination of oil with water, forming sludge. The straight oil–water interface in the reservoir is pushed and tilted. The tilted water–sludge–oil interface creates a hydrodynamic trap. A hydrodynamic trap is rare and also not of much commercial importance.

3.10.3.4   Permafrost Trap

Permafrost is a frozen rock layer at variable depth below the earth surface in cold areas. Permafrost containing methane hydrates forms excellent traps. The effectiveness of a permafrost trap as a barrier to fluid flow depends on many factors such as the buoyancy pressure of the fluid, the thickness of the trap, minerals along with their porosity and permeability and capillary pressure across the pores of the trap.

3.11   PETROLEUM OIL/GAS FIELD

Petroleum oil and gas field may be included as a final part of the petroleum system. A petroleum field is defined as 'The oil and gas fields are the area from where numbers of producing wells operate to draw hydrocarbons from the subsurface accumulation'.

Natural petroleum oil/gas fields are located at on-shore or off-shore areas where subsurface hydrocarbons are accumulated. The accumulated hydrocarbons from underground wells are drawn to the surface by production equipment set up at the surface. A well is a bored hole in the ground to draw the oil/gas.

The oil/gas from different reservoirs and wells in the same field generally possesses similar properties. s. An oil-rich producing area is known as an oil field. A gas-rich accumulation area is called a gas field. The petroleum area may extend to several hundred square km. A series of producing wells as well as exploratory wells are spread throughout the whole field area. The outer wells are used to demarcate the boundary of the field. In addition to the operating wells, the site consists of a pipeline network for oil/gas gathering and transportation, the metal structure of rigs and pumps, support services, electricity, water installations and housing. The petroleum field also includes the initial surface treatment, separation and processing equipment for making safe and saleable oil/gas. Crude oil is separated into oil, gas, condensate and water at the surface installation.

The produced excess gas that cannot be utilized economically is flared up, away from well. The flaring of gas requires a combustion system, furnace, chimney, etc. Nowadays flaring should be done according to environmental regulations. The flaring of gas at the oil/gas field gives a pleasing sight to watch from a distance, especially at night. In fact, a petroleum field gives an impression of an illuminated small town.

3.12   PETROLEUM MIGRATION

The process of migration involves a stationery solid rock mass and migrating fluid (oil/gas/water) through different pores of sedimentary rock. A pore is a small hole in rock filled with fluid (oil/gas/water) and a means for the migration of fluid in the rock. The subsurface fluid (oil/gas) flows under the influence of geophysical, geochemical and geological factors. Petroleum is formed in deep source rock, but it is found in other shallower reservoir rock, located in a slanting upward direction. The flow of fluid within the source rock and outside in the carrier bed (rock) is called petroleum migration. Petroleum migration is not a continuous or uniform flow of oil/gas. The rate of fluid flow from the source/carrier bed rock is not uniform; a stop-and-go situation may exist. The total path of migration of fluid from the source rock to the accumulation reservoir rock may be a few meters long or several hundred kilometers. The flow pathway from source rock and accumulation is described by three types of migration:

- **Primary migration** is the flow of generated oil/gas from the solid kerogen, movement of fluid within the source rock and expulsion of oil/gas to the adjacent rock of same porosity and permeability as the source rock.
- **Secondary migration** is the movement of expelled fluid from source rock to a coarse-grained porous and permeable carrier rock. Simply any movement of the fluid outside the source rock in the carrier bed is a secondary migration. The end of secondary migration is the formation of a fluid

reservoir pool by sealing with trap rock. If no trap rock is encountered in the migration path way, secondary migration continues in an upward direction, till the appearance of fluid at the earth surface as 'seepage'.

- **Tertiary migration** is the flow of fluid from already formed accumulation. It is the remigration, or redistribution or flow within or outside the reservoir rock through a leak.

3.13 PRIMARY MIGRATION

The migration of fluid starts from the center of source rock and involves two materials, namely rock and rock pore fluid (oil, gas and water). The role of both materials is important for effective migration. Rock geological characteristics affect the migration. Rock properties vary from one source rock to another and also within the rock. Temperature, over burden pressure and pore size are variable. The pore fluid quality and quantity also change in source rock. Accordingly the expulsion and migration of fluid (oil/gas) are adjusted in the changed environment.

All these source rock and fluid parameters, along with some other factors affecting the movement of oil/gas within the source rock and ultimately fluid expulsion from the rock, are described as below.

3.13.1 ROCK PORE DIAMETER AND MOLECULAR SIZE OF HYDROCARBON

A proper combination of pore diameter of the rock and molecular size of fluid is needed for easy movement of the hydrocarbon molecules in subsurface condition. A pore is void space between the grains of the rock. The pore may assume multiple shapes and dimensions. It may be mentioned that the pore diameter of the shale source rock is between 5 and 20 nm with comparatively less porosity (<10%).

3.13.1.1 Rock Pore Diameter

Pore diameter is used to express the void volume between the rock grains. Proper distribution of pores and their sizes in the source rock is an important factor for the expulsion of fluid from source rock. Larger pore sizes than the molecular diameter of the fluid facilitate the migration. A pore diameter greater than 5 nm and a porosity of 5–10% are a good combination for effective migration from the source rock. The shape, size, diameter and height of the rock pores vary and depend on location, type of minerals and geological environment. Normally, a pore is defined by the size of its diameter, considering it to be circular, but the fact is that a pore may assume any shape, circular, square, rectangle or flat. The shape may be a capillary or narrow pore-throat or wider cavity. Rock pores, grains size and shape are constantly changing during geological time. Pore diameter decreases with depth. At greater depths, only light hydrocarbons, especially gases, are found.

3.13.1.2 Molecular Size and Expulsion Efficiency

The molecular size of the fluid is expressed as its molecular weight or as its molecular diameter (nm). The molecular size is one of the main factors controlling migration.

The size of a fluid molecule is related to the molecular weight or simply the number of carbon atoms in the molecule. The higher the carbon numbers, the higher the molecular size. It is expected that the concentrations of hydrocarbons that migrated from the source rock to the reservoir rock contain lower molecular weight compounds than high molecular weight (ring structure) compounds. Lighter paraffin and cyclo-paraffin hydrocarbons migrate more easily from source rock to reservoir rock than the higher molecular weight aromatic ring compounds. The high molecular diameter substance is retained in the source rock, whereas the lower diameter molecules are allowed to migrate.

Quantitatively the migration is expressed as 'expulsion efficiency'. The fluid expulsion efficiency from the source rock depends on molecular size for a given porosity and geological condition. The fluid expulsion efficiency in defined as 'The amount of migrated fluid expressed as the percent (%) of the total quantity in the undisturbed source rock'. The fluid expulsion efficiency is the highest for lower molecular weight hydrocarbons compared to those with higher carbon numbers (more than 50). The expulsion efficiency decreases with carbon numbers. Aromatic ring compounds are 1–3 nm in diameter and asphaltene's molecular size is 5–10 nm, whereas hydrocarbon gases and lower molecular oil have a 0.34–0.48 nm molecular diameter size.

3.13.2 PORE WATER AND PETROLEUM MIGRATION

Rock pore water plays an important role not only in primary migration but throughout the entire migration of oil and gas. The water content of the pores of the rock decreases with depth. The pore water content of shallow and young rock is about 80%; with depth, water is squeezed out. In deeper rock the water exists only as absorbed layers on the pore wall. Water is completely absent after the metagenesis stage. Pore water is saline. The water contains dissolved minerals. The saline water is heavier than fresh water. The water properties are modified depending on the concentration of dissolved minerals. More saline water has greater density and hydrostatic pressure. The hydrostatic pressure gradient for fresh water is 100 bar/km and for saline water it is 120 bar/km. It means that hydrostatic pressure depends on the salinity of water. The salinity increases with depth. The vertically upward moving water from deeper rock loses salinity due to membrane filtration. The interaction of water with clay minerals is interesting. Water plays a duel role in clay minerals. Clay is an efficient adsorbent for a variety of substances including water. The surface area of the clay is many times greater than most of the other minerals. Two types of water exist in the pores of clay. One is flowing or stationery water. The other type is adsorbed water on the clay walls of the pores. Adsorbed water adheres to the pore wall as multilayered liquid or solid ice and becomes an integral part of the wall, reducing the pore diameter. It is possible that oil and gas migrate independently as separate phases. Even in this particular case, water affects the migration direction and distance covered by migrating oil and gas. Water exhibits unique properties, because of hydrogen bonding (polymeric structure). It is liquid because of hydrogen bonding. Other elements of the oxygen group in the periodic table form gases with hydrogen, for example hydrogen sulfide. Oil and gas are transmitted through water replacement in the water-filled pore, or with some other type of interaction with water (details to follow).

3.13.3 LITHO-STATIC-GEO PRESSURE-DRIVEN MIGRATION

Pressure at a certain point of depth is governed by the following factors:

- 'Litho-static pressure' is the weight of the overlaying sediments known as 'overburden pressure'.
- 'Hydrostatic pressure' of the stationery pore water is the weight of the water column for a given height of water in a pore.
- 'Hydrodynamic pressure' is the result of fluid flowing in the pore.

Together overburden and water pressures constitute a combined pressure, known as 'litho-static-geo pressure'. Litho-static-geo pressure-driven expulsion of fluid from source rock is the most important cause of primary migration. With increasing depth of sediment burial and geological time, pressure and temperature are increased. Pressure is created by the weight of overburden rock.. The hydrostatic pressure of the water in also increased with depth. The fluid flows faster in sandstone and limestone reservoirs than in the clay/shale strata. Therefore hydrodynamic pressure in the reservoir rock is much more effective than in the source rock. The reservoir rock is more permeable to fluids.

Pressure in a dynamic state continues to change due to geological factors. Normal pressure maintains the system in equilibrium. With the increase of litho-static-geo pressure, the porosity of the source rock decreases. From the high litho-static-geo pressure in the rock formation, fluid is driven out of the pore to the lower pressure rock zone. The lower pressure zone of the rock has more porosity due to either wider pores or fracture. The zone may be in the source rock or a different, adjacent carrier bed.

3.13.4 TEMPERATURE/PRESSURE-DRIVEN MIGRATION

Gaseous and liquid hydrocarbons behave differently with temperature and pressure. Solid, liquid and gas expand to different extents with a rise of temperature. Gas expands enormously, followed by liquid, with the least expansion in solids. A rise of temperature leads to an increase in pressure due to the expansion of fluid in the pores of shale rock. The geo temperature gradient changes from field to field as does the geo pressure gradient. The variations in both the temperature and pressure gradients are due to rock compactness, burial depth, gravity effect of overlaying sediments, water circulation and geo-physico-chemical factors. The extent of average temperature variation from surface to deep zones (say 5000 m) may be between 25 and 200°C. The average geothermal gradient is reported as 25°C/km, and the average pressure gradient (the rate of increase of pressure) is about 2–3 psi/meter depth or 2000–3000 psi/km (133–200 bar/km). The pore fluid expands under the influence of increased temperature and exerts pressure to drive out the fluid from the pore of the shale rock. Rock thickness, depth, pressure, temperature, compactness, porosity and permeability are inter-dependent; all affect pore saturation and migration.

3.13.5 Depth-/Compaction-Driven Migration

With depth, burial and compaction, the sediments and rock become denser and harder. The density increases and the pore size, porosity and permeability decrease. Several physical and chemical changes occur with compaction and depth. The changes depend on the types of sentiments. Clay and sandstone are comparatively chemically inert. Both of these rocks undergo physical alteration. Limestone and evaporite rocks are chemically active. Both rocks are subjected to physical and chemical alteration. Pore water along with oil and gas are expelled by the compaction pressure. In deeper rock, severe compaction and overburden lead to lithification, cementation and crystalliza-tion of the minerals with rapid expulsion of residual fluid, mostly water and gases.

Paraffin and isoparaffin hydrocarbons, including gases, have the lowest molecule size and so are found in deeper rock, even in small pores of the rock. Greater depth means a small pore diameter. Asphaltene and aromatic ring compounds have the highest molecular size, so gradually their concentrations decrease with depth com-paction. They are retained and trapped at lower depths. At a depth greater than 2000 meters, the pore diameter in rock is reported to be less than 5 nm. The migration of all of the crude oil components is possible except the asphaltene molecules (molecu-lar diameter '5 nm). Asphaltenes make a homogenous solution with light hydrocar-bons; homogeneous solutions move easily. Any un-dissolved asphaltene remains in the source rock as solid material. The relative retention and movement with increas-ing depth and decreasing pore size are related to the molecular size of the hydrocar-bons. The larger sized molecules are progressively trapped (retained) with increasing depth and compactness of the rock. Asphaltene molecules are of a higher molecular size, cyclo-paraffin and normal paraffin hydrocarbons are lower in size and aromatic ring compounds are of intermediate molecular size.

3.13.6 Hydrocarbon-Driven Migration

Hydrocarbons are generated in all three phases of matter, gas, liquid (oil) and solid (asphaltene), in different stages of kerogen thermal maturation in source rock. The three phases of hydrocarbons are derived from a single solid phase known as 'bitu-men/kerogen'. They undergoes thermal cracking, splitting into gas, oil and asphal-tene hydrocarbons and residual kerogen. The gas, oil and asphaltene fraction of hydrocarbon takes place in migration. The simple and low-weight molecules migrate first, followed by intermediate and high molecular weight compounds. Gases take the lead, followed by the oil, and the solid asphaltene lags behind in the migration process. Asphaltene migrates in dissolved form.

Most hydrocarbons are generated during the 'oil window' temperature range 60–150°C in the catagenesis stage. Enormous pressure is generated in the pores of the source shale rock. High pressure creates fractures in the rock with the release of hydrocarbons. First to come out are gases which go into the pores of carrier rock. Gases are followed by oil and then higher molecular weight hydrocarbons. After the release of hydrocarbons and the establishment of equilibrium pressure, the fracture point in the source rock is repaired by the overburden pressure in due course of time.

3.13.7 MIGRATION BY DIFFUSION

Diffusion is a physical property exhibited by gases. Gases diffuse from regions of high concentration to regions of low concentration or diffuse into a void space to form a uniform mixture of gases. This is the dispersion of gases from one place to another, to equalize the concentration. The relative rate of diffusion of hydrocarbons decreases with the increase of molecular size. The diffusion rate of methane is the most rapid, followed by ethane, propone and butane petroleum gases. Apart from methane and ethane, the diffusion rate of liquid hydrocarbons is very low and their migration by diffusion is negligible.

3.13.8 OIL/GAS SINGLE PHASE MIGRATION

All the three phases of matter (g, l, s) are pressure and temperature dependent. Most of the expulsions and migrations take place between 2000 and 5000 meters depth and in a temperature range between 60 and 150°C, by various methods employed for migration. Homogeneous solution of gas, liquid and solid by single phase. Single-phase flow is more easy through saturated rock pore. Saturation is defined as the fraction of the pore space that is occupied by a fluid phase (oil, gas, water). The high saturation of pores with oil or gas favors continuous flow in the absence of water or little adsorbed water. In fact, a water wetted pore wall facilitates the oil flow.

Briefly it may be stated;

- Saturated pore with oil/gas favors migration.
- Water wetted pore facilitates migration.
- Homogeneous solution as single phase of oil/gas support migration.

3.13.9 MIGRATION OF OIL/GAS AS SOLUTION

Single phase flow is reemphasized here. True solution migration is an example of single phase flow. The migration of oil and gas in water solution phase is related to the solubility of different hydrocarbons under subsurface conditions. Normally lower molecular weight compounds, within the same hydrocarbon type, are more soluble in water than heavier molecules.

The solubility of lower molecular weight hydrocarbons in water is appreciable. Gases have lowermolecular weight in addition to solubility and because of eases of diffusion, lower molecular are widely dispersed in subsurface sedimentary rock. The solubility increases rapidly with depth under pressure. An increase of temperature reduces the solubility of methane in water; however the effect of increased temperature on methane solubility is negligible compared to pressure. The solubility of methane in water is reported to be 25 ppm at normal temperature and pressure, whereas the solubility of methane is 250,000 ppm under an overburden pressure of 7000 psi and temperature 150°C. Total dissolved salt (salinity) in water reduces the methane solubility, but a rise in pressure compensates the salinity effect. The migration of gases, especially methane, is well established through water solution migration. However, the migration of oil in complete solution form is limited. The underground solubility of hydrocarbons is a function of depth, temperature, pressure and salinity; depth and pressure increase and temperature and salinity decrease the solubility.

3.13.10 Migration of Oil/Gas as Emulsion

By the interaction of water and oil, under certain subsurface geological conditions, an oil-in-water emulsion is formed. The particle size of a solute in an emulsion is bigger than in true solution. Gases in water form foam, not an emulsion. Oil is hydrophobic so water does not dissolve oil in significant quantities. An emulsion is a forced solubility. Naturally occurring surfactants in crude oil (hydrocarbon containing O, N and S), under overburden pressure and temperature, act as emulsifying agents to make a colloidal solution or emulsion of oil-in-water. In the initial stage of hydrocarbon generation in the narrow shale pore, the emulsion is likely to be a contributory factor for migration. But the oil-emulsion and gas-foam do not seem to take a major part in hydrocarbon migration.

3.13.11 Migration, Oil Suspension and Gas Foam

Suspension is the dispersion of oil-in-water by vigorous agitation. Suspensions of oil are the small oil drops (globules) forcefully dispersed in water. Suspensions break easily. Foam is the dispersion of gas in liquid. Under pressure, gas dissolved in water creates foam. Under reduced pressure, gas is released as bubbles from the foam.

The migration of oil and gas through water-saturated pores of fine-grained shale rock, under higher pressure and temperature, takes place as globules and bubbles. The sizes of oil globules and gas bubbles are important. If the pore diameter is greater than the sizes of the globules and bubbles, there is no problem in migration. Narrow pores create capillary pressure that opposes the forward movement of the globules and bubbles. Extra pressure is needed to overcome the resistive force for globule and bubble transportation in narrow pores, or a fracture of the rock may help the oil and gas in migration. Oil and gas migration as globules and bubbles is a fairly established phenomenon in compact rock.

3.13.12 Migration of Oil as Micelles

Aggregates of colloidal particles formed in water are known as micelles. On the basis of polarity, hydrocarbon molecules are divided into two types. One type is polar and the other is non-polar molecules. The majority of the hydrocarbons are neutral or non-polar molecules and contain carbon and hydrogen atoms. Polar hydrocarbons, in addition to the carbon and hydrogen elements, contain hetero-atoms nitrogen, sulfur and oxygen. The heteroatoms form a functional group in the molecule. The functional group creates a charged pole at the one end of the molecule (water loving). The other end of the molecule assumes a non-polar character (water hating). Thus two ends, one hydrophilic (water loving) and the other hydrophobic (oil loving), are created in the molecules. The molecule functions as the detergent or surfactant or emulsifying agent. The surfactant is joined by oil molecules at one end, and at the other end with water molecules to form an aggregate of many molecules of different nature. The aggregate is termed 'micelle'. The micelle is dispersed into the water, forming a colloidal type of solution. Obviously the size of a micelle is much bigger than the suspension/colloidal particles.

The migration of oil as a part of the micelle seems to be possible. However the formation of micelles is limited. The major reason is the limited quantity of hetero molecules in the petroleum. It cannot account for the large commercial-scale migration of oil.

3.13.13 SEDIMENT FACTOR

All three major sedimentary minerals play their due and specific role in migration because all have different physico-chemical properties under different geological conditions. Sedimentary rock containing clay/shale with a minor quantity of sand is the main source for the generation of oil/gas from preserved organic matter. The generation and migration of oil/gas are related to the composition and structure of the rock. The source sedimentary rock contains alternate zones of fine-grained shale rock and coarse-grained sand rock. This creates a litho-logical pressure gradient in the rock. Likewise organic-rich zones and organic-lean zones in the rock generate different quantities of hydrocarbons. Therefore a pressure differential is created by organic matter and hydrocarbons. The litho-logical and hydrocarbon pressure differential acts on the fluid to force it to migrate from a higher pressure zone to a lower pressure zone in the sedimentary rock, and possibly out of the source rock.

3.13.14 TECTONIC PRESSURE FACTOR

Tectonic movement brings about many changes in the sedimentary rock, for example compression, contraction, fracture, deformation and expansion. Tectonic movement establishes high pressure zones in the rock, resulting in rock fracturing and the release of fluid from the source rock.

3.13.15 ROLE OF JOINTS, FAULTS AND UNCONFORMITIES IN MIGRATIONS

So far oil and gas migration through rock pores under different conditions has been discussed. In addition to this, oil/gas also migrates through rock joints, faults, fractures, fissures and geological unconformities.

The initial role of pressurized fracturing for expelling oil and gas from source shale rock has been discussed above. High capillary pressure exists, especially in the narrow pores of shale rock. Therefore high displacement pressure is required for the displacement of oil/gas from shale sediments. Sufficiently high pressure is developed, creating a high-pressure compartment in the rock. The high fluid pressure exceeds the capillary displacement pressure, to cause the fracturing of the shale.

The fracture provides a path way for oil/gas migration. Joints, faults, fractures, fissures and unconformities are caused by tectonic and high-pressure compartments. They may provide a path way for migration. The upward vertical movement of fluid through faults and fissures is more common than through the porosity/permeability of shale rock.

3.13.16 DISCUSSION

The primary migration and expulsion of fluid from source rock are related to the maturation and thermal cracking of kerogen during the catagenesis and metagenesis stages. At the shallow depth up to 1500 meters, minor amounts of hydrocarbons are generated in the diagenesis stage. The hydrocarbons generated are of lower molecular weight compounds or gases. They are unlikely to take part substantially in migration and making a reasonable pool of oil and gas. A major portion of hydrocarbons is generated during the catagenesis process from the kerogen at a depth of 1500–5500 meters in the oil window zone. At these depths, geological conditions such as higher pressure, higher temperature, lower porosity, lower permeability and extreme compactness prevail, which help the expulsion of oil/gas in primary migration.

With further burial and depth beyond 5500 meters, the pressure and temperature, density and compactness further increase. These conditions favor the severe thermal cracking of kerogen into small molecular weight compounds; methane gas is especially dominant.

It is said that about 70% of hydrocarbons are expelled from the source rock. The un-expelled 30% oil remains attached to the kerogen and pore wall surface of sedimentary source rock. With this, the primary migration is terminated, and the beginning of secondary migration in wider porosity and permeable rocks of sandstone and limestone occurs.

3.14   SECONDARY MIGRATION

After crossing the source rock in primary migration, the fluids (water, oil, gas) enter into carrier rock to start the secondary migration. When carrier rock is sealed at the top by a trap rock, it turns into a reservoir rock and secondary migration is terminated. The geophysical conditions in carrier rock are different than those in source rock. In the carrier bed, wide pores and more space are available with greater porosity and permeability; those result in the release of rock pressure. The release of pressure facilitates phase separation into gas, liquid and water. The multiphase movement of fluid through carrier rock takes place. At the same time, incoming nearly homogenous fluid from the source rock is split into oil globules and gas bubbles. Migration from the carrier bed to the reservoir rock is normally in an upward slanting direction. The rate of migration of fluid depends on a number of parameters such as the oil/gas ratio, quality and quantity of oil/gas, rock pore size, water properties, buoyancy of fluid, permeability, joints, fractures, etc. The details of the factors that control the secondary migration are given below.

3.14.1 BUOYANCY RISE OF OIL AND GAS

Buoyancy is the upward force exerted by fluid. It is a natural property of a liquid. An object floats at the surface of water because of the upward buoyancy force. The upward pressure of the fluid is enhanced by the difference of specific gravities of gas/ oil/water in fluid-saturated pores. The specific gravities of gases are between 0.1 and 0.55, of oil between 0.55 and 0.99 and for water the range is 1.0–1.25. The specific gravity of water depends on salinity (salt content). The higher the gravity differences, the higher the buoyancy pressure. A gas/water system has a higher buoyant rise than an oil/water system. The buoyant pressure continues to work until a less permeable trap rock is encountered. Upward buoyant pressure is balanced by trap rock, where pores' hydrostatic pressure prevents the further rise of the fluid. Thus a reservoir is created. The underground pore water may be mobile or immobile. Mobile hydro- dynamic pressure enhances the buoyancy, whereas immobile hydrostatic pressure retards this buoyancy pressure.

3.14.2 SURFACE TENSION AND CAPILLARY PRESSURE

The surface tension controls the formation and stability of emulsion and foam and the detergent property of the oil. Surface tension may also be related to the recoverable oil from the reservoir, as the viscosity is correlated to the production oil from well.

Surface tension is the cohesive/adhesive forces of attraction acting on the bound- ary between liquid and gas molecules. The force acting at the liquid–gas interface is known as interfacial tension. A molecules at the surface of the liquid are attracted only inwardly by the molecules below. The surface molecules tend to go inside the center of the liquid. The liquid surface is under tension and tries to contract into the smallest volume. A drop of liquid in air assumes a spherical shape due to the inward contraction of the molecule. A sphere is formed due to the inwards pull of the molecules. The sphere is the smallest space for a given volume of a liquid. Due to the inward pull experienced at the surface of the liquid, the liquid is under constant inward pull. The liquid surface is under tension. The tension is the inward pull of the molecules. The surface tension of a liquid is defined as 'the force acting along the surface of a liquid at right angle to a line of one centimeter expressed as dynes/ cm'. The surface tension of water is too high (72.75 dynes/cm at 20°C). The surface tension of hydrocarbons is very low (benzene's surface tension is 28.5 dynes/cm at 20°C). Surface tension is associated with wetting ability, capillary rise and contact angle, which are related to oil/gas migration.

3.14.2.1 Wetting Ability

Wetting ability is the tendency of a liquid to adhere or not on a solid surface. High and low surface tension correspond to high and low surface wetting ability. Water has a higher surface tension and high wetting ability, so it adheres to the pore wall surface preferentially. Gas, with the least wetting power (negligible surface tension), occupies the center of the pore, far away from the solid pore wall. Oil occupies the middle of the pore, because oil possesses intermediate wetting ability and surface tension. The water wet pore wall supports oil flow and a dry wall creates friction. The adhere water keeps the flowing oil away from the wall. Thus eliminating the friction factor. The water preferentially adheres to the wall of the pore minerals. This is most true for sandstone mineral. Oil in a sandstone reservoir rock is at the center of the pore and water away from the center, at the wall of sand grains. Therefore sandstone is more 'water wetting' than the oil. On the other hand, other common reservoir rocks of limestone and dolomite are preferentially "oil wetting" fluids. Oil tends to adhere to the grain surface rather than water. Therefore it is expected that oil recovery from a sandstone reservoir is more than from the limestone or dolomite reservoir rocks. An oil wet surface retards oil movement. The oil flow from the center of the pore is frictionless and produced easily. On the other hand, oil that is adhering to the grain surface encounters more friction and requires more force to displace. Consequently that adversely affects the flow rate and production quantity. The relative wetting ability of fluids (g, l and s) in a magnified cross section of the pore is illustrated in Figure 3.4.

3.14.2.2 Capillary Rise

Water or any other liquid rises to a certain height in a capillary (narrow pore). This is due to the surface tension force that is acting along the inner circumference of the

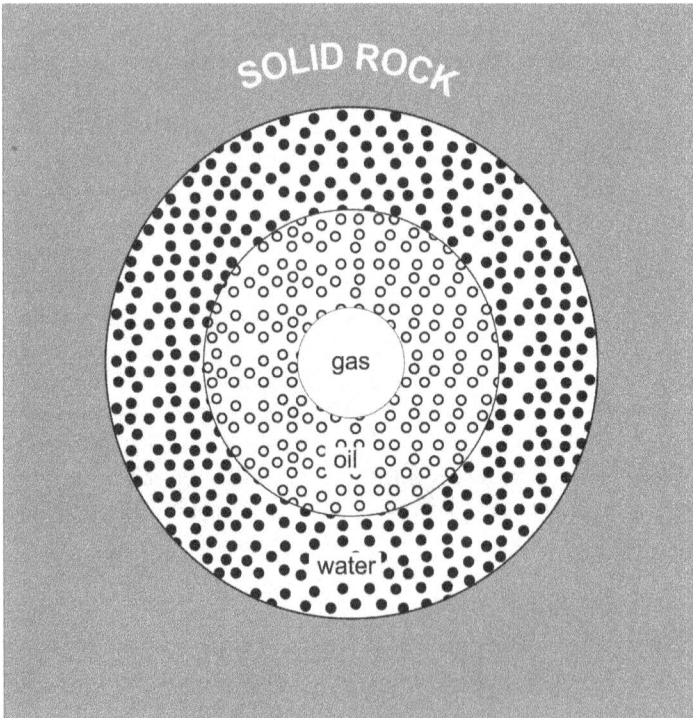

FIGURE 3.4 Magnified cross section of rock pore containing gas, oil and water.

capillary pore. The capillary rise of water depends on the diameter of the pore and density of water. At equilibrium the capillary upward water pressure due to surface tension is equal to the downward hydrostatic pressure of the water column. At equilibrium conditions:

Upward capillary pressure = downward hydrostatic pressure

$$2Л\gamma\curlyvee = hЛ\curlyvee^2 \, w$$

Surface tension (γ) = h$\curlyvee$ w/2

$\curlyvee$ = radius of the capillary pore, h = height of the water column in the capillary pore, w = weight of the water column.

For any movement of the oil and gas through a pore, the upward capillary pressure of the fluid must exceed the downward hydrostatic pressure for replacing the water column with oil or gas. The liquid surface in the capillary pore is contracted downward, making a concave shape.

3.14.2.3 Contact Angle

When two immiscible substances, liquid–liquid or liquid–solid, are in contact with each other, the shape (angle) of their interface is controlled by the adhesive and cohesive attractive forces of the molecules of both substances. The resultant of those forces determines the contact angle(Θ), wetting ability of the liquid and surface tension.

The wetting ability of the liquid is related to the contact angle (Θ) as follows:

Contact angle (Θ) = 0 ° strongly water wetting
Contact angle (Θ) = <90° water wetting
Contact angle (Θ) = >90° oil wetting
Contact angle (Θ) = 180° strongly oil wetting

The wetting ability or the contact angle is affected by the type and composition of the fluid, pore, minerals, microbiological activities, temperature and pressure existing in the rock matrix. Two different contact angles are generated by two different liquid drops. A spherical oil drop, when placed on a glass plate, maintains its spherical shape; the contact angle between the oil drop and glass surface is 90–180°. On the other hand, a water drop adheres to the glass plate surface; the contact angle is 0–90°. The spherical shape of water is distorted.

3.14.2.4 Pore Throat

Pores may be interconnected through channels or may be isolated. The narrower channels connecting the pores are known as the 'pore throat'. The pore throat is the narrowest portion of the pore. Normally the pore throat is a hindrance in fluid flow. Oil and gas are slightly soluble in water but can be emulsified in water as small droplets (globules) and gases as foam bubbles. Globules find difficulties passing through the 'pore throat'. Oil droplets cannot pass, without being distorted. The interfacial surface tension resists the distortions. At the equilibrium state the hydrostatic downward pressure and upward buoyancy pressure are equal and the oil/gas cannot move. This is a case of water entrainment. Below is the oil/gas and above is the water in the pore, all in rest position. To break the equilibrium in an upward direction, an extra buoyancy pressure is required to distort and force the globules through the pore throat by overcoming the globule distortion force. The extra buoyancy pressure is generated by the fresh arrival of fluids (oil/gas/water). The height of oil/gas in the pore is further lifted and provides extra pressure to force the globule into the pore throat.

3.14.3   Hydrodynamic, Capillary and Buoyancy Pressures

Rock pore water exists in the subsurface conditions either in a dynamic (flow) or static (stationery) state. The static water needs a displacement pressure in order to make space for oil migration through the pores. Flowing water conditions in the pore have been mentioned earlier in connection with hydrodynamic traps. How do hydrodynamic pressures affect the secondary migration? After the expulsion from the source rock the fluids assume an upward inclined (ramp) pathway in the carrier bed. The upward path may be resolved into horizontal and vertical components.

- Fluid buoyancy pressure supports the vertical movement. The buoyant force contributes little in the horizontal movement of the fluid;
- A hydrodynamic pressure gradient across the boundary affects upward vertical as well as horizontal movement. If the driving gradient forces are not sufficient, the water remains static, there is no displacement of water or oil transfer.
- The vertical movement is opposed by the capillary pressure. The water-filled rock pore creates a capillary barrier for the oil/gas migration. With the increase in water column in the pore, the opposing capillary pressure increases.
- Fresh arrival of fluid increases the buoyancy pressure and hydrodynamic force.
- The net result among the three forces, hydrodynamic, buoyancy and pore capillary pressure, determine the oil migration, its direction and flow rate as shown in Figure 3.5.

3.14.4   Oil–Water Contact Line

The contact lines between gas and oil, oil and gas and oil and water are subjected to different pressure in different parts of the reservoir. High capillary pressure is expected at the top of the reservoir rock, because of the vicinity of the cap rock. Therefore, stationery conditions of contact lines between gas and oil are established. The subsurface bottom water is under varied hydrodynamic conditions. The water

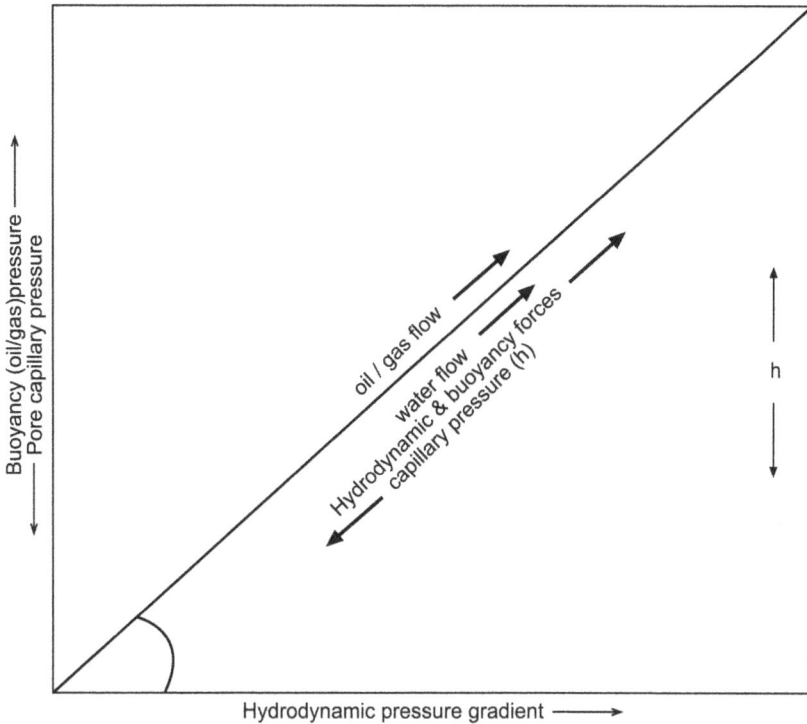

FIGURE 3.5   Magnitude and direction of fluid flow is dependent on the buoyancy rise, capillary and hydrodynamic pressure.

may be fast flowing or slow flowing or in a stationery state. The distinct contact line between oil and water in the reservoir is established according to varied hydrodynamic conditions. Under a high water fluid flow rate and high hydrodynamic pressure, the oil–water line drags farther ahead in the direction of flowing water with increased inclination of the oil–water line. The trapping of oil may be disturbed and may lead to the formation of an almost new horizontal contact line and new accumulation. At a low water flow rate, the sharp oil–water contact line slightly drags in the direction of flowing water but remains almost horizontal, trapping the oil/gas in the reservoir. In conditions of no water flow, the oil–water contact line is a perfect horizontal line and the reservoir is in a stable state.

3.14.5   JOINTS, FAULTS, FISSURES AND UNCONFORMITIES IN MIGRATION

The roles of joints, faults, fissures and unconformities in secondary migration are not common in high-porosity limestone and sandstone rocks; rather these are prevalent in compact shale rock. Joints, faults, fissures and un-conformities in shale rock are formed due to abnormal high pressure. Small pore size and large amount of generated oil create abnormal high pressure in shale rock. This situation hardly arises in in high-porosity and -permeability limestone and sandstone rocks.

The secondary migration is terminated with the accumulation and formation of an oil/gas pool called a 'petroleum reservoir'.

Important phases of migration are summarized below:

- Major factors that affect the migration are pore type, sediment type, compaction, salinity of pore formation water, permeability and geological environment, and the most important is the quality and quantity of hydrocarbons available in source rock for expulsion.
- Pure hydrocarbons hardly form true solution of oil in water. Oil solubility in water is negligible. Only low-quality compounds containing heteroatoms are soluble to some extent in water. Hydrocarbon compounds containing heteroatoms form a small portion of the total crude oil. Migration of the crude oil as solution is insignificant.
- The emulsion/suspension of oil and gas in water forms small globules and bubbles. The breaking of emulsion and suspension gradually begins in the carrier bed/reservoir rock due to availability of more space. Tiny drops aggregate and coalesce to form bigger drops. Bigger drops respond more to the upward 'buoyancy force' in water. In the same way with repeated action, during migration oil/gas is collected over water in reservoir rock.
- High molecular weight and heterocyclic hydrocarbons are left behind. They settle down. The heterocyclic hydrocarbons are adsorbed in the rock minerals.
- Migration continues in upward ramp directions along with the separation of gas, oil and water according to the difference of density, till the fluid is trapped by a cap/seal rock.
- Hydrodynamic, buoyancy rise and capillary rise pressures control the direction and rate of flow of oil/gas.
- The roles of joints, faults, fissures and unconformities in migration are limited to compact shale formation.
- The bulk of off-loading of oil and gas from the carrier bed occurs in a trapped reservoir rock forming a petroleum accumulation.

3.15   TERTIARY MIGRATION

Fluid possesses a mobile tendency, it hardly rests in one position. Similarly petroleum in reservoir rock is not static. There is a redistribution of petroleum in the reservoir. In fact, there is nothing static in nature with respect to geological time. Even an equilibrium state tries to find its own way to further adjustment. The remigration of petroleum may be within reservoir rock or it may continue outside of the rock till seepage appears at the surface.

The main factor for tertiary migration is geological tectonic processes, such as folding, faulting, bending and uplifting of strata. The other factor is that when a

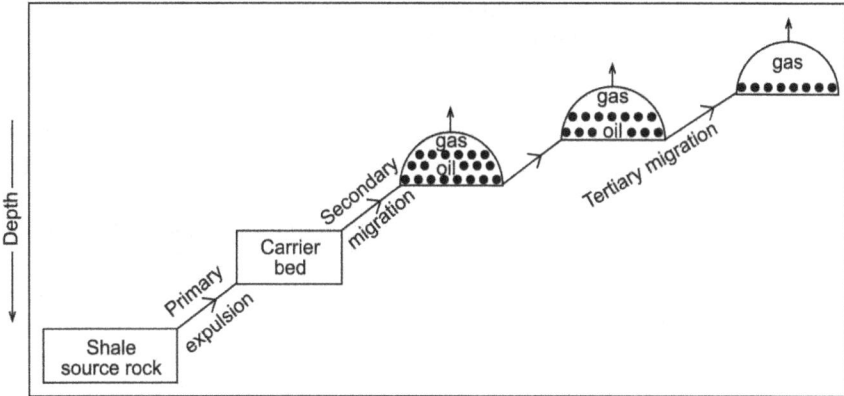

FIGURE 3.6 Migration (primary, secondary and tertiary), separation and formation of new oil/gas reservoir.

reservoir rock is full to capacity, extra oil migrates out from the sides of the reservoir to a shallower reservoir trap. The gas bubbles float up from liquid, and oil migrates from the side of the reservoir. The remigration may result in the partial or complete separation of oil/gas in a newly formed reservoir. This depends on the extent of oil/gas remigration. Gas bubbles normally prefer to migrate in an almost vertical direction, whereas oil migrates in a slanting lateral upward direction. The resultant of the preferred migration of oil and gas determines the composition of the new reservoir. If the vertical escape of gas bubbles is faster than the lateral oil flow, then the gas occupies a shallower reservoir above, devoid of oil or with minor quantities of oil. Primary, secondary and tertiary migrations and formations of accumulations have been summarized as shown in Figure 3.6.

3.16 FORMATION OF PETROLEUM ACCUMULATION

A petroleum accumulation or pool is a combination of reservoir and trap rocks along with formation fluid (gas/oil/water). Most reservoir rock is formed by sandstone and limestone minerals followed by dolomite and fractured shale. Reservoirs are also formed by the uplifting of basement metamorphic and igneous rock into sedimentary rock. The petroleum accumulation is a semi-porous reservoir rock, suitably trapped by impervious cap and seal rocks. It is possible that a sedimentary rock fulfills the requirement of a reservoir but is without petroleum deposit. This is said to be a 'barren reservoir rock'. The most commonly found commercial reservoir is a uniformed anticline (dome) structure rock. An example of a stratigraphic type accumulations is unconformity trap rock. A salt dome is an example of a combination trap formed by halite and anhydrite sediments.

A typical accumulation is characterized by the presence of gas, oil and water separated by clear, distinct and horizontal contact lines between gas and oil and oil and water, under hydrostatic equilibrium conditions. In these conditions, the downward capillary pressure is equal to the upward buoyancy pressure. In the absence of balanced hydrostatic equilibrium pressure, the fluid contact lines are disturbed, overlapping each other.

The pores of the reservoir are occupied by the migrated oil/gas/water replacing the original water. However within the reservoir rock, some pockets may exist, where the pore water could not be replaced or was partially replaced by hydrocarbons. This is due to very narrow pore size, non-interconnected and non-uniform void space and pores. Very high capillary pressure is generated in this situation. The buoyancy pressure could not balance and overcome the capillary pressure. Water remaining in the pore is produced along with crude oil and separated at the surface installation.

A reservoir may be completely filled or partially filled with fluids. A completely filled reservoir is called a 'saturated pool'. An 'unsaturated pool' is partially filled by fluids. Saturated and unsaturated pools have different patterns of hydrocarbon production and amounts of saline water. A well may have only gas without pore water or oil; such a reservoir is called a 'dry gas hole' producing 'dry gas'. If condensable liquid is associated with the produced gas, the well is called a 'wet gas hole' and the gas is 'wet gas'.

3.17   DISTANCE, DIRECTION AND TIME OF MIGRATION

The distance traveled by the migratory fluid varies considerably. It may be a meter or several thousand meters from the source rock. It depends on the structure of the carrier bed, its size and permeability and the quantity and quality of oil/gas. The amount of fluid and its composition are other important factors that affect migration. Higher quantities of generated oil in the source rock at high pressure have high mobility and higher distances of migration. As expected, thin oil is more mobile than thick oil and covers more distance in migration. Gas moves faster and longer distances in migratory path. Faults, fractures and joints in rock facilitate the quick migration.

After expulsion from the source rock, about 60% of the total expelled oil and gas migrates in the vertical component direction; the remaining 40% of oil and gas migrates as a horizontal component. Perfect horizontal movement is highly unlikely as tremendous hydrodynamic pressure is required to drive the fluid. Upward migration of oil and gas is quite natural, because of higher buoyancy and lesser density, than the saline formation water. Upward movement facilitates gas, oil and water separation and reservoir formation. The fluid flow rate is highly dependent on the permeability of rock. Slight contamination of sandstone rock with less permeable clay sediments greatly reduces the flow rate.

3.18   PETROLEUM SEEPAGE

Seepage is the leakage of oil and gas from underground to the surface of the earth. Seepage fluid should be visible at the surface. If leakage occurs in an underground rock and the fluid is lost in strata, and does not appear at the surface, it is not considered as seepage. The oil/gas seep is defined as 'the actual and visible presence of oil and gas at the surface of the earth that originated from subsurface deposits'. Seepage is the slow escape of fluid through a small opening in the rock. The seepage/leakage appears on land as well as on the surface of the aquatic bottom. The terms 'seepage' for oil and 'leakage' for gas may be used.

The seepage of oil/gas is an example of underground fluid migration. Although the seep may be a part of primary, secondary and tertiary migration, it is considered as tertiary migration. If the fluid finds a proper opening, the seepage can originate from any component of the underground petroleum system source, reservoir rock or from the carrier bed. In fact every component of the petroleum system leaks to some extent and can hold oil/gas to some extent. Seepage can be minor or major. Minor seepage is detected by sensitive instruments. Major seepages are visible to the naked eye. Minor seep occurs from minor underground deposits, and major seep is due to a commercial reservoir. Rock containing oil/gas continues to leak and disperse oil/gas to the adjacent strata. The leaked oil/gas follows the permeable path way through proper openings, faults, fractures and fissures in a high-porosity and permeable enough formation. Under proper path way conditions, the oil/gas appears at the surface; otherwise it is lost in the strata. The duration of seepage and the amount of seepage oil/gas depend on many factors, namely high subsurface driving pressure, quantity of oil/gas and porosity/permeability. There is no perfectly permeable (reservoir) or impermeable (seal) rocks, and there is thus no permanent opening for leakage. Porosity and permeability continue to change with geological time. The seepage can be for a short duration or for a longer period, depending on the conditions of porosity. The original porous structure changes or is modified by the overburden pressure, continued fresh sediment depositions and the movement of neighboring strata. The modified porosity/permeability affects the migration/seepage of oil/gas. Some very small pores in tight rock remain undisturbed and oil/gas is trapped.

The seepage of oil/gas to the earth's surface is a geological process and has been historically known to mankind since ancient times (1300 BC). Ancient civilizations were using asphalt for building and construction work. Asphalt was the residual product after the evaporation of volatile seeped oil at the surface. Asphalt possesses adhesive properties which are needed in the construction industry. The underground gas leaks were utilized for illumination purposes, especially in places of worship. Historically the appearance of oil/gas at the surface was the most important geological evidence used in prospecting for underground hydrocarbon, and petroleum exploration and drilling were concentrated in the oil/gas seep areas. Major oil fields were discovered nearby the seep region. With the development of technology and sophisticated instruments, seep-based exploration has diminished. It is assumed that only about 15% of the total oil generated in source rock is available in reservoir rock for commercial production. The rest (85%) is either held up in source rock, carrier beds and reservoir rock or lost as dispersed fluid in the adjacent strata. Only a fraction of the dispersed oil appears at the surface as seepage. Some factors responsible for seepage are given below:

3.18.1 Geological Activities and Seepage

Major causes for the appearance of underground oil/gas at the earth's surface from underground accumulation are the continuous geological forces (plate tectonic movement, volcanoes, magma thrust and earthquakes) acting on the accumulation. The uniformity and continuity of the subsurface strata undergo changes and dislocation, causing the opening of faults, fractures, joints, cracks and erosion between rock layers. The disturbed oil and gas try to find new equilibrium by adjusting their flow through openings. Through the opening seepage of oil/gas occur first in the subsurface; afterward, it may appear at the surface.

3.18.2 CYCLIC PRESSURE AND SEEPAGE

The obvious cause of the seepage is the pressure-driven oil/gas migration from the subsurface source. Pressure is developed due to the geological forces of compression and compaction in the uniform substrata. If the high-pressure zone is well established in the subsurface accumulation and has an opening to the earth's surface by a permeable path way, the driven oil/gas appears at the surface of the earth.

High pressure inside the deposit causes fractures (openings and pores) and gives passage to oil/gas leaks. On the other hand, a fracture is repaired and blocked by sediment mixture and water under overhead pressure. The sediments along with water are forced into the opening (fracture, pore) by overburden pressure, fill the gap and stop the leak. It may be noted that the pressure sometimes creates a fracture in the rock and other time repair by filling the rock opening. The seepage from the fractured rock is not a permanent process. There may be a stop-and-go situation. In some conditions there is a leak and in other conditions the leak is stopped. The leak occurs in cycles; it opens and closes.

3.18.3 VOLCANO ACTIVITY AND SEEPAGE

Volcanic activity originates from the igneous and metamorphic rocks due to the eruption of underground magmatic lava. Volcano activity supports seepage. When pressure builds up due to the volcano activity to the extent that a fracture is enacted in the accumulation, the oil/gas is forced out as seepage/leakage. The seepage fluid follows the volcano activity of periodic 'eruption' and 'quietness'. The cyclic volcano activity continues the process of fracture and repair again and again, the leaks open and close periodically. A volcanic eruption is accompanied by the spreading of particle-sized lava called 'flying ash' in a round area.

3.18.4 WATER SPRING AND SEEPAGE

Water springs are common, especially in hilly areas. In most cases water springs are associated with oil/gas seepage. Subsurface water flows with high pressure and emerges at the earth's surface. A water spring is an underground water leakage or seepage at the surface. If the underground water is associated with subsurface oil, then the oil comes out along with water at the surface. At the surface under atmospheric conditions, the oil separates from the water and forms a thin film at the top. In due course the oil is drained out and oxidized.

3.18.5   OUT-CROPPING OF SUBSURFACE ROCK AND SEEPAGE

The uplifting of underground rock to the earth's surface is also known as the 'out-cropping' of rock. Surface out-cropping is a common geological feature. This is a part of attaining equilibrium due to underground geological forces acting on rock. If any oil/gas is associated with the out-cropped rock structure, that is also carried to the surface. At the surface the uplifted rock structure is exposed to atmospheric conditions of oxygen, water and sunlight and undergoes erosion, abrasion, weathering and oxidation. The rock is gradually eroded with the creation of fractures and openings, allowing any associated oil/gas to drain out as seepage.

3.18.6   MUD VOLCANIC AND GAS LEAKAGE

One of the interesting examples of gas leakage is the eruption of mud volcanos. A mud volcano is a leakage through the fracture of gas accumulation under pressure at shallow depth. The mud volcano emits mud with force. A mixture of mud–water–gas forms a cone-shaped 'mud volcano'. Mud volcanos are found in various parts of the world in argillaceous (clay) sedimentary rock basins. The cone shape of a mud volcano mound is similar to the igneous magma volcano, but its size is much smaller. The mud is emitted out from the mouth, a crater in the middle of the cone, with different eruption speeds. The mud spreads around a wide area depending on the mud/gas volcano pressure. The eruption of mud is not a continuous process. The volcano activity is sandwiched between eruption (active) and quietness (sleeping). The eruption of mud is associated with gases. Methane is the main gas in the emission of mud volcanos. Gas bubbles can be seen coming out from the mud. In case the mud is thick and viscous, the gas bubbling is stopped. When enough pressure is developed, gas bubbling again starts. The amount and duration of gas bubbling are different from time to time and from one volcano to another. A view of mud volcano eruption and bubbling gas is depicted in Figure 3.7.

FIGURE 3.7   A view of a mud volcano and bubbling gas.

3.18.7 REVIEW OF GEOLOGICAL FACTORS AND SEEPAGE

- A shallower oil/gas accumulation seeps more than deeper deposits.
- Seepage from deeper reservoirs is possible, but the oil/gas is dispersed in substrata and does not reach up to surface.
- The connection of a permeable carrier bed up to the top surface leads to seepage.
- Seepage occurs due to erosion, fracture, rapture, folding and faulting from reservoir rock.
- Oil/gas associated with out-cropped rocks appears at the surface.
- A tectonically active earthquake area contributes to seep.
- A shallower trap rock affected by diapir-rich (uplifting minerals) strata facilitates seep.
- Oil/gas seepage on the earth's surface is subjected to weathering. Oil is washed away by water or lost by the evaporation of volatile hydrocarbons, leaving asphaltic residue.
- Seeps do not give a positive indication of a commercial-sized underground deposit of oil/gas.

3.19   OIL AND GAS BALANCE IN A PETROLEUM SYSTEM

The balance sheet of oil and gas from their generation in source rock to the accumulation in reservoir rock via carrier bed and final production at the surface could be speculated. The numbers quoted below do not bear any practical verification, but definitely serve the purpose. The balance sheet elaborates the distribution of subsurface oil/gas. The basis of the estimation is that all the oil/gas generated from kerogen in the source rock is taken as 100%. The distribution of the oil and gas is as follows.

- 70% oil/gas expelled from source rock. The 30% remaining adsorbed to kerogen and rock grains.
- The 40% from the expelled oil/gas was lost in secondary migration due to adsorption, dispersion, leakage, alteration and seepage of oil/gas.
- From the carrier bed and at the end of secondary migration, 30% of oil/gas is initially accumulated in the reservoir.
- During tertiary migration, the same processes of adsorption, dispersion, seepage and fluid alteration continue to reduce the accumulation by 15%.
- The net deposit of oil and gas in the reservoir pool was only 15%.
- Finally only 10%, out of 15% oil/gas in the pool, is recovered at the surface installation, through production operations.
- The 30% oil/gas that remained in source rock is subjected to metamorphism. The increasing burial, overburden pressure, temperature, compactness and geological time bring about modification to form graphitic carbon.

3.20   CONCLUSION

The process of migration and accumulation of underground fluid involves a stationery solid mass and migrating fluid. Source rock, carrier bed, reservoir rock and trap rocks are stationary solid masses. Fine- grained shale/clay source rock of very low porosity and permeability is able to retain the organic matter over long periods of geological time for proper maturation and conversion. Sandstone or limestone reservoir rock has enough porosity and void space to accumulate large quantities of oil/gas. Carrier bed rock is of similar characteristics as reservoir rock but without the trap rock, so oil continues to flow. Trap rock is of very low permeability so that it can prevent the escape of fluid.. The primary migration starts from the center of the source rock. Fluid expulsion from narrow pore source rock to wider pore carrier bed takes place mainly through pressure compression in primary migration. Molecular size, pore diameter, sediment type, temperature, pressure, tectonic, type of sediment, fluid phase, fluid solution, dispersion, suspension, emulsion, micelle and geological structural features, such as faults, fractures, joints and unconformities are the main contributory factor for primary migration. The water plays an important role in migration.

Oil–gas–water separation initially begins in wider pores of the carrier bed, during secondary migration outside source rock. Secondary migration is finally completed in a trapped reservoir rock. The buoyancy rise, capillary pressure and hydrodynamic pressure control the direction and rate of flow of oil/gas in secondary migration. Surface tension, wettability and contact angle are elated to capillary pressure. Reservoir fluid consists of water, oil and gas. Components of fluid separate according to the buoyancy and density differences in an accumulation. Gas occupies the top, water at the bottom and oil in the middle position of the reservoir. The movement of oil/gas, due to tectonic activities, in or out of reservoir rock is termed as tertiary migration. The seepage/leakage of oil/gas occurs throughout the petroleum system, but only those are counted as seepage/leakage that appear on the earth's surface. The main causes of seepage are underground geological activities, volcano eruption, water springs, mud volcanos and subsurface unconformities. Varied distance is travelled by the oil/gas, from a few meters to thousands of kilometers, during migration. It is speculated that of the total hydrocarbon generated in source rock, only 10% oil/gas is produced from the reservoir rock. Major portion of the earth crust is covered by sedimentary rock. Further geological information of sedimentary rock and crust is given in the next Chapter.

4 Petroleum Geological Survey

4.1 INTRODUCTION

A geological survey is an initial inspection of an area so as to draw a map to get a general view of the earth's structure. A petroleum geological survey collects data on a particular block of earth through a number of techniques and measurements and reports the data through a geological map or model for interpretation. The petroleum geological survey has become the back bone of petroleum prospecting, due to the integration of various disciplines of science and technology. The purpose of the petroleum geological survey is to locate a possible underground rock for the presence of oil/gas petroleum.

The first petroleum well, 'Drake Well', was drilled in 1859 in the USA. The site for the well was identified by a nearby oil seepage. At that time, oil/gas seepage was the only clue for finding petroleum in a particular field. But soon it was established that oil seepages do not always give positive information for an oil discovery. However subsurface oil and gas seepages on the surface of the earth are useful evidence for petroleum. Oil seepage is also an indication of underground geology. The appearance of subsurface oil and gas at the earth's surface as seepage provides information that there is enough driving pressure provided by the geology of the underground location. Geological processes continue to act on underground sedimentary rock. Fissures, faults, the displacement of rock strata and movement of fluid continue to occur according to changing subsurface geology with time.

For the last 150 years, the petroleum industry has expanded manifold, mainly due to advances in the fields of petroleum geological survey, exploration, drilling, production, refining, petrochemical and marketing innovations. The terms petroleum geological 'survey' and 'exploration' seem to be interchangeable. However they may be treated separately to get a better understanding of the subject. Even then, overlapping between these two terms, survey and exploration, cannot be ruled out. Primarily the distinction may be visualized in that the findings of a geological survey are more academic, scientific, theoretical, general and indicative in nature, whereas the results obtained by explorations are practical and close to reality. That is why, in most of the cases, exploratory boreholes drilled are converted into production wells. The geological survey is the initial step on the journey towards the discovery of petroleum, so its importance cannot be minimized. Almost all the petroleum producing wells are found in the sedimentary rock of the earth's crust. There are a few exceptions; some oil fields are found in basement igneous or metamorphic rocks. However, these rocks are located near sedimentary oil reservoir rock, and oil might have migrated from this to igneous or metamorphic rocks. Therefore a petroleum survey begins with the study of sedimentary rock in terms of its origin, structures, constituents, characteristics and other factors that contributed to the gradual buildup of the sedimentary rock over geological time and history.

Preliminary work, prior to surveying, is carried out by public/private organizations to formulate the permission/license/concession documents and demarcate the land to be surveyed. Later, the activities of survey/exploration, drilling, production and sale of petroleum are covered by additional legal documents.

Geological survey is conducted to gather the maximum information of the earth's crust and to differentiate between promising and non-promising areas for petroleum potential, so that expensive drilling operations are avoided. The initial study of the earth's surface by the geologist is conducted by traveling on land or flying by airplane to get a feeling of the surface and taking samples for laboratory examination. Under sea bottom surveys and sample collection are carried out by sea divers. Air and satellite photographic surveys are common and convenient nowadays. The data so obtained are represented by contour mapping (lines on a map showing points of equal character/property surface data). The collected earth surface data are correlated with the geological characteristics of the underground structure. The geological survey is comprehensive data presented by mapping. The petroleum survey encompasses the geological, physical, chemical and biological aspects of the rock supported by physics, computer and engineering. The work is multidisciplinary and enormous. A survey for petroleum prospecting comprises the following three studies:

- Petroleum geological survey
- Petroleum geophysical measurements
- Petroleum geochemical analysis

The last two subjects are discussed in later chapters of the book. The present chapter is concerned with the petroleum geological survey. The study of the earth's crust is the main activity of the survey.

4.2   EARTH CRUST AND ROCK

The earth crust is the distinct outer most solid shell of earth globe. Beneath the crust is the earth's mantle. The earth's crust is composed of hard rigid stone rock as well as soft sediment rock. Rock is a mass of naturally occurring conglomerates of loose or consolidated grains/particles of the same or different minerals. The average thickness of the crust is 17 kilometers.

The natural process of rock formation and erosion is continuous, occurring since the origin of the earth. The rock cycle depicts the origin, formation and erosion of rock. Upwelling hot magma from the inner earth's core passes through the earth's mantle and cools down and crystallizes to form igneous rock. The lava from the earth's mantle extrude into the above earth's crust. At crust condition the lava forms igneous rock. The earth's crust was formed from the igneous rock. The igneous rock leads to the formation of metamorphic and sedimentary rocks. The igneous rock was formed by underground magma at high temperature and pressure. The minerals of igneous rock cannot remain stable and in their original shape in atmospheric conditions prevailing in the earth's crust. The minerals of igneous rock are altered, changed and modified to attain equilibrium with the atmospheric environment. Under atmospheric conditions the igneous rock cools down and is subjected to weathering and erosion, creating the raw material needed for sedimentary rock. The sedimentary rock is formed by the slow deposition, compaction and consolidation of fragmented particles of igneous and metamorphic rocks. The sedimentary rock in geological time sinks down into the earth and metamorphoses (geological modification) to metamorphic rock under conditions of overburden pressure and high subsurface temperature. Ultimately the metamorphic rock sinks down further into the earth's mantle and core to become part of the hot magma, thus completing the rock cycle.

Initial testing of rock is carried out by taking a small piece of rock sample from the visible outcrop or core and examining it under a microscope. The color, texture, particle grain size, geometry, mineralogy and rock types are determined. The data become the basis for determining the types of rock and identifying the subdivision of the rock into individual litho-stratigraphic units. The geological branches that study rock are 'petrology' and 'litho-logy'. The difference between petrology and lithology is that the former focuses on and studies microscopic information and the latter on the macroscopic description of the rock. Both describe physical characteristics, origin, composition, constituents and their distribution and rock structure.

4.2.1 ROCK TYPE

Rock type is determined on the basis of the dominant mineral present in the rock. Minerals are the fundamental units of rock. From the minerals the whole structure of the rock is built up. A mineral may be defined as a 'basic unit of the rock'. A mineral is naturally occurring abiogenic (inorganic) solid, mostly crystalline with a fixed chemical composition. The study of minerals is known as mineralogy. An ore is a commercial mineral from which one or more valuable elements are obtained by chemical treatment, for example hematite iron ore. Oil/gas, coal and subsurface water are not considered either minerals or ores, because of their distinct characteristics. On the basis of mineralogy, the whole of the earth's crust consists of three different types of rock:

- Igneous rock
- Metamorphic rock
- Sedimentary rock

The earth's crust is composed of igneous, metamorphic and sedimentary rocks. The mode of formation, characteristics and function of each type of rock are different. The main features of the three types of rocks namely igneous rock, metamorphic and sedimentary rocks along with their mode of formation, characteristics and examples of each rock, are summarized in Figure 4.1. The position of sedimentary rock along with igneous and metamorphic rocks in the earth's crust surrounding the earth's mantle is explained in Figure 4.7. Additionally the earth's crust may be defined as

```
                          ┌─────────────────┐
                          │      Rock       │
                          │ minerals composit│
                          └─────────────────┘
```

Igneous rock

Formed by cooling
  and solidification
  of magma

Sedimentary rock

Formed by sedimentation
  of weathered rock, organic
  residue and mineral
  by precipitation

Metamorphic rock

Formed when igneous
  and sedimentary rocks
  are subjected to intense
  heat and pressure

Characteristics

-Crystalline
-Rate of cooling
  determines whether
  its course/fine grained

Characteristics

-Broken inorganic
  sediments
-Chemically formed
-Organically formed
-Layered appearance
-May contain fossils
  or crystals, depending
  on how they were formed

Characteristics

-Crystalline
-May have banded
  or layered structure

Examples

-Granite
-Basalt
-Rhyolite
-Quartz latite

Examples
-Sandstone
-Shale
-Limestone
-Coal

Examples

-Marble
-Gneiss
-Quartzite
-Slate

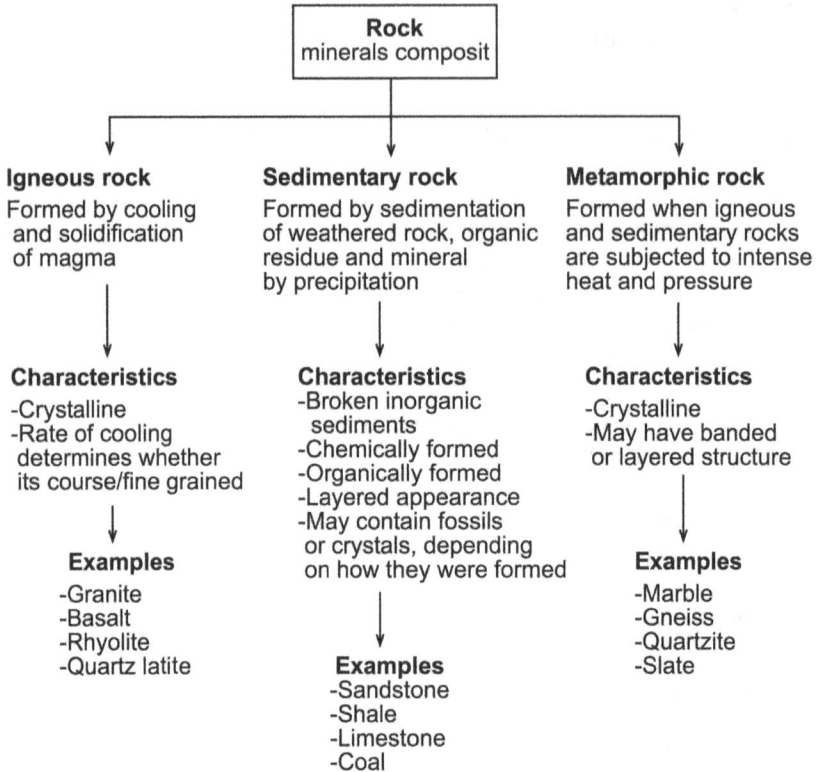

FIGURE 4.1   Rock formation, types, characteristics and examples.

composed of sedimentary, igneous, metamorphic rocks, hydrosphere, atmosphere and biosphere. The earth's crust is floating above the asthenosphere. The asthenosphere is the upper part of the earth's molten mantle.

4.3   IGNEOUS ROCK

Igneous rock is composed of crystalline solid minerals formed directly from the earth's hot molten magma from the earth's mantle. The majority of the earth's crust, about 90%, is composed of igneous rock. But it is mostly covered by metamorphic and soft sedimentary rock. At the inner end of the crust or the border line between the hot mantle and the crust, the rock materials are almost igneous. Igneous rock is formed from the cooling and fractional solidification of hot magma. With cooling, the dissolved minerals crystallize according to their melting points and separate out as solid crystals. Different locations and cooling conditions give rise to two types of igneous rock.

- One is the 'plutonic igneous rock' that is formed below the earth's crust in the outer portion of the earth's mantle. The upwelling hot magma from the earth's core and mantle is surrounded by cooler crust materials. Crust material forms a thermal insulation. Therefore the crystallization of minerals takes place under a very slow process of cooling. The conditions favor the formation of large perfect crystalline substances.

- The second type is 'volcano igneous rock'. It is formed by the rapid cooling of the ejected molten lava at the surface conditions of temperature and pressure. Lava is the basalt silicate ejected from the mantle to the earth's surface. The rapid cooling of lava results in rapid solidification and crystallization. The crystals formed are of poor quality, small size and ill-defined geometry.

4.3.1 Mineralogy of Igneous Rock

Silicate groups of minerals are rock-forming material and constitute about 99% of the molten magma of the earth's mantle and igneous rock. Silicate mineral consists of anions of silicon and oxygen atoms $(SiO_4)^{-n}$ together with metallic cations of Al, Na, K, Fe, Ca, Mg and some other metals.

Silicates are polymeric compounds composed of individual metal oxides linked together through various covalent and coordinate bonds, forming enormous numbers of silicate minerals. The oxides are the basic units (monomers) of the macromolecule silicates. The chemical formulae of silicates are complex, but can be simplified to 'metal oxide – aluminum oxide – silicon oxide'. These oxides are naturally arranged in different ways to produce a variety of stable neutral molecules forming different groups. Crystal and chemical structures of silicates are discussed in Chapter 8. On the basis of their similarity in mineralogy, the total silicates in magma may be conveniently placed into seven mineral groups. Brief descriptions of each silicate mineral group are given below.

4.3.1.1 Silica or Quartz Mineral Group

Pure silica (SiO_2) is an amorphous whitish power and has variable properties. The silica is not found, as such (pure stae), in igneous rock. It occurs abundantly in many polymeric forms. The most common is quartz. The quartz silicates contain only silicon and oxygen atoms. The fundamental monomer unit of the polymeric quartz minerals is SiO_2. The SiO_2 units combine themselves to produce a continuous framework of tetrahedral $(SiO_4)^-$ crystalline structure. Each oxygen atom of unit $(SiO_4)^-$ is shared among other tetrahedrals $(SiO_4)^-$ giving a net chemical formula SiO_2. During cooling and crystallization, the prevailing conditions of temperature, pressure and composition determine the number of SiO_2 units that combine and arrange to form macromolecule quartz silicate. Pure silicon dioxide polymerized macromolecule minerals beside quartz are latite, tridynite, flint and chert. Sand is made up of tiny granules of quartz minerals that have been broken by wind and water. Trigonal quartz is the second most abundant mineral, after the feldspar mineral in the earth's crust. Pure quartz is colorless, but tinted due to impurities. It is the hardest of the common minerals. Quartz is also associated with beach sand grains and limestone. Many silica group minerals are semi-precious stones and are used in jewelry.

4.3.1.2 Feldspar Mineral Group

Feldspar is the most common of all silicate minerals, constituting about 75% of the total silicates, followed by quartz, having the general formula $[MNO_8]$. M represents sodium, potassium, calcium and barium, N is silicon or aluminum. All feldspar contains aluminum and silicon. The general chemical formula can also be written as $[M(Al, Si)O_8]$. The first groups of feldspar are the silicates of potassium $[K(AlSiO_8)]$, and barium $[Ba(Al_2Si_2O_8)]$ is mostly present in magma and igneous rock. Groups of feldspar silicates containing the lighter atoms sodium and calcium are mostly found in metamorphic rocks. Anorthite feldspar is rich in calcium, and albite feldspar is rich in sodium.

4.3.1.3 Feldspathoid Silicate Group

The feldspathoid silicates resemble feldspar, but have lower silica contents than alkali-aluminum silicates. The group is formed in magma that is rich in alkali and deficient in silica. The group consists of seven types of silicates. The most well-known feldspathoid silicate is leucite.

4.3.1.4 Pyroxene Silicate Group

Pyroxenes are a large class of igneous rock-forming minerals containing calcium, magnesium and iron. The group is also known as ferromagnesian silicates. The minerals crystallize in both orthorhombic and monoclinic crystal systems. Pyroxenes have the general formula $[XY(Si,Al)_2 O_6]$, where X = Ca, Na, Fe(II), Mg and Zn, and Y = Cr, Al, Ti, Fe (III) and Co. Chemically it is analogous to amphibole silicate except that the hydroxyl group is absent in pyroxene. The group contains 12 types of silicates, typically in the form of prismatic crystals.

4.3.1.5 Amphibole Silicate Group

The amphibole silicate group is rich in magnesium atoms and occurs with feldspar and quartz minerals. The amphibole group obeys the same chemical system as the other silicates. That is, the metal oxide – aluminous oxide – silicon oxide system. Additionally, the hydroxyl (OH^-) functional group is attached to the amphibole silicate. Occasionally 'OH' in replaced by a fluorine (F^-) ion. Most of the amphibole silicate series are monoclinic crystal. However some silicates of the group exhibit amorphous characteristics. The anthophyllite and hornblende are well-known silicates of the group. Amphibole silicates are multi-colored from dark to light shades.

4.3.1.6 Olivine Silicate Group

Olivine silicate is an olive-green or grey-green or brown mineral in igneous rock, containing varying percentages of bivalent metals iron, magnesium and others. Olivine silicates are represented by the general formula $(Fe, Mg)SiO_4$. Their crystal structure is orthorhombic. Olivine is used as a gem stone and is said to be associated with healing properties.

4.3.1.7 Mica Silicate Group

Mica is the most common silicate and has a shiny texture. The mica group silicates are layered and sheet-forming materials. All have a monoclinic crystal structure.

The mica breaks along the cleavage plane, forming very thin elastic slices. The basic unit is the SiO_4^- tetrahedral ion. Each tetrahedral ion is linked by three corners to the neighboring tetrahedral forming a sheet structure. The crystal structure of mica is monoclinic. The most common mica silicates are biotite and muscovite. Mica has various colors from white to brown to black. Muscovite produces colorless transparent slices. 'Biotite' refers to the dark mica series. The mica group silicates possess a high density range between 2.8 and 3.1 g/cm^3. Mica is used as a thermal and electrical insulator.

4.3.1.8   Igneous Composite Rock

Common components of igneous rock are course-grained granite and fine-grained basalt rocks. The major components of granite rock are quartz, feldspar and amphibole silicate minerals. Quartz imparts a light color to granite whereas the other minerals give it a colored tinge. A small quantity of iron mineral imparts a reddish or pinkish color in granite. Granite is a useful construction material. The next most important component of igneous rock is a fine-grained basalt igneous rock. It is a composite of olivine, magnesium-rich pyroxene and calcium-rich feldspar silicate minerals. The thermal conductivity of basalt is very high. It cools rapidly. While cooling rapidly during volcano eruption, gas bubbles are trapped in basalt that look like holes.

4.3.2   Crystallization of Silicate Minerals

Hot magma inside the earth's core is almost in a homogenous state due to very high temperature, forming a uniform solution of silicate minerals. The separation of silicates begins in comparatively less temperature prevailing in the earth's mantle The separation of minerals from the magma is a fractional crystallization process according to their melting points inside the earth. During crystallization, some interaction between minerals also takes place to produce new varieties of silicates. With the cooling of the magma, the olivine silicates are the first to crystallize/solidify from the solution at about 1200°C. As the temperature drop further, the olivine silicates interact with silica minerals, to produce pyroxene silicates. The pyroxene starts to crystallize at some lower temperature (<1200°C), followed by amphibole minerals. Fractional crystallization of the silicates continues from 1200°C and ends at about 750°C. The order of crystallization (precipitation) of some silicate minerals from the hot molten magma is as follows:

Olivine (first) → pyroxene → amphibole → feldspar (calcium-rich anorthite → feldspar sodium-rich albeit) → quartz (last)

4.4   METAMORPHIC ROCK

The dominant components of the metamorphic rock are silicate minerals with various structural formulae. Metamorphism means change in shape of a mineral from one to another. It is a change of a solid into another solid form with minor or substantial chemical alteration. Metamorphism produces new or modified minerals, originally present in igneous or sedimentary rocks. Metamorphism tries to establish a new chemical equilibrium in the changed environment, to produce modified and stable minerals. Metamorphic rock is a changed version of the igneous or the sedimentary rock. Igneous rock, while cooling under the lower temperatures and pressure near the earth's surface and undergoing erosion, forms metamorphic rock. Sedimentary rock, while buried deeper into the ground at higher temperatures and pressure, transforms into metamorphic rock. Metamorphism is a subsurface process under varied temperature, static-dynamic pressure and geochemical conditions. Static pressure reduces the volume and increases the density of the mineral. The dynamic pressure changes the shape and geometry of the mineral. Subsurface conditions of temperature, pressure and geochemical environment cause the re-crystallization and modification of the mineral within the solid state. The geochemical nature of minerals and the chemical environment, namely the presence of carbon dioxide and acidic materials, all affect metamorphism. The presence of active fluid (acidic water) or other chemical species may facilitate chemical conversion by adding or removing certain material from the original mineral during metamorphism. A mineral may be soluble under certain pressure-temperature & chemical conditions but may assume a new shape under different conditions. Few examples of metamorphism are as follows:

• Limestone metamorphosed to marble.
• Quartzite mineral is a metamorphosed version of sandstone.
• Mudstone and shale metamorphosed to form slate rock.

Metamorphic and igneous rocks were mentioned as 'basement rock' in connection with oil and gas production. Basement rock does not exist as a separate type of rock. Normally the basement rock refers to metamorphic or igneous rocks located below sedimentary source rock. Due to some geological processes, for example tectonic, the sedimentary petroleum source rock moved below or merged with the igneous/metamorphic rocks. That is known as the sub-cropping or in-cropping of sedimentary source rock into the basement (metamorphic/igneous rock). The sub-cropped sedimentary rock may carry oil/gas accumulation.

4.5 SEDIMENTARY ROCK

Sedimentary rock is important from the point of view of petroleum. Sedimentary rock is a thin layer of solid material (sediment) overlaying the igneous and metamorphic rocks of the earth's crust. Sedimentary rock is composed of different organic and inorganic materials and minerals and are formed by gradual deposition, accumulation and sedimentation of the organic and inorganic matter (sediments) on land surface and in aquatic (sea, lake) bottom. The deposition of the large amount of sediments forming rock takes geological time. Strata of sediments lie on top of each other. The deposition and preservation of organic matter, its conversion to petroleum, migration and accumulation occurred in the subsurface sedimentary rock.

The earth's crust was primarily formed from igneous and metamorphic rocks. The exposed, weathered and eroded igneous/metamorphic rocks provides the necessary raw material (sediments and mineral particles) for making the sedimentary rock. The sedimentary rock is generated by the slow deposition of fragmented particles of igneous and metamorphic rocks at comparatively lower temperature and pressure, in the presence of air, wind and water currents prevailing at the earth's surface or in shallow water. One may expect that the distribution of minerals in sedimentary rock should be similar to that in igneous rock, since the sedimentary rock is formed from the weathered fragments of rock. Ingenious rock is fairly homogenous, and sedimentary rock is heterogeneous. Igneous rock contains few minerals, whereas sedimentary rock is a composite of many minerals. Mineral distribution in sedimentary rock is wide and non-uniform. The period between the weathering of igneous rock and sedimentary rock formation is very long (geological time) and subjected to several variable geological conditions, that not only affect the mineralogy but also drastically change the physical and chemical characteristic of the minerals, however stable mineral retains their originality.

Different types of sediments/minerals form different types of sedimentary rocks. Each rock has its own distinct characteristic according to the sediments and mineral type. On the basis of the dominant minerals, three main types of sedimentary rocks have been identified, namely shale, sandstone and limestone rocks (Chapter 3). Minerals are heterogeneously distributed in the rock. Even within a particular type of rock, characteristics differ from one point to another considerably.

A sedimentary rock is mostly characterized by its thickness, which varies from a millimeter to many hundreds of meters. The rock differs in uniformity and thickness. A rock layer may be thick and broad at one end and thin and sharp at the other end. The characteristics vary continuously from one end, through the middle, to the other end or over lapping may occur. Sedimentary rock covers most of the land and some portion of the sea bottom. Some portion of the earth's crust is without sedimentary cover. It is especially true for the bottom of ocean where igneous and metamorphic rocks occupy the earth's crust and are devoid of sedimentary crust. A thick succession of sedimentary rocks is found in the ocean's continental margin; the thickness may range up to 7 km. The average thickness of the sedimentary rock is 1.5 meters (approximately) of the whole earth's crust. The thin layer covers about 72% of the earth's surface, but quantitatively sedimentary rock forms a small percentage of the crust compared to the igneous and metamorphic rocks.

On the basis of the mode of formation of minerals and their incorporation into the sedimentary rock, three basic forms of sedimentary rocks are well known.

- Mechanically formed clastic sedimentary rock
- Organically formed sedimentary rock
- Chemically and evaporite sedimentary rock

But considering the content of organic matter in the form of fossils and the composite nature of sedimentary rock, the following two additional forms are included:

- Fossiliferous sedimentary rock
- Composite sedimentary rock

4.5.1 Mechanically Formed Clastic Sedimentary Rock

Clastic sedimentary rock is also called detrital sedimentary rock. Clastic and detritus refer to broken stone pieces of all kinds. Loose, broken and detached pieces of existing rock are known as fragments/particles/grains. The particles/grains are of various sizes and shapes. Size and shape are an important subject for the petroleum geologist and help in understanding the underground 'petroleum system'.

The stone pieces are generated usually in high mountain areas by the natural process of weathering. Weathering is a slow process of the breaking of rock by the mechanical forces provided by air, wind, water current, wave, ice and gravity. The stone pieces are carried to faraway places by erosion and ultimately settle down in low lying regions in stagnant aquatic basins (sea, lake) or land depressions. Erosion is another form of weathering. It is accompanied by substantial mechanical movement. The erosion removes the pieces from the location of their origin (mountain area) and takes them to areas of settlement usually a land surface depression for accumulation and deposition. Along with continuous deposition, the overburden pressure of the sediments causes the expulsion of trapped water in the stone pieces (particles/sediments). Further compaction and cementation of the sediments result in a solid mass of sedimentary rock. The compaction and cementation together are known as 'lithification'. The lithification leads to the formation of clastic sedimentary rock. Clastic sedimentary rock consists of silicate, clay, sand, sandstone and minerals. Some specific examples of clastic rocks are as follows:

- Quartz-sandstone is a medium-grained clastic rock that contains more than 90% quartz grains (silica).
- Greywacke is a sand grain clastic rock surrounded by dark fine-grained clay matrix.
- Shale is a fine-grained clastic sedimentary rock composed of silt and clay grains. It is a layered rock and can be fissile (broken).
- Siltstone is a harder and courser-grained rock than the shale.
- Clay-stone consists of clay-sized grains and is hard and not fissile.
- Mudstone is composed of silt and clay-sized grains, compact and massive.
- Conglomerate sedimentary rock is formed under pressure, and is composed of clastic rounded gravel/pebbles/grains cemented together by a fine-grained matrix.

4.5.2 Organically Formed Sedimentary Rock

Calcareous (containing calcium) and siliceous (containing silicon) materials of biological origin composed of calcite ($CaCO_3$) and chert (SiO_2) minerals contribute to the formation of organic sedimentary rock. Calcium carbonate and silica are the constituents of calcareous and siliceous shelled marine living organisms (zooplankton). They build their shells through the assimilation of dissolved calcium and silicon compounds. The remaining materials of the dead zooplankton continue to be accumulated in the aquatic bottom forming layers of sedimentary rock With the passage of geological time, overburden pressure increases, the accumulated material is converted to a calcite and chert sedimentary rock. Additionally chemical alterations of calcite rock take place in the presence of subsurface saturated magnesium solution. A double salt mineral, known as dolomite ($CaCO_3MgCO_3$), is formed.

4.5.3 CHEMICALLY FORMED AND EVAPORITE SEDIMENTARY ROCK

Under-ground water may be stationery or running depending on the environmental condition. The water running through subsurface rock formations dissolves some of the constituents of the rock. With time and distance traveled, the dissolved minerals are built up to saturation point in the water. At the saturation stage, less soluble minerals, limestone and dolomite, precipitate out and are redeposited and accumulate in a new environment to produced calcite and dolomite rocks. The chemical rock is also formed from naturally occurring minerals or by chemical precipitation. Precipitated calcium carbonate is formed by the interaction of carbonic acid with calcium ions present in the soluble saturated solution. Dolomite is the double carbonate of calcium and magnesium. At shallow depth, calcite is metamorphosed into oolite sedimentary rock. Oolite is a spheroidal grain of lime. Gypsum ($CaSO_4$) is a mineral of salt rock; it is precipitated and deposited from water containing sulfate ions.

More soluble minerals, halite, anhydrite and gypsum, remain in solution, until all of the water is evaporated, leaving these minerals as residue to form evaporite rock. Salt rock and gypsum rocks are in fact true evaporite sedimentary rock.

4.5.4 FOSSIL/FOSSILIFEROUS SEDIMENTARY ROCK

Fossils are organic constituents of some sedimentary rock. Fossils are also building blocks (sediment) of sedimentary rock like inorganic minerals/sediments. Fossils are the mineralized remains of dead plants and animal organisms in the rock. Fossilized sedimentary rock may be called 'fossiliferous rock'. On equating the sedimentation with fossilization, one can find that the origins of sediments and fossils are different. Their process of incorporation into sediment agglomerate is different, but in the end, both become constituent and integral parts of the rock. Fossils from 30 to 500 million years old have been recorded.

Living organisms consist of organic and inorganic substances. Inorganic refers to metallic or metalloid materials forming bones, skeletons, teeth, shells, skulls, vertebrates and exoskeletons of different organisms. The salt and water contents of organisms are also inorganic but they are not fossilized. Tissue, hair, feathers and skin are organic substances. The inorganic parts are less reactive (more stable). Fossils originated from inorganic parts are abundant. The organic substances are more reactive (less stable) and yield to the environmental conditions and decay with time accordingly. However, plant branches, stems, roots and trunks, though organic, have been found as fossilized and preserved components in sedimentary rock.

4.5.4.1 Fossilization Process

Fossilization is a process of incorporation of fossils into the sedimentary rock. The process of fossilization is the incorporation of organic matter into the sediments and sedimentary rock. The remains of dead organisms are associated with different sediments. The associated parts undergo settling, accumulation, compaction, contraction and cementation along with sediments in different layers of the sedimentary rock with different quantity and quality. The processes of preservation, fossilization and finally mineralization of fossils depend on the types of organisms and environments. The ultimate end of all fossils is incorporation, mineralization and metamorphosis under temperature and pressure into the rock matrix. Three fossilization processes are as follows:

- The mineralization of the fossils of the dead organisms is known as the 'per-mineralization process'. The soft parts (tissue) of the dead organism are replaced by inorganic sediment minerals. Sediment grains find their way into the pores and cavities of the dead body of the organism through saline water. This mostly happens in aquatic conditions, where hardening and compaction take place. With compaction and the passage of time, water is expelled, leaving behind the inorganic sediment with preserved and fossilized organic remains.
- A natural 'mold and cast' are formed by the hard-shell structure of the animals, buried in sediment. Slowly and gradually the outer surface of the shell is replaced by minerals. Ultimately the original walls of the shell are replaced by hard sediment. The hard sediment retains the original shape of the mold (shell). The original pattern of the shell is available but made of a new re-crystallized agglomerate of sediments.
- The fossilization of organic parts take place by the 'carbonization process'. The carbonization process is the conversion of organic matter into carbon residue. Fossil containing major portions of hydrocarbons, particularly from terrestrial plants, undergo catagenesis/metagenesis conversion and the reduction of complex molecules, leaving almost pure carbon residue or carbonaceous residue. The carbon residue is stable. The residue is adsorbed and compacted on the mineral grains surface with time. The process is known as 'carbonization', resulting in the fossilization of carbon grain. When the rock is broken and split into parts, one portion of the rock appears with a layer of carbon and the other with only the 'impression' (print) of carbon grains, without the fossil carbon layer. The formation of the carbon film is due to the 'compression' in the presence of high pressure and temperature conditions. The carbon film represents an outline of the original geochemically altered state of the fossil. If the clay has adsorbed carbonaceous film it turns black. The black shale represents the presence of fossilized carbon in the rock.
- The animal shell fragments are compacted and preserved as calcareous and siliceous minerals.

4.5.4.2 Factors for Fossilization

Favorable conditions for fossilization are initially similar to the conditions for the generation of petroleum source rock. Dead organisms that are rapidly buried in the aquatic bottom, and also rapidly covered by enough sediment, are likely to be fossilized properly. The dead remains of organisms are prone to decay and destruction, by natural processes of erosion and chemical decomposition. The dead remains are also consumed by living organisms. These processes are more rapid and destructive in land conditions; therefore their remains vanish, leaving almost nothing to be preserved and fossilized. Most fossils are found in aquatic conditions.

The process conditions between the death and burial of the organism into the sediments are important for favorable conditions for fossilization. A favorable scenario is rapid burial without much decay and consumption, and a high rate of sediment deposition. These conditions are met in quiet and stagnant aquatic environments. Flowing water and harsh conditions on land facilitate the rapid destruction of organic matter, therefore the contribution to fossilization is minimal.

4.5.4.3 Types of Fossils and Fossiliferous Rock

Types of fossil depend on the type of the organism from which the fossil was generated in the sedimentary rock. Chemically the fossils are carbonate/silica minerals and carbon residue. The majority of fossils were formed by zooplankton rather than phytoplankton. Several types of fossils are known, namely body fossils, microfossils, trace fossils, index fossils and biochemical fossils.

A 'body fossil' is the preserved hard and bony parts of the dead organism in the sedimentary rock. Chemically body fossils are either calcareous (carbonate) or siliceous (silica) substances. Most known body fossils are of corals and clam species. Corals are marine invertebrates. The clam is a common name for bi-valve mollusks. The animal species are found in coastal belts or at shallow depths, where they get their food from land-derived sources. The shells of organisms are mainly composed of limestone calcareous minerals. On being compressed and cemented, the rock generated from these bony skeletons is known as 'calcareous rock' or 'calcareous ooze'.

'Microfossils' are found in the deep ocean bottom. The fossils contain siliceous matter. The microfossils were produced from the remains of algae bacteria microorganisms. The remains of the microorganism, on being compressed and cemented, become constituents of the rock. Chemically the microorganisms composed of siliceous compounds (SiO_2 units) form "siliceous rock" or "siliceous ooze". Siliceous ooze is a soft deposit of fine-grained siliceous remains along with some clay. Various types of microfossil exist.

'Trace fossils' are not fossils themselves. They represent the past activities of some animals. The animal can be traced back by observing certain marks left behind by organisms on sedimentary rock. The marks left are foot prints, bores, barrows, tracks and trails and bite marks. The life of the animal can be interpreted by examination of these trace marks.

'Index fossils' give indications of the rock associated with petroleum. Such index fossils are geochemical and retain their original characteristics of organic matter from which the petroleum was generated. Presence of geochemical is also related to the origin of petroleum oil. Such geochemical is known as 'biomarker'.

'Biochemical fossils' are the biochemical indicators, left behind by the organisms, after their death. Solid execration (feces) and inorganic residue of liquid (urine, blood, etc.) are biochemical indicators, and may throw some light on the life of the dead animal.

Fossils have both theoretical and practical applications. The study of fossils is a useful tool for the study and age determination of past life and events. The representative samples of fossils are drawn from underground strata at different depths from rock cuttings and core samples. The study of fossils has become a specialist subject called paleontology.

4.5.5 Composite Sedimentary Rock

Any demarcation on the basis of the form (mode of formation) or type (dominant mineral) of sedimentary rock is not 100% correct. All rocks are formed and composed in widely varying proportions of broken pieces of several kinds of inorganic minerals/organic sediments derived from clastic by mechanical, chemical and organic means as elaborated in Section 4.5 above. Among the three types of sedimentary rock, limestone rock is distinct from the others on the other hand similarities and overlapping exist in shale and sandstone rocks. Limestone minerals exist in substantial quantity in sandstone rock. Likewise sandstone minerals are found in appreciable quantity in limestone rock. Both sandstone and limestone minerals may form composite sedimentary rock.

4.6 SEDIMENT AND SEDIMENTATION PROCESS

Sediment is a term used for the accumulated and buried fragments/particles of solid materials consisting of both inorganic rock minerals and dead organic substances. Sediments move from one place to another on the surface of the earth, ultimately settling down in a depression leading to the formation of sedimentary rock.

The sedimentation process is defined as, 'The deposition, accumulation, compaction and distribution of sediments loaded in water, wind and ice into the basin for possible rock formation'. Sedimentation is a sequential deposition and conversion of sediments into rock. Sedimentation is a complex physical, chemical, biological and geological process involving the erosion, transportation and deposition of sediments and ending with the formation of rock. A related term is 'sedimentology', that is the knowledge of sediment that sinks to the bottom. The sedimentation process begin with tiny particles and lead to the formation of a solid compact sedimentary rock. Sedimentation is slow process; it can only be expressed in a geological time scale. The solid particles in suspended form in air and water settle down on the bottom of the basin. The assemblage of sediments forms a solid rock layer over geological time under proper geological conditions. The weathering, erosion and decay of inorganic and organic materials (rock & dead organism particles) are the main factors of the sedimentation process, which is brought about in the presence of the hydrosphere and atmosphere. Broadly sedimentation is the interaction of earth crust materials (sediments and mineral particles) with the hydrosphere and atmosphere.

The sedimentation process follows the 'law of superimposition' or the 'law of succession'. The deposition of sediments one by one on top of each other forms a series of layers (strata), with older layers down near the bottom, and new layers above nearer to the earth's surface. The most recent layer occupies the top of the already accumulated sediments (rock).

The sediments are continuously transported and preserved in the basin over geological time. More and more sediments continue to be deposited in the basin, increasing the 'litho-static overburden pressure'. Recent and loose sediment particles are soft and unconsolidated. With continued sediment deposition, and the increase of overburden pressure and burial time, the sediments are sedimented, compacted, cemented and become hard rock.

The void pore spaces between sediment particles are normally filled with water. Ground water plays very important role in sedimentation and the whole petroleum system from the generation and accumulation of organic matter to the production of oil and gas. Ground water is actually rock pore water. With burial time and over-burden pressure, the sediments are compacted, pore sizes are reduced and water expelled. However, regardless of compaction, there will be some pore space filled with water. Under these conditions some dissolved salt in the water is precipitated out, and adheres to the walls of the sediment grains, further reducing the pore void space. The coating of the walls of the grain with salt grows steadily and continues to reduce the pore diameter. The coating and adhering of salt to the pore wall is known as the 'cementation' process. Limestone, quartz and halite are common cementation agents. Cementation and pressure expulsion of water add to the solidification of sediments into a hard, rigid rock.

4.6.1 MINERAL STABILITY AND SEDIMENTATION

The mineral particles in sedimentary rock are mostly derived from the weathered igneous rock. Igneous rock was formed long before, during the cooling of hot magma inside the earth. Obviously the presence of mineral particles in sedimentary rock is dependent on their reactivity/un-reactivity or stability/instability over all geological time.

About 92% of the sedimentary rock consists of silicate minerals. Among the silicates, the most abundant are feldspar and quartz. Other prominent minerals are calcite and evaporite. The following is a list of minerals from the most stable/un-reactive down to the most unstable/reactive.

Quartz $(SiO_2)_n$ is the most stable and un-reactive. It retains its original chemical form and is present only as detrital fragments of the mineral in sedimentary rock.

Feldspar minerals are stable but undergo changes over geological time. The conversion of feldspar to other forms of silicates is known. Feldspar is a large group of silicates forming 60% of the earth's crust.

Mica is the next group of silicates which are moderately stable but undergo conversion with time to form new types of mineral, under aquatic conditions.

Amphibole, pyroxene and olivine silicates are less stable and convert into secondary minerals.

Clay minerals are secondary silicates. Clay minerals are formed from other silicates, by the action of air, water and carbonic acid. Clay silicate minerals undergo hydrolysis, in the presence of water, to produce aluminum hydroxide and silicic acid. Both aluminum hydroxide and silicic acid are reactive and produce other types of silicates. Most of the clays are hydrated minerals. The grain size of clay is very small (<0.004 mm). Clay can form gels and colloidal solutions. There are many forms of clay minerals, depending on the chemical formula and their structure. Important clay minerals that make up the sediments are kaolinite, chlorite, illite and montmorillonite. All these clay minerals are soft, white with slight tinge and with layered structures.

- Mud is a mixture of clays and other silt-sized minerals, namely quartz and calcite.
- Shale minerals are formed from the compaction of mud. The shale mineral is fissile, soft and laminated clay.
- Mudstone is hard. Mudstone and shale are the tertiary products of clay; they are stable in atmospheric conditions.

Calcite and dolomite are carbonate. The minerals are stable in the absence of aqueous medium. The minerals undergo physico-chemical change in the presence of moisture. Biological processes play a vital role in their formation and conversion to a more stable form of calcite. Dolomite is formed by the interaction of calcite and magnesium ions in solution.

Evaporite and halite are important minerals of sedimentary rock, and are stable under atmospheric conditions but highly soluble in aqueous medium. They interact with other soluble salts and convert from one form to another.

4.6.2 SOIL AND SEDIMENTATION

Soil is the upper soft solid layer of the crust. Soil formation and sedimentation are closely related to each other. However, the changes occurring in the soil take place in a short time compared to geological time. Soil is composed of the constituents formed by the weathering and erosion of rock and brought by flowing water, rainfall, rivers and streams, floods, wind and rock sliding. Later on, the same factors cause soil erosion as well, coupled with natural and human activities such as natural climate change, plantation, agriculture, deforestation, urbanization, road, bridge, dams, mining, industrial, commercial and economic activities, etc.

Soil is a mixture of mineral particles derived from solid rock, organic matter and water. Mineral particles include sand, silt, clay and others. The relative proportion of these minerals determines the fertility of the soil. Sand is made up of coarse-grained particles and is unable to retain water; therefore pure sand soil is not fit for agricultural purposes. Clay is a fine-grained mineral and adsorbs water and leads to mud formation, therefore it is also unfit for vegetation. On the other hand, fine-grained slit particles associated with sand and clay particles provide a fertile land.

4.7 PHYSICAL FEATURES OF SEDIMENTARY ROCK

The majority of sedimentary rock is of shale minerals, followed by sandstone, limestone and evaporite rocks. The approximate share of each type of rock in the world's sedimentary rock is given below:

Clay/siltstone/mudstone/shale rock	=70%
Sandstone rock	=15%
Limestone rock	=10%
Evaporite/anhydrite rock	=5%

Sedimentary rock is made of minute particles of minerals and sediment particles called 'grains'. A small sandstone detrital rock piece contains multiple grains. When small parts (detrital pieces) of reservoir rock are examined under a microscope, the following components can be easily distinguished in the rock matrix:

- Detrital (broken) rock. It is a small broken piece of rock. It has appreciable porosity. It contains numbers of mineral grains (for example sandstone grains). The grains may be touching each other or separated to create void space (pore).
- Matrix. This is the small detrital rock piece in which mineral grain, filler and cementing agent are embedded.
- Grains. The grain is a small mineral particle that makes up all rock. It is the framework of the rock, for example sand grain. Many mineral grains exist in detrital rock pieces.
- Fillers. These are the pore-filling solid materials, for example clay, silt and sand.
- Cement. This is the cementing material between the grains' void space, for example limestone, quartz and halite.
- Pore. This is the void space in rock or between the grains. The pore may be empty or filled with a fluid (water, oil, gas). The void space is called interstitial space or pores.
- Pore fluids are water, oil and gas. Water preferentially adheres to the pore rock wall, making a water film. Gas occupies the center of the pore, and oil is between the water film and the gas.

4.7.1 GRAIN SIZE CLASSIFICATION OF SEDIMENTARY ROCK

Sedimentary rock classification is based on many considerations. Classifications of rock based on the nature/type of minerals and mode of formation have already been dealt with earlier. Now the classification is considered on the basis of mineral grain size. Not only is the nature of the mineral components of sedimentary rock important, but also equally the size (diameter) of the mineral grains plays

TABLE 4.1
Sedimentary Rock Grain Size

Particle/Grain	Size Diameter mm	Sedimentary Rock
Boulder (large stone)	>256 mm	–
Cobbles (paving stone)	64–256 mm	Conglomerate
Pebbles (small stone)	4–64 mm	”
Granule (particle)	2–4 mm	”
Very coarse sand	1–2 mm	Sandstone
Coarse sand	0.5–1 mm	”
Medium sand	0.25–0.5 mm	”
Fine sand	0.125–0.25 mm	”
Very fine sand	0.0625–0.125 mm	”
Course silt	0.031–0.0625 mm	Silt
Medium silt	0.016–0.031 mm	”
Fine silt	0.008–0.016 mm	”
Very fine silt	0.004–0.008 mm	”
Clay	<0.004 mm	Clay/shale
Carbonate	0.003 mm (approx.)	Carbonate

a vital role in the underground petroleum system. On the basis of grain size, the sedimentary rock is classified into five types, namely conglomerate, sandstone, silt, clay and carbonate rocks (see Table 4.1). Sandstone rock consists of sand-sized grains (2–0.0625 mm). Shale rock is composed of clay-sized grains (<0.0004 mm). Similarly, other rocks are classified by their grain size. Pore volume depends on both the grain size and geometry. The rock's previous geological history could be imagined from the size and shape of the mineral grains. Larger grains give rise to larger void spaces and pores than the fined-grained rock. The fluids get more space for easy flow in large-sized pores and course-grained rock. This type of rock forms good oil and gas accumulation (reservoir). Grain sizes are classified as:

Fine grained = < 0.1 mm
Medium grained = -0.1 – 2 mm
Coarse grained = > 2mm.

4.7.2 TEXTURE OF THE ROCK

Texture defines the appearance or consistency of the surface or print impression of the sedimentary rock. The texture depends on the size, geometrical shape and structural arrangement of the grains. The crystallization processes of minerals and weathering and erosion of rock affect the form of the texture. The rock's previous geological history could be imagined from their surface texture. They give information about the erosion process to which the rock has been subjected. They describe how the rock has been broken, and how the present shape of the minerals, rounded or angular or spherical, was attained during transportation under different weathering conditions. A well-rounded shape has minimum angular bends and edges. The well-rounded grain has a minimum and uniform surface. An angular grain may have several edges and an irregular shape. Spherical grains in the rock create more void space than the angular and irregular grains. Spheres placed together have minimum contact points, thereby creating more void space among them.

4.7.3 Sorting (Grain Size Distribution)

Sorting defines the grain size distributions in the sedimentary rock. Sorting is a natural process during transportation and sedimentary rock formation. A rock may have or may not have a uniform particle size distribution. A rock may be composed of several different grain sizes. The rock with uniformly sized particles is known as 'well-sorted rock' and the rock with non-uniformly sized particles is called 'poorly sorted rock'. The weathering, transportation and deposition of sediments are accompanied by the simultaneous sizing and sorting of sediment grains. In addition to the size, sorting also determines the void space or porosity of the rock. Fine-grained rock is poorly sorted; the rock creates less porosity. In poorly sorted rock, the smaller grains occupy the void space between larger grains. A good reservoir rock is course-grained, well-sorted rock. It can hold and transmit more fluid.

4.8   STRUCTURAL FEATURES OF SEDIMENTARY ROCK

Structural geology is the study of the alteration and distribution of sedimentary rock units as a result of geological and environmental forces. Broadly two different kinds of structural changes take place as follows:

- Surface changes (visible structural features and alterations of sedimentary rock)
- Underground changes (structural deformation of sedimentary rock)

The causes of structural changes in both cases are different; some factors may support alteration in both cases. The portion of the rock that is exposed to the atmosphere or to the ocean bottom undergoes visible structural changes.

The underground structural deformation of sedimentary rock is mainly due to two reasons, first by sediment unconformity and second by compressional (pushing) and tensional (stretching) stress caused by magma movement and tectonics.

It may be noted that the terms 'alteration' and 'deformation' may be used for any change in sedimentary rock structure. However it seems appropriate to use deformation for any drastic structural change caused by underground geological stress and sediment unconformity (discussed in Section 4.10).

4.9   SURFACE STRUCTURAL FEATURE (VISIBLE) OF SEDIMENTARY ROCK

Physical features give useful information about sedimentary rock. However structural features are not only distinct, but also give detailed information about the depositional and accumulation environment. Structure is the geometrical representation of the rock. The factors responsible for visible structural features of sedimentary rock are weathering, erosion, mechanical, physical, chemical and biological conditions prevailing in the environment. The visible structural changes of the rock are based on the following processes that also contributed to the formation of the original rock.

- Weathering and erosion alteration
- Mechanically and physically altered structure
- Chemical and biologically sedimentary structure

4.9.1   WEATHERING & EROSION ALTERATION

Weathering and erosion bring about significant visible alteration in sedimentary rock. Weathering is the breakdown of the solid rock into small or big portions caused by mechanical force (air, water and sliding) or dissolution of the rock by chemical activity. Sand rock is hard and rigid and resists weathering and dissolution. On the other hand, limestone is reactive and soluble and undergoes deterioration in aqueous medium. Clay is soft and yields to weathering even under mild conditions. Erosion is related to the transportation of rock particles, from their original location to some other place, usually to a depression basin. The rise and fall or the low and high tides of the sea alternately submerge and expose the coastal land to the atmosphere, causing weathering and erosion simultaneously. Other causes of erosion and weathering are climate change, rain fall, wind, land sliding and icebergs.

4.9.2   MECHANICALLY & PHYSICALLY ALTERED STRUCTURE

Weathering and erosion, responsible for structural alterations of sedimentary rock, are associated with mechanical phenomena. The physical sedimentary structure is related to transportation, mode of transportation and deposition that result in stratification or layered structural rock and surface structural alteration. Stratification is the single most important aspect of structural geology.

4.9.3   CHEMICAL & BIOLOGICAL ALTERED STRUCTURE

Chemical processes are responsible for the creation of certain types of surface structure in the rock by the precipitation of minerals on top of each other. The precipitation process also brings about stratification. Biological processes or biogenic processes are initiated by living organisms. The processes generate tracks and trails in the rock, known as 'bioturbation sedimentary structure'. Bioturbation is the reworking of sediments by living animals and plant organisms which disrupts and alters the original structure. Animal habits and activities including taking food and passing feces also affect the sediment structure.

Most of the visible structural changes brought about in sedimentary rock by the above-mentioned weathering and erosion, and mechanical, physical, chemical and biological phenomena are described as follows.

4.9.3.1   Sole Mark/Mechanical Impact

'Sole mark' is an important structural feature of sedimentary rock. The sole mark in the rock is created by mechanical impact so it can be termed as 'mechanical mark'. The sole mark is created by a strong fluid (water, wind) current. Fluid current may be accompanied by other heavier materials such as dust, sediments, and pieces of rock and ice. The shape, size and geometry of the mechanical mark can be related to the history of the fluid current prevailing in the locality. A sole mark that has been created once can be destroyed by a current carrying sediment. The sole mark remains distinct in spite of filling by recent sediments. Another example of a sole mark is small circular holes created on the soft surface by the action of falling rain drops.

4.9.3.2   Soft Sediment Structural Change

Soft sediment structure is the initial step before sedimentation, lithification and rock formation. Soft sediment structural alteration may be mild or drastic. In the initial stage sediments are loose, soft and mobile, with minimum contact between sediments having large porosity. The system is not in thermodynamic equilibrium. The unstable system yields to geological forces (earthquake etc.). The soft sediments arrange themselves into a new orientation and undergo many changes, including from un-consolidation to the subsequent consolidation stage.

4.9.3.3   Loading by Gravity Differential

Loading is a process in which heavy sediments settle down at the bottom and the lighter sediments rise up. It is like 'sorting' but not on the basis of grain size, rather on the gravity differential. The loading phenomenon arises from a recent sedimentary rock. The recent sedimentary rock contains sediments of different weights (heavy and light) under unstable conditions. During the consolidation process, heavier sediments in recent rock sink down replacing light density sediment in the lower layer due to the difference in gravity. Heavier sediments tend to go down in the bottom layer and lighter sediments to the top layer. The lower density sediments rise up to the upper layer. The loading phenomenon consolidates the structure. Recent sediments before loading are unstable. After loading, compact and stable sediments are formed. New rock stratification is created when the heavy sediments go down and the underlying ligh sediments go up. Loading is more prominent when water-saturated sediment flows like a semi-viscous material on an inclined plane. The degree of loading depends on the angle of inclination and flow rate of sediments.

4.9.3.4   Cross Bedding

Cross bedding refers to horizontally inclined sedimentary rock layers. Cross beds are groups of inclined layers. They are also known as 'cross strata'. Cross bedding is formed during deposition on an inclined surface, in the presence of suitable sediments and flowing water or blowing wind.

4.9.3.5   Graded Bedding

Graded bedding strata is defined by coarse-grained heavier sediments at the lower base and fine-grained light sediments at the top. Graded bedding rock is created by the principle of 'sediment size and gravity sorting'. Bedding rock is formed by flowing turbid water. The turbid water is saturated (loaded) with sediments. When the motion of the turbid water (turbidity current) drops, the unloading (settling) of sediments takes place. The sequence of settling follows the order of the size and weight of the sediments. The largest and heavier sediments are the first to settle down, followed by the medium sized, and finally the smaller sediments separate out and settle down. The bottom bed contains larger particles, the middle bed has medium-sized particles and the top bed light particles.

4.9.3.6   Ripple Mark

Ripple marks represent a sedimentary structure. The ripple marks are formed by the blowing wind and flowing water. Ripples are produced by the action of the fluid current and waves on the surface sand sediments. Sand ripples appear after water recedes in coastal areas. In a desert, the blowing wind produces big ripples known as dunes. Coastal and desert environmental conditions differ considerably. The humid coastal conditions differ from the dry desert sand, which affects the ripple formation in the two locations. Two kinds of ripple marks are known as follows:

- 'Symmetrical ripples' are generated by the to-and-fro motion of the blowing wind or flowing water waves often witnessed in the desert and at beaches. The to-and-fro of the water waves creates perpendicular ripple marks with sharp crests and rounded troughs. There are three different types of symmetrical ripple marks; the first is continuous ripple series, the second is overlapping ripples and the third is discontinuous ripples.
- 'Asymmetrical ripples' are formed by flowing water or blowing wind in one direction. They are found on riversides or in the desert. The asymmetric patterns lead to a determination of the direction of water and wind current at the time of ripple formation.

4.9.3.7   Mud Crack

See Figure 4.2. Mud cracks are formed on clay-rich (mud) sediments, where an aquatic and warm environment is available. With time, the mud dries and shrinks. Shrinkage generates cracks. The patterns of the cracks may be rectangular, rounded or polygonal. Cracks are filled with different types of sediment. The original shapes of the cracks are retained by well-marked boundaries.

FIGURE 4.2 Mud crack in clay-rich sediment.

4.10 STRUCTURAL DEFORMATION OF SEDIMENTARY ROCK

Sedimentary rock is originally deposited in a basin as a horizontal layer, one layer on top of the other. Over geological time the original layers are disrupted and deformed. Deformation is a severe structural change in the original deposited horizontal layers of the sedimentary rock. There are two types of deformations; one is caused by sediment unconformity, and the second kind of deformation is due to underground geological stress.

4.10.1 UNCONFORMITY DEFORMATION OF SEDIMENTARY ROCK

Unconformity is a subsurface erosion process between two different rock layers. Unconformity may be defined in many ways and has a number of causes.

Unconformity deformation is the absence of a regular pattern in a rock structure. An unconformity is a contact line between young and old strata. Unconformity is the buried erosion surface of strata. Unconformity is a break in the otherwise continuous geological record of the strata. An unconformity is a deformation of the underground rock strata caused by the non-uniform and irregular deposition of sediments. An unconformity is caused by a period of sediment erosion and the absence of further sediment accumulation followed by new sediment deposition.

The unconformity deformation is divided into major and minor categories as follows:

- Discontinuity (major)
- Non-conformity (major)
- Angular unconformity (major)
- Buttress unconformity (minor)

4.10.1.1   Discontinuity Unconformity

A discontinuity is an unconformity deformation between parallel layers of sedimentary rocks. Different sediments, shale, sand, limestone and basement rocks, create interface discontinuities.

A discontinuity deformation is witnessed between old (consolidated shale rock) and recent (unconsolidated shale rock) sedimentary rocks. The surface between the old and recent rocks is eroded or represents non-deposition and non-preservation of sediments. The older compact rock is beneath the recent soft rock.

4.10.1.2   Non-Conformity

A non-conformity is a deformation between different types of rocks. A non-conformity is an unconformity between sedimentary rock and igneous or metamorphic rock. The sedimentary rock is deposited above the eroded or broken surface of the old igneous or metamorphic rock. The interface between the two types of rocks represents a region of non-conformity.

4.10.1.3   Angular Unconformity

An angular unconformity occurs when sediments are deposited on a tilted eroded rock. An angular unconformity is a deformation between the underlying eroded tilted rock layers and the overlying parallel horizontal sedimentary layers.

Moreover upper, middle and lower parallel layers in sedimentary rock are disturbed and continue to change due to geological forces over geological time. The parallel layers are tilted and deformed. A favorable site for angular unconformity is in coastal marine area. Initially horizontal sedimentary strata are deposited in the marine bottom due to water movement. Alternate to–fro movement of water exposes the land surface to atmosphere. Weathering at the land surface couple with the to-and-fro motion of water creates angular unconformity tilted layers.

4.10.1.4   Buttress Unconformity

A buttress in geology defines a protruding rock structure resembling the buttress of a building. A buttress is a steep slope structure to support the wall of a building. A buttress is a minor unconformity between old and recent rock layers. The younger protruding layer deforms the old rock structure.

4.10.2   STRUCTURAL DEFORMATION OF SEDIMENTARY ROCK BY GEOLOGICAL STRESS

Structural alteration of underground sedimentary rock is severe deformation due to geological forces. Geological tectonic and magma movement processes generate deforming forces that produce folding and faulting in the sedimentary rock. Sedimentary strata are deformed by the tensional and compressional stress acting on them. The stress comes from magma movement or tectonics, creating compression, contraction and expansion of strata, affecting the continuity or conformity of the underground structure. Different forms of deformations caused by geological factors in subsurface strata are given below.

4.10.2.1   Homocline (Monocline) Deformation

A homocline is an inclined layered structure in which layers of rock dip uniformly in one direction. It is a simple, one-sided uniform inclination (dip) of the rock layers.

4.10.2.2   Fracture Deformation

A geological stress force that exceeds the cohesive forces among the rock particles creates cracks and deformation. Any mechanical break (deformation) in the rock is known as a fracture, joint, fissure or fault depending on the type of deformation.

Fracture. The fractures are perpendicular or oblique to the direction of the layers. They divide the rock into various pieces or blocks. The fracture depends on the type of minerals and defines the type of rock.

Joint. This is any fracture without visible splitting of the rock; the fractured block has not moved. It only demarcates the rock into various blocks.

Fissure. This is a deeper and narrow fracture. It leads to the appearance of a breach in conformity. The fissure can be small or with a long opening filled with minerals. It can create perforation in the rock. A fissure may be linear or branching.

4.10.2.3   Faulting Deformation

Faulting is the most important and severe fracture in which relative movement of two adjacent rock blocks take place. Faulting is a crack in the earth's crust resulting from the displacement of one side of the rock with respect to the other. A fault can also be defined as a planer fracture or discontinuity in a rock across which there has been significant displacement of the rock. A fault causes the parts of the rock to move vertically, horizontally or diagonally. Fault is generated during earthquakes and accompanied by movement of rock blocks. The stability is attained, after the passage of geological time.

The generated geological deformation/unconformity provides more porosity and permeability that support the subsurface migration and formation of oil and gas trap rock. Excellent traps are formed by fault lines. Specifically faulting occurs due to two reasons, one in which the crustal rock layers are subjected to tensional forces (pulling force)and the second in which the crustal rocks are subjected to compressional forces (pushing). The rocks that are subjected to tensional forces are pulled apart and stretched to either side, resulting in a fault. Compressional forces may cause a fault or fold depending on the geological characteristics of the rock that is flexibility or brittleness. Types of relative moments of the rock may assume different geometry, giving different types of faults and geological structures. Modified geological structures and the classification of faults as a result of the deforming movement of two rock blocks along with related terms are explained as follows.

Vertical dip-slip faulting deformation. Vertical up-and-down movement of two adjacent blocks of the rock along the fault line (linear perpendicular) is known as a 'vertical dip-slip fault'. As a result of a vertical dip-slip fault one block goes up and the other down. A vertical dip-slip fault divides the rock 'side by side' into two rock blocks. The two rock blocks move along the fault line relative to each other in up and down directions.

Horizontal strike-slip faulting deformation. The horizontal movement of two adjacent blocks of rock along the fault line is termed a 'horizontal strike-slip fault'. Horizontal strike-slip faults divide the rock into two rock blocks, one is above the other. If the upper rock block has moved to the right side, this is called a 'right lateral fault'. The rock bock that has moved to the left side is called a 'left lateral fault'.

- *Oblique dip-slip faulting deformation.* The fault line is neither vertical nor horizontal; it is slanting. The oblique dip-slip fault is formed by the movement of the rock blocks between horizontal (x-axis) and vertical (y-axis). Usually the angle between the fault line and horizontal line is less than 45°. The slanting fault line divides the rock into two blocks named the 'hanging' and 'foot' walls. Oblique dip-slip faults are of two kinds, one is the 'normal oblique fault' and the other is the 'reverse oblique dip-slip fault' as illustrated in Figure 4.3 and described below:
- A normal oblique fault between the two rock blocks is created by pulling apart (tensional force) geological stress. The two blocks separate away from each other by the tensional forces acting upon them on either side of the fault line. The block that goes above the fault line is known as the 'hanging wall', and the portion of rock that goes below the fault line is known as the 'foot wall'. The hanging wall moves downward relative to the foot wall.
- Reverse oblique dip-slip fault. The two blocks are separated due to compressional (pushing) forces. Some sections of the two blocks overlap on the fault line. The hanging block goes up relative to the foot block.
- Reverse thrust fault or reverse thrown fault. This is also a reverse oblique dip-slip fault. When the hanging and foot walls are thrown apart, the fault line is known as a 'thrust fault'. In this case compressional forces exert

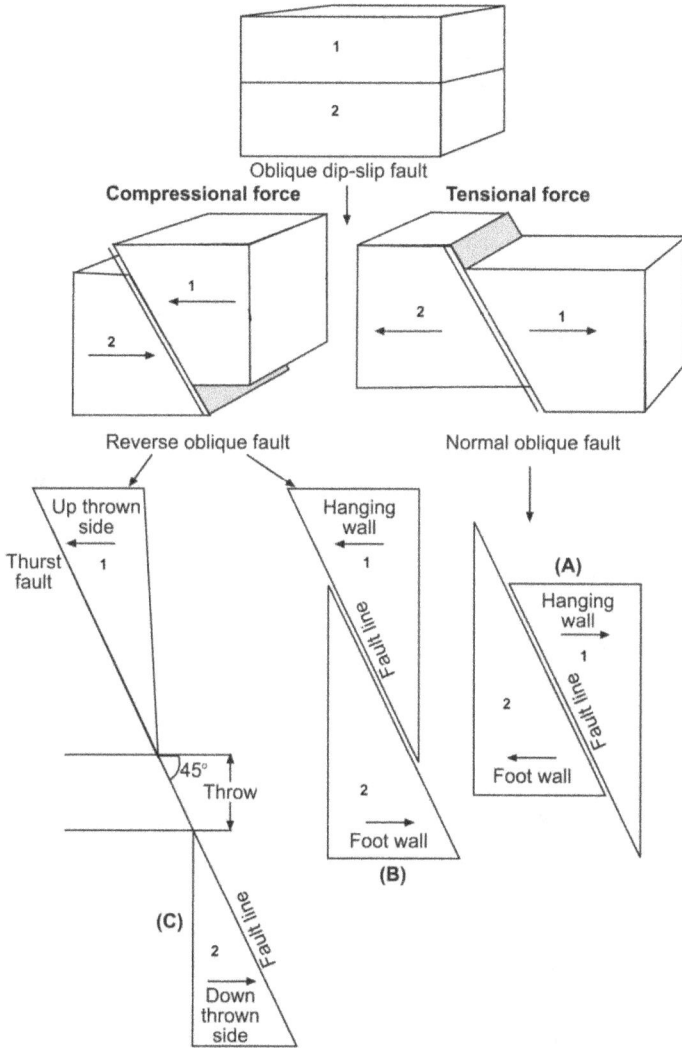

FIGURE 4.3 Types of oblique dip-slip fault: A = normal oblique fault, B = reverse oblique fault, C = reverse thrust fault. Source: Modified from J. Hyne Ph.D., *Non-Technical Guide to Petroleum Geology, Exploration, Drilling and Production, Penn Well Book* (e-book), Oklahoma, USA, 1995 (Figure 5–16, page 80).

more pressure on the two rock blocks. The hanging wall portion of the broken rock block goes over and above the other side (foot wall). The hanging and foot walls are completely separated. The distance between the two parts is known as the 'throw', that is, how much the blocks are thrown away from each other. The upper hanging wall has been thrown (thrust) over the lower foot wall.

4.10.2.4 Horst and Graben Deformation

Horst and graben are the elevated (hill) and depressed (valley) blocks of crust that lie between normal faults. A graben is also known as a 'rift valley'. Both occur side by side. Horst and graben structure is formed by tensional stress and are dip-slip faults. A horst is rock up thrown by faulting and a graben is a block dropped down between faulting lines. A horst is the hillock between two grabens. Both horst and graben vary greatly in size and area. A graben may form a petroleum basin supported by horst-associated trap rocks.

4.10.3 FOLDING DEFORMATION

Folding deformation is a major structural alteration. Sedimentary layers are bent or curved during deformation geological stress. Folding deformation may be minor or major. The fold may be of microscopic or mountain size. The horizontal layers of sedimentary rocks are disrupted and deformed by compressional and tensional forces. The layers of rock bend or fold when compressed, and the rock layers slide along the fault line under tensional force. It may be pointed out that the sedimentary rock is hard, rigid, brittle and devoid of flexibility. Such materials do not bend, fold, elongate or compress without breaking (yielding). But under long geological conditions of temperature and compression, the rock is gradually deformed and folded into various geometrical shapes. The imported folded structural shapes of the rock are as follows.

4.10.3.1 Anticline Folded Rock

An anticline rock is a dome-shaped folded rock in which the sides of the rock bend downward from the middle crest. Anticline dome-shape folded rock is formed when the sedimentary layer is compressed by geological processes in such a way that two ends (limbs) of the layer are folded in opposite directions and raise the middle portion to form a crest. The crest is formed by up-thrown rock block from the middle portion of the layer. An anticline rock is like an arc or dome. It resembles an inverted bowl. Both sides of the rock structure push down and dip away from the crest. The upward projected rock (crest), being near to the earth's surface, may be exposed to erosion. Sometimes the raised crest dome structure emerges out at the earth's surface and looks like a hill. Anticline rocks are usually large and form good reservoir rock (see Figure 3.2).

4.10.3.2 Syncline Folded Rock

Syncline folded rock looks like an open bowl or trough basin. The rock simulates a concave shape, whereas an anticline assumes a convex structure. Syncline is a large downward arch, created by the downward bending of the middle portion of the rock layer and in which the sides (limbs) of rock layer go upward. A syncline rock is a trough (valley) of folded strata. The presence of anticline and syncline rock structures indicates that the area has been subjected to geological compressional stress. The rock plays no role in underground petroleum systems.

4.11   PETROLEUM BASIN

The earth's crust can be divided broadly into rigid and soft blocks. The rigid block is of relatively low mobility is called a 'platform', and a soft, mobile block is known a 'geosyncline basin'. A platform is a part of land that is covered by sedimentary rock underlain by igneous and metamorphic basement rocks. Both igneous and metamorphic often crop out at the surface. A geosyncline basin is a large-scale depression in the earth's crust. It is an open concave elongated trough (valley) on the earth's surface. A platform block is unfit for petroleum, and the geosynclines basin is a favorable location for petroleum prospecting. A petroleum basin is a portion of a geosyncline basin. A petroleum basin is an open, shallow depression of greater width and length than depth and filled with sedimentary materals. Petroleum is generated and generally found in the sedimentary rock. Although sedimentary rock covers about 75% of the earth's crust, not all regions of rock are suitable for oil and gas formation. Petroleum gas and oil fields are restricted to some selected areas of sedimentary rock in the form of land depressions or basins. Oil- and gas-bearing sediment basins are mostly located in the coastal and sea continental shelf area of the earth's crust. Many off-shore petroleum fields are discovered and operated in these regions. A petroleum basin may be an oil-bearing or gas-bearing basin, or both oil and gas.

Oil and gas are found in different structures of the sedimentary rock. Some sedimentary structure favors the presence of oil and gas. Therefore, a detailed geological history, along with a sketch and map showing the general distribution and features of land and sea, greatly help in surveying for petroleum. A thick sedimentary rock is more likely to produce oil and gas than a thin rock. Absence of sedimentary rock means an absence of oil and gas.

4.11.1   BASIN FORMATION

The formation and alteration of basins (depression) and mountains (uplift) are regular features throughout geological time. There are many geological factors in the creation and modification of basins:

- Movement of hot and molten magma. The major factor for the creation of a petroleum basin is the deformation of the earth's crust, due to the underground geological movement of hot and molten magma. The slow movement of the hot magma exerts stress on the above earth crust. The crust tries to adjust itself to the new thermodynamic conditions. Crust basement rock (igneous and metamorphic) is affected by the upward thrust of the magma. The basement rock may compress or expand or protrude into the above sedimentary rock. In the former case a land depression is created that may be filled with water creating a lake or the depression on the land may remain dry. The expansion or protrusion of basement rock into the earth crust and sedimentary rock creates a mountain range by uplifting the surface sediment layer.

- Subsidence of basement rock. A basin is created by the subsidence (contraction) of igneous or metamorphic basement rock, followed by downward bending (depression) of the earth's crust. The depression may be filled by sediments and water to form a basin.
- Folding and faulting. The downward bending of rock is further modified by folding and faulting during sedimentation. Graben basin is formed along one side of the faulting line.
- Basin modification. Geological compressional forces acting both on the basin and mountain range continue to work during geological time. New mountains and basins are continuously formed with new sediments. Simultaneously erosion and weathering continue to affect and continue to modify the basin and mountain during geological time.
- Basins are formed in low-lying valley areas not far from mountain ranges. The mountains erode and rain water carries the sediments below to form a new basin.
- Mountain areas are not fit for petroleum generation. Mountains are fully exposed to atmospheric and climatic conditions. There are greater possibilities of opening, leakage, sliding and loss of deposited matter including oil/gas. They are subjected to weathering and erosion. They cannot preserve and accumulate the organic deposition for a long time, and there is no fresh supply of sediments.

4.12   STRATIGRAPHY OF SEDIMENTARY ROCK

Stratigraphy is one of the many structural features of sedimentary rock. Stratigraphy deals with the classification, nomenclature, correlation and interpretation of the stratified sedimentary rock. The stratigraphy defines the rock layering or strata. The plural of the word 'stratum' is strata; stratigraphy tells about the layers sequence of the rock. Stratigraphy and sedimentology are related geological subjects. Overlapping between the two exists as evident from the following description of stratigraphy. The formation of strata or layering in a rock is known as the stratification process; a stratified sedimentary rock is shown in Figure 4.4.

The subject of stratigraphy may be applied to igneous and metamorphic (volcanic) rocks on a limited scale. The detailed discussion here is related to the stratigraphy of sedimentary rock that is closely related to petroleum system. Stratigraphy is useful in defining the rock strata, and their relative position and distribution in the rock. It reveals the sediment composition and their depositional behavior that ultimately give an interpretation of the relative age of the rock strata and geological history. Stratigraphy study helps to identify differences in grain size, changes in composition, color, shade and concentration of individual sediments, for example particles, pebbles, fossils, etc. On the basis of sedimentary facies, several kinds of stratigraphic subjects have been developed, for example litho-logical-stratigraphic, geo-chronological-stratigraphic, biostratigraphic and magneto-stratigraphic units, etc. Some aspects of litho-logical stratigraphy are given below.

FIGURE 4.4   A view of stratified sedimentary rock.

4.12.1   Litho-Logical Stratigraphy

The litho-logy of a rock unit is a description of its physical characteristics, rock type, origin, color, texture, grain size, mineral nature and composition, constituent distribution and rock structure. Litho-logy stratigraphy is based on subdividing rock sequences (layers) into individual litho-stratigraphic units. Originally the term 'litho-logy' was used for rock types. The three major rock types are sedimentary, metamorphic and igneous rocks. Stratigraphy distinguishes rock formation on the basis of litho-logy. A 'rock formation' is a distinct unit of the whole rock. A rock formation has particular litho-logical characteristics that differentiate the formation from other formations of the rock.

Stratigraphic rock is formed by sediment deposition in any suitable basin. The repetition of the sediment depositions, forming layers on previous layers, over geological time, generates a stratified rock. The following observations are noted for the stratification process:

- Slow and sustained and periodical depositions of sediments facilitate the formation of well-stratified rock.
- Rapid, violent, quick and storming deposition leads to poor stratification.
- Stratification is the superimposition of young sediments over the older deposits.
- The sequence of strata has distinct facies, for example fine- or coarse-grained layers.
- Almost all sedimentary rocks are stratified to a lesser or greater extent. The layers may or may not be distinguishable.
- At the beginning of deposition, the layer tends to reflect the pattern and irregularities of the bottom surface. With further depositions and time, the sedimentary layer assumes a horizontal flat shape. The original flat

horizontal shape of sediments continues. The far ends of the layer may witness irregularity and erosion.

- The present geometry of strata may change to another stratigraphy after compaction and consolidation.
- The layer may be tilted to a certain angle (cross stratification).

The rock strata are formed according to the 'law of gravity' and the 'law of superimposition', mentioned earlier. The sediments are filtered and deposited according to the size/weight/ gravity of the sediments in a stagnant aquatic environment. Sediment flowing with river water are first discharged the sediment at a shallow marine depth and then towards deeper zones. The sediments are of different sizes and types. The succession of sediments flows down in the inclined bottom, from the river to the ocean. The heavier sediment particles first separate from the river water and then progressively settle down to the bottom. Lighter and fine-grained particle sediment occupies progressively higher layers above, forming distinct rock layers. The most recent layer superimposes itself on the older deeper layer.

The stratigraphic rock spreads in all directions on the flat bottom surface. The spreading is called 'lateral rock continuity'. But there is a limit; the continuous extension of the rock is not possible. Breaks and separation are observed in rock layers. There are a number of reasons that may cause interruption in layer continuity. Interruption is due to the types of sediment, grain size of sediment, composition of the sediments, shape of the basin and geological environment. Course-grained rock strata show an earlier break than the fined-grained strata. A basin with a flat wide long bottom supports discontinuity to a greater extent than a narrow basin. One of the useful aspects of stratigraphy is that the strata can form a reservoir trap. A stratigraphic impermeable rock is tilted by the subsurface moment in such a way as to form a bended trap. A stratigraphic rock in combination with the other structural features can also form a trap rock, known as a combination trap.

4.12.2 Primary and Secondary Stratified Structure

Many factors are involved in rock stratification as in sedimentation process. The stratified structure is variable with time and depends on the processes involved throughout the geological history of the rock. Primary stratification begins with sediment transportation, settling, deposition, accumulation, compression, compaction, consolidation, lithification and rock formation. Stratification is a time-dependent process; the rock structure continues to change in the presence of new environments. The rocks are built up by physical, chemical and biological phenomena. Physical processes, such as sediment transportation, settling, accumulation and compaction, are certainly the main contributors to the stratification process that result in good layering. Chemical processes, for example the precipitation of $CaCO_3$, produce good stratified rock. But chemical contribution to stratification is minor compared to physical processes. Biological processes do not produce stratification, but fossils, tracks, trails and burrows help to identify different rock strata. Secondary structure or modified structure continues to build up with geological time by secondary forces. Most of the factors that are responsible for the sedimentation and consolidation of rock give rise to various forms of stratification, given as follows.

4.12.3 Parting Plane

A plane is a horizontal flat surface. A parting plane is a border line between two layers. A parting plane defines the separation (splitting) of individual layers in a stratified sedimentary rock. Planes of parting are termed stratification planes. Planes of parting are horizontal when sediments are deposited on flat surfaces but are inclined when the deposited surfaces are slanting. A parting plane is a demarcation space between two layers, for example a layer bonded above by a thinner layer and below by thicker layers. A parting plane separates and distinguishes a thin and thicker layer, for example a thinner shale layer bordering thicker limestone rock layer. The parting can very small, less than 1 centimeter, or as large as more than 100 centimeters. The parting of a stratified plane (separation) is created by erosion, weathering, type of sediments and cohesive forces among the rock grains and environmental conditions.

4.12.4 Plane Stratification

Plane stratification defines the plane horizontal strata. Plane stratification results when an interior rock layer plane is parallel to the horizontal bottom plane or the internal layer plane is parallel to the top horizontal surface. In practice, the bottom layer, interior layers and top surface layer are not exactly parallel; they are close to horizontally parallel. Discontinuity in the bottom surface is reflected in the bottom layer and in the successive upper sedimentary rock layers.

4.12.5 Cross Stratification

Cross stratification, also known as cross bedding, is layering within stratified rock at an angle to the bottom surface not parallel to the bottom or top layer. The strata are inclined relative to the horizontal bottom and the top layer at the surface. Cross bedding is formed on the inclined bottom surface, during the period of sediment settling, accumulation and consolidation.

The geometrical shape of the cross bedding differs widely. Cross bedding is an indicator of specific environmental conditions prevailing at the time of bedding formation. The shape of cross stratification depends on the mode of transportation and the nature of the sediments along with the direction and speed of wind blowing and water flowing. The shape and size of the cross stratification are used to find out the nature of wind and water current during the formation of the cross bedding. The direction and flow rate of water and wind can be predicted by noting the dip angle (inclination) of the cross stratification and the length and height of the bedding. This information also helps us to find the size distribution of sediments in the bedding. Examples of cross stratification are given below.

4.12.5.1 Ripples Cross Stratification

Ripples are the indicator of cross bedding on the surface. Ripples are produced by the action of the fluid current and waves on the surface sediments. After water recedes in a coastal area, sand ripples appears with stratification. In the desert, the wind produces big ripples known as dunes. Coastal and desert environmental conditions differ considerably. The coastal conditions of water, humidity and nearby plant roots significantly modify the cohesive forces of sediment compared to the dry desert sand, that affect the ripple formation on the beach and in the desert. The 'symmetrical ripples' are generated by the to-and-fro motion of the blowing wind or flowing water waves often witnessed in desert and beaches. To-and-fro water waves create perpendicular ripple marks with sharp crests and rounded troughs. There are three different types of symmetrical ripple marks; one is continuous ripple series, the second is overlapping ripples and the third is discontinuous ripples. The 'asymmetrical ripples' are formed by flowing water or blowing wind in one direction and have asymmetrical ripple patterns. They are found on riversides or in deserts. The asymmetric patterns lead to the determination of the direction of water and wind current at the time of ripple formation.

4.12.5.2 Mud Cracks Cross Stratification

See Figure 4.2. Mud cracks are also surface indicators of cross stratification; they show horizontal and vertical cross stratification. Mud cracks are formed and appear on the land surface in the presence of aquatic and warm environments by the evaporation of water from clay-rich mud, leaving dry cracked clay. With time, the mud dries and shrinks. Shrinkage generates cracks. The patterns of the cracks may be rectangular, rounded or polygonal. Cracks are filled with different types of sediment. The original shapes of the cracks are retained by well-marked boundaries.

4.12.5.3 Favorable Location for Cross Stratification

Cross stratified beddings are formed by to-and-fro movement of water and also from the wind–sand system in the desert. Favorable sites for cross stratification are lake, lacustrine, river channel, coastal beach tidal wave and desert wind areas. Some locations are reported below:

- *River channel and delta.* A delta is formed at the discharge point of a faster flowing river into comparatively slower or stagnant water. This happens when a river meets an ocean, sea or lake. A river flowing from a high-altitude area to a lake or sea brings along with it a large quantity of sediment. The sediments are of various sizes, from particles (small) and pebbles (medium) to gravel (large). The flow rate of water gradually slows down as the water reaches the ocean or lake. The kinetic energy of water decreases, and it is unable to hold the suspended sediments. First to separate and drop are the heaviest gravels and pebbles, followed by lighter and finer sediments, on the delta river bank and on the bottom of the water. The deposits build up. A cross bedding is gradually formed by the repeated settling of sediments from flowing river water and interruption by the advancing of sea water into the land. Two kinds of beds are formed. One bed is on the land side, known as 'fore-set bedding', and the other layer on the aquatic side, formed by the deposition of sediments between flowing and calm water, is known as 'fore-most bedding'. The delta basin is rich in various types of sediments brought by the river from wide and far places. The bedding is rich in nutrients, good for farming and agriculture. Most of the world petroleum reservoir occurs in deltaic zones.

- *Glacier.* During the formation of glaciers, sediments are trapped in the freezing water. Sediments become an integral part of the glacier. Later on, the glaciers with sediments move and slide down. When the glacier melts, the water leaves the sediment and moves on. The sediment deposits by glaciers are known as 'moraine' rock.
- *Coastal/tidal wave.* Tidal zones lead to the formation of cross bedding. The upcoming tidal wave forms upward sediment deposits on the sea side. A shorter inclined bed is formed on the land side by the receding tidal wave. The elongation of the bedding and its thickness depend on tide (low andhigh), water flow rate and the kinds of sand grains. The land-side layer is small compared to the sea side sediments. Between the land-side layer and sea-side layer, there is a sediment peak or crest. The height of the crest depends on the water flow rate and tidal frequency.
- *Aeolian (wind) zone.* Dust, dirt and sand are moved rapidly during dust storms and sand storms. Surface rocky materials are broken down into small particles by wind. The particles' sizes are reduced by collisions with each other. When the wind stops or slows down, the airborne sand particles are deposited to form cross bedding that appears as sand ripples or sand dunes. The extent of bedding depends on the wind speed and direction.

4.12.6 TABULAR/TROUGH STRATIFICATION

The tabular or trough stratification is cross bedding formed in a deep small basin. There is not much difference in the breath and length of the tabular basin. The stratified strata are curved according to the bottom shape of the trough of the basin.

4.12.7 LOADING STRATIFICATION

Loading is a sediment consolidation process, according to the gravity of sediments, with time in a given basin. The process of the rising up and sinking down of the sediment on the basis of gravity is known as 'loading'. It is a time-dependent process. Actually this is gravity separation. Recent stratification occurring in rock is not stable. In recent rock there is uneven distribution of sediments and water that results in uneven distribution of layer densities and grain size. It is possible that an agglomerate of heavier density occupies an upper layer above the lighter sediment layer. Naturally there will be a tendency for the higher sediment to sink down and the lighter sediment to rise up. An example of loading is the sinking down of water-saturated sands into the underlying mud.

4.12.8 Gravity Stratification

Gravity stratification is related to water flow loaded with sediments of different gravity on the inclined sea bottom of the continental margin. The stratified sedimentary layers are formed by the downward flow of water carrying sediments. Strata of rock are formed according to the 'law of gravity'. The sediments transported by water from land are distributed according to the sediments' gravity in the sea continental shelves. The water and the accumulated sediments flow with different rates on the inclined continental shelf bottom surface. The flow rates depend on the nature, geometry and weight of the sediments. Such flow is called 'gravity flow'. The larger and heavier sediments move faster. They form the 'head' of the moving sediment mixture. The finer and lighter sediments lag behind and form the 'tail' of the moving sediment stream (see Figure 4.7).

4.12.9 Turbidity Stratification

Turbidity in water is caused by the presence of suspended, dispersed and colloidal particles in water. It is also possible that the turbid water contains completely dissolved particles. Turbidity indicates the saturation of water with sediments. Turbid water is opaque, and light practically does not pass through it. After the settling down of the suspended solute particles, the water becomes clean and transparent. A water column saturated with sediments behaves in three different ways.

- The soluble sediments, mostly salt minerals, make a homogeneous solution with water and are difficult to separate from the water column under normal conditions.
- Colloidal dispersed salute particles remain in the water, but can be separated under mechanically or chemically forced demulsifying conditions.
- The insoluble particles are known as suspended materials in water. The flowing water loaded with suspended sediment is 'turbid water/turbidity current'. The phenomenon of holding insoluble sediments in water is known as 'turbidity'. How long does the turbidity remain or the water retain the sediment as suspension? It depends on the physico-chemical environment and the nature of the sediments. Rounded and larger/heavier gravels separate first and settle down at the bottom surface from the water suspension. Next the sand-bearing sediments settle, and finally the clay particles slowly trickle down the water column. Together all the sediments form a stratified rock with successions of well-identifiable layers with some overlapping (see Figure 4.7).

4.12.10 Thickness of the Stratified Sedimentary Rock

A stratified rock layer of the sedimentary is distinguishable from the adjacent layer. A layer is a sheet of uniformly and continuously distributed sediment. To distinguish different layers, the thickness of each layer is cited. The thickness of a layer as well as of the stratified rock is an important aspect of sedimentary rock. Classification of the stratified rocks is carried out on the basis of their thickness. Broadly, stratification with larger layer thickness (>100 mm) is classified as 'bedding' and with smaller layer thickness (<100 mm) is known as 'lamina'. The lamina is the minimum recognizable thin layer of sediments. The bedding rocks are mostly formed by thick course-grained sediment particles. The lamina is formed by fine-grained sediment particles. A layer/lamina is formed by the deposition of one set of sediments at a time. A layer/lamina corresponding to the size (diameter) of the set of the sediments. One set of sandstone sediments forms a mono layer of thickness 0.1–2.0 mm to correspond to the diameter of the sediments (0.1–2.0 mm). One set of fine-grained shale sediments will form a mono lamina of thickness of 0.004 mm according to the diameter of the shale sediments. In the same way another set of sediments will form another lamina corresponding to the thickness (diameter) of the second set of sediments.

A non-uniformity/discontinuity in stratification is developed when a large amount of sediment is deposited in geological time to make thicker layers. The rapid deposition of sediment creates an observable discontinuity in the successive layer.

4.13   BIO-STRATIGRAPHY

Litho-stratification discussed above defines rock by its litho-logical distributions, sediment types and degree of consolidation during geological time. On the other hand, bio-stratigraphy identifies and describes the distribution and arrangement of fossils in the strata of sedimentary rock. A bio-stratigraphic rock unit is a collection of fossils, independent of the types of sediments and minerals in the specific rock strata.

On the basis of fossil contents, the basic unit of the bio-stratified rock is termed the 'bio-zone'. The bio-zone of the rock varies considerably both in thickness and extent. There may be non-bio-zone area that may be devoid of fossils. Such an area without fossils is known as a 'barren zone'. A study of bio-stratigraphy and the fossil content of the strata leads to the determination of the age of the rock. The nature and type of sediment may differ even if they are of same age because the strata might have been formed by shale minerals or with sandstone or limestone minerals. But fossils of the same kind and class found in strata correspond to the same age. The fossils must have been deposited and incorporated in the strata at the same time. Fossils give the following information:

- Some fossils are rock-forming; they form fossiliferous biogenic sedimentary rock along with other minerals (litho-logy). Higher flora and fauna, upon fossilization in sedimentary rock, form carbonaceous, calcareous and siliceous rocks (coal, diatomite and coral reef).
- The most common fossil is known as the 'fossil index' that refers to a period of geological history of the earth. Trilobites (extinct marine arthropods) and foraminifers are two examples of the fossil index.

- Certain fossils help in understanding the 'theory of evolution of species'. The emergences, growth, life style, sustainability, decay and finally extinction and the replacement by new improved species, through geological time, are the fundamental principles of the theory of species evolution.
- A combination of biology and geology leads to the emergence and development of subjects like paleontology and palynology.

4.13.1 PALEONTOLOGY (OLDER)

Paleo (old)-onto (being)-logy (study). Paleontology is the study of old fossils originating from animal species forming body fossils, for example animal skeletons and the teeth of large animals. Trace fossils are the animals who left their marks and excretions on the sediments. Paleontology studies the marks on the sediments and excretions and correlates the findings to the behavior, habit and life style of past dead animals. Index fossils are used by paleontologists for rock age correlation and sometimes for petroleum prospecting. Paleontology is the scientific study of life on the earth since it first appeared about 3 billion years ago.

4.13.2 PALYNOLOGY (RECENT)

Palynology is the branch of bio-stratigraphy that studies small living species and their fossils. The main focus is on 'pollens' and 'spores'. Pollens are fine to course powder-like materials and are produced by land plants. They occur in the seeds and flowers of a plant which produce a male gamete (sperm cell). Spores are asexual and uni-cellar organisms, for example lower organisms like algae and fungus. These spices have a microscopic structure (5–500 nm). They are chemically stable and decay very slowly. They are not only abundant in sediments and sedimentary rock but also spread in the vast area of the atmosphere and hydrosphere. Pollen and spores in the presence of oxygen in the atmosphere are not as stable as in the preserved sedimentary deposits in the absence of oxygen. Pollen and spores exhibit polymorphism. They exist in many forms. Pollen and spore fossils are important tools for the study of bio-stratigraphy and palynology of sedimentary rock. Palynology covers the recent geological period known as the 'Quaternary period of Cenozoic era', spanning from 2.58 million years ago to the present age.

4.14 MAGNETO-STRATIGRAPHY

The magneto-stratigraphic unit of sedimentary rock can be viewed as a litho-stratigraphic unit and a bio-stratigraphic unit. A litho-stratigraphic unit is a part of the rock formation possessing particular sediment characteristics. A bio-stratigraphic unit is a part of the rock formation possessing biological characteristics (fossil content). On the basis of these characteristics, the litho unit and bio-stratigraphy of the rock are distinguishable from the other units or parts of the rock. Just like a litho-stratigraphic unit and bio-stratigraphy, magneto-stratigraphy defines the earth's natural magnetic field for a magneto-stratigraphic unit of sedimentary rock. Just like a magnet magnetizes an iron bar, in the same way ferromagnetic materials (magnetic sediment) present in the sedimentary rock are magnetized by the earth's natural magnetic field. Magnetic property is introduced in the sedimentary rock and creating a magnetic zone. Magneto-stratigraphy divides the rock strata on the basis of magnetic properties of the sedimentary rock. The magnetic properties of given strata are affected in the presence of magnetic sediments, at different stages of deposition, thickness and compactness. When measurable changes in magnetic properties occur from layer to layer (stratigraphic-ally) that defines the 'magneto-stratigraphic unit'. A layer in which the magnetic parameter is uniform is known as the 'magneto zone'. The technique is used to date the sequences of strata.

4.15   VOLCANIC ROCK STRATIFICATION

The terms 'volcanic rock' and 'basement rock' are used for igneous and metamorphic rocks. Both igneous and metamorphic rocks are comparatively less heterogeneous than sedimentary rock; their mode of formation is under severe conditions of temperature and pressure and stratification has limited application. The study of stratification can be applied to volcanic rock. Volcanic rock originates from the igneous and metamorphic rocks due to the eruption of underground magmatic lava. A volcanic eruption is the spreading of particle-sized lava called 'flying ash' in the nearly sea bottom and land atmosphere. Flying ash is composed of mostly magmatic materials. The gravity and the size of the lava particles along with wind affect the 'distribution' and 'sorting' of the particles accordingly. Lava falling from the atmosphere, settling and cooling creates -well-sorted' layers of igneous rock on the ground or water bottom. The repeated eruption, falling and settling of flying ash generate well-stratified igneous rock. Volcanic rock is formed at the land and marine bottom surfaces.

4.16   GEO-CHRONOLOGY

Chronology is the record of events in the order of their time of occurrence. Geo-chronology is the science of the determination of the age of rock and related geo-events and the arrangement of them in time sequence. The geological events which a sedimentary rock has gone through can be compiled from the study of litho-logy, litho-stratigraphy, bio-stratigraphy and magneto-stratigraphy.

4.16.1   CHRONO-GEOLOGICAL TIME UNIT/CHRONO-STRATIGRAPHIC TIME UNIT

The geo-chronology record is subdivided into two groups: (1) the chrono-geological time unit is an expression of the time interval in which a litho-logical (rock type, origin, color, texture, grain size, mineral nature, composition, constituent distribution and rock structure) event occurred, and (2) the chrono-stratigraphic time unit is the record of the formation of particular rock strata in a particular time interval (see Chapter 1).

4.16.2 BIO-CHRONOLOGY

Bio-chronology relates biological events to geological time with the help of the fossil content of rock. The bio-stratigraphic zone or simply the 'bio-zone' of the sedimentary rock is defined as 'the rock layers in which the remains of organisms (fossil) are preserved'. The bio-zone represents the location (layer) where the organism has died. Organism moved from their place of birth to another.

4.16.3 MAGNETO-CHRONOLOGY

A magneto-chronologic unit is a section of strata having the same magnetic field (polarity) that allows the unit to be differentiated from other strata. Precisely it is called the 'magneto-stratigraphic polarity unit' or simply the 'polarity zone'. Each polarity zone is related to a specific time called 'magneto-zone'.

4.17 SEDIMENTARY ROCK AGE DETERMINATION

Age is determined in relative or absolute terms. It is common to use relative age determination in sedimentary rock strata and arrange the strata in time sequence. Absolute age is determined by the radioactive element dating method.

4.17.1 RELATIVE AGE DETERMINATION

The principle of relative strata age determination is based on chrono-geology and chrono-stratigraphy using the litho-logical, biological, fossil and magneto characteristics of the rock. Relative age tells how much older or younger a strata is than other. Naturally the bottom layer is older than the strata above. The topmost stratum is the youngest in a sedimentary rock. Sometimes geological events disturb the sequence of strata and bring the old to the top and the younger to the bottom. Similarly the occurrence of rock structural features such as faulting, folding, intrusion, extrusion and deposition may also be relatively dated. As a natural corollary, an event that has occurred on a rock must be younger than the rock itself. For example a fault that has been found below a rock stratum is older than the deposited younger layer above the fault. A biological species (organism) has a very short life duration between its birth and death compared to the time scale of rock. The age of the fossil corresponds to the age of the rock. The presence of fossils in sedimentary rock is related to a particular geological time because the organism lived and died in a particular period of time. On the basis of fossil study, the rock strata are dated in geological time scale. Strata sequences are arranged accordingly. Over time, fossils continuously undergo alternation. There is a clear distinction between an older and a younger fossil. The relative age of the two rocks could be estimated by examination of the fossil content of the deposit. Fossils are incorporated into the sediments simultaneously during the sedimentation process. The particular stratum in which the fossil is found is of the same age because of the simultaneous deposition of sediment and fossil. By following the dating procedure, the whole vertical section of the strata from bottom to the top can be dated relatively.

4.17.2 Absolute Age Determination

Absolute age of rock is determined by the radioactive element dating technique. Chemical elements are of two kinds, stable or unstable. The unstable are radioactive elements. The radioactive elements have an imbalanced proton/neutron ratio in the nucleus. The radioactive elements are called isotopes. The isotope atoms contain the same number of protons and a different number of neutrons in the nucleus, that is, the atomic number is the same but the atom weight is different. The atomic nucleus of an unstable element (parent) disintegrates to form stable and smaller atomic nucleus elements (daughter), simultaneously releasing energy in the form of atomic particles and radiation. This spontaneous breaking of an unstable atomic nucleus with the release of radiation is termed 'natural radioactivity'. On the other hand, some stable element nuclei, when bombarded with high-speed atomic particles and radiation such as neutron particles and gamma (γ) rays, produce unstable radioactive nuclei. The phenomenon is known as 'artificial or induced radioactivity'. The absolute age of a rock is determined with the help of radioactive atoms. Out of over 115 elements in total, about 50 elements are classified as radioactive, but most of them occur in traces naturally. They either decay very fast or very slowly, rendered them unfit for dating. Only four to five radioactive elements are significant for geological dating. A radioactive element, called a parent element, is unstable and continuously decays and disintegrates until a stable daughter element is formed. The decay of radioactive elements is a spontaneous nuclear reaction. The reaction ends with the formation of the stable daughter element along with the release of radiation (α and β energy particles and γ electromagnetic photons).

4.17.2.1 Half-Life of Radioactive Elements

The half-life of a radioactive element is the time it takes in years for one-half of the total atoms to decay into stable daughter atoms. The half-life of radioactive elements differs considerably. It can be as small as a few seconds for isotopic polonium (Po^{212}_{84}). Tritium, an isotope of hydrogen, has a half-life of ten years. On the other hand samarium, a radioactive element, has a half-life of over 100 billion years. One daughter atom is formed by the decay of one parent atom. The concentration of the parent radioactive element decreases, and correspondingly the quantity of the stable daughter element increases with time. By measuring the residual concentration of the parent radioactive element and the daughter atoms generated, the age of the mineral rock in which the radioactive element occurs could be computed.

4.17.2.2 Decay of Radioactive Elements and Absolute Age

The technique measures the age of a radioactive element in the rock. It does not measure the age of the rock. The age of the element is correlated with the age of the rock. It is supposed that the radioactive element has been incorporated in the sediment at the time of deposition. Uranium, potassium and thorium radioactive elements have half-lives between 0.7 and 14 billion years; this decay period is a significant time window for the determination of the age of the rock. Two sets of radioactive atoms are quoted below for rock dating:

- Potassium–argon, parent/daughter (K–Ar) radioactive dating is most common. The radioactive potassium (parent) is found abundantly compared to other radioactive elements in rock formation. Argon (daughter), being a stable and inert element, is not associated with either with potassium or any rock mineral. It solely originates from the disintegration of K^{40}_{19}. The point of 50% disintegration of potassium and 50% growth of argon corresponds to a 1:1 ratio and half-life of potassium of 1.3 billion years. If the potassium/argon ratio in a deposit is 1:1, the age of the

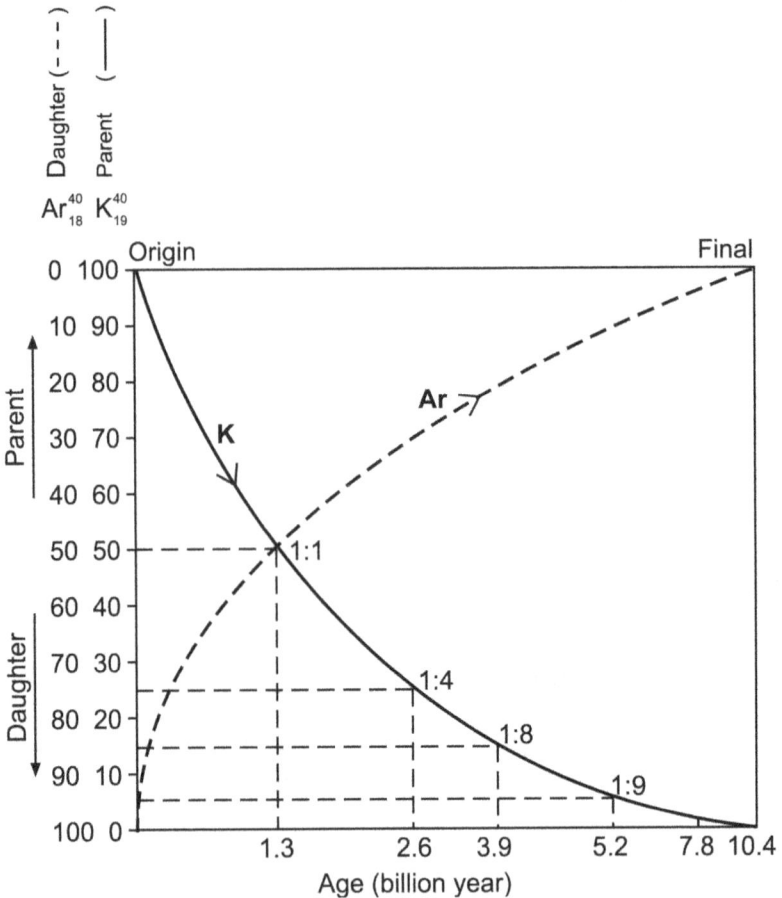

FIGURE 4.5   Radioactive decay curve of K^{40}_{19} isotope and growth curve of Ar^{40}_{18} element: decay of potassium parent isotope (—). Growth of argon daughter element (...). Source: Modified from J. Hyne Ph.D., *Non-Technical Guide to Petroleum Geology, Exploration, Drilling and Production, Penn Well Book* (e-book), Oklahoma, USA, 1995 (Figure 4–3, page 54).

deposit is 1.3 billion years. Similarly, the age of the rock could be computed for other ratios of potassium/argon. At the final point (10.3 billion years) parent radioactive K all vanishes and 100% stable Ar is generated. The radioactive decay curve of K^{40}_{19} (parent) and growth curve of Ar^{40}_{18} (daughter) are shown in Figure 4.5.

- Thorium–lead (Th–Pb) technique: the parent radioactive thorium (Th^{230}_{90}) decays to form the stable daughter element lead (Pb^{208}_{82}). The half-life of radioactive thorium is 14.0 billion years, corresponding to the time period that has elapsed since the Big Bang.

4.17.2.3   Drawback of Radioactive Element Dating

Sedimentary rock cannot be used directly for age dating, as it is not rock dating, it is radioactive element dating. The radioactive element is dated and correlated with the age of sedimentary rock. The method uses a radioactive element that has been supposed to be incorporated at the time of the initial sedimentation process. But the fact is that radioactive elements may have been added into the rock in various stages of sedimentation. Most of the trace radioactive elements originate from igneous rock. Sedimentary rock is formed by raw sediments coming from the weathering and erosion of igneous rock, so the radioactive elements may be incorporated or may not enter into the rock during the sedimentation process.

4.18   TOPOGRAPHIC AND GEOLOGICAL MAPPING

The presentation of geological features in the form of mapping is a well-established technique. Broadly the earth's surface (land and sea bottom) and underground structures are studied separately. The study of the earth's surface is called topography. The data are presented by 'topographic mapping'. The geological subsurface data indicating the underground features are processed and presented by graph and 'geological mapping'. The topographic mapping and geological mappings are different kinds of presentations but are related to each other. Topographic mapping deals with visible surface features and objects, whereas geological mapping is concerned with hidden underground features.

4.18.1   TOPOGRAPHIC MAPPING

Primarily topography is the science of describing the natural and artificial physical features of an area. In geology topography refers to the natural and artificial characteristics of the earth's surface. Topography mapping is a detailed representation of the features using different methods, for example drawing a contour map. Additionally topography finds applications in geography and a wide range of subjects of science, environment, engineering, medicine and defense studies.

Topographic maps depict different forms of land, raised structures, middle terrain and low depressions, along with the earth's land features shape, surface, elevation, dip, slope, location and surroundings. It clearly delineates (marks with limits) natural features such as mountain, hills, valley, basin, forest, river, lake, creeks, glacier, tectonic land, and underwater bottom and artificial man-made structures.

The major factor that is shown in the topographic map is the elevation of the earth's surface. The map replicates observation as witnessed from a point above the land in the air (top view). An area is selected and demarcated for topographic surveying operation with a specific purpose, for example exploration for natural resources. The various features shown in the graph are represented by standard signs, symbols and color. A topographic survey records the elevation (heights) of the sampled points.

4.18.1.1 Topographic Relief

Topographic relief shows the amount of topographic variation within a particular terrain, and relief map show the elevation and depression by different shades. A 'local relief' shows variation in elevation between two adjacent points in the topographic elevation (hill) and depression (valley). The 'total relief' is the difference between the lowest and highest points in the terrain map.

4.18.1.2 Contour Line and Map

A contour line joins all the points of subsurface strata height/elevation of equal value on the map. The base level contour line on the map is sea level. A numbers of contour lines are drawn on the map, each line showing a particular height. Any point on the contour line is above sea level. The difference in elevation between two adjacent contour lines is called the 'contour interval'. After every five contour lines, a line is selected and designated as the 'index contour'. The index contour is marked by a bold (darker) line. These facilitate reading and quick identification of elevation at any point. Elevation changes more rapidly on an inclined surface than on the nearly flat or less sloped surface. In a very steep slope structure, the contour lines are closely packed and their contour intervals are small. In a less sloped object, the contour intervals are large. The shapes of contour represent the object, such as a hill or valley or plain. A slope covers both an elevation and depression in the broader sense. A vertical erected wall side has a maximum slope of 90° and maximum inclination, represented by single contour line and a flat surface has a minimum slope of 0° and minimum inclination; 45° indicates middle inclination. The shape and the number of contour lines drawn in the map gives the details of the selected area.

Figure 4.6 illustrates the principle of topography and correlation with a contour map, by showing the topographic features of land elevation, depression and slope. Simplified correlations between the contour map and the terrain characteristics are given below:

- Contour maps are drawn below the hill with rapid and gentle slopes and broad and sharp hills, from the left to the right of the figure. The shape of contour maps below the figure represents the characteristics of the above three hills.
- The contour map on the below right-hand side of the figure represents the contour map of two hills and a connecting valley.
- The contour map on the below left-hand side of the figure represents the contour map of an elevated hill with rapid and gentle slopes.

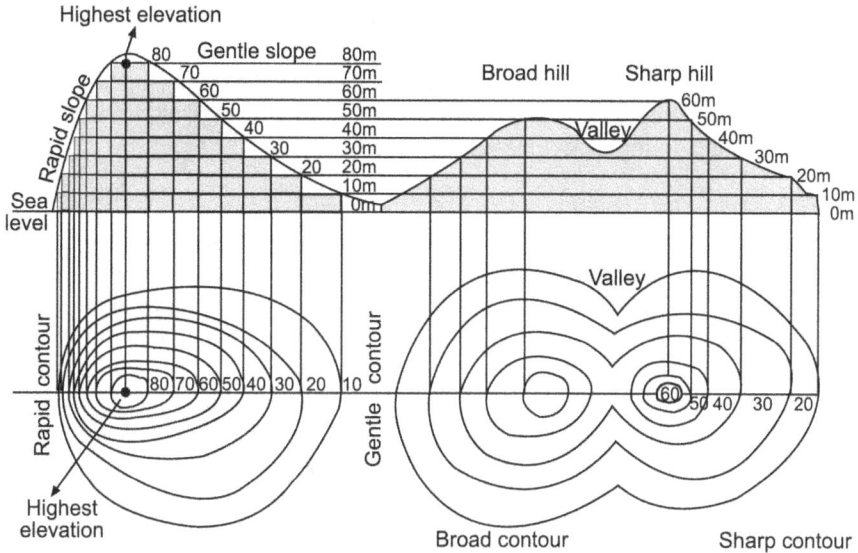

FIGURE 4.6 Contour maps (below) of different types of topography: elevated hill with rapid and gentle slopes (top left side). Broad and sharp hill and valley (top right side). The contour map on the below right-hand side represents two hills and a connecting valley. The contour map on the below left-hand side represents a hill with rapid and gentle slopes. Source: general, generic modified.

- Contour lines are concentric circular, not divided or split, and form a closed circuit. Each point on the contour is indicated by its elevation.
- The summit of the hill (peak) is indicated by an encircled center point of the map.
- Uniform and evenly placed contour lines indicate a symmetrical slope.
- The rapid or steep slope (left side of the figure) is indicated by closely placed contour lines.
- The gentle (less) slope is indicated by widely drawn contour lines, that is a large contour interval.
- A flat surface (angle of slope inclination zero) is indicated by widely spaced contour lines.
- A highly steep surface (a wall is highly steep surface inclination angle 90°C) is indicated by very close or touching contour lines.
- Unevenly spaced contour lines indicate closely located hills, ridges, valleys and pits.

4.18.2 GEOLOGICAL MAPPING

Geological maps are used to show rock features such as underground structures and processes that brought changes in structure, location, outcrop of rock, folds, fault, arrangements of various types of rock, sediment deposition patterns, etc. A map gives the material and structure information along with the formation history of the rock.

A particular 'rock formation' is selected for geological mapping. A rock is divided into several rock formations. Each formation or stratum is marked with a specific color and symbol to highlight the rock structural features.

4.18.2.1 Outcrop Mapping

An outcrop of rock is an uplifted part of the underground formation that is exposed and is visible at the earth's surface. Two special features of the outcropped rock are worth noting, the 'dip angle' and the 'strike'. The dip angle measures and indicates the angle in which the rock is dipping down into the earth (going down). Precisely the dip of a rock layer is the angle and direction in which the rock goes down into the subsurface. The strike represents the direction in which the rock is outcropping at the surface (going up). The movement of the outcropped rock is expressed as a vertical component. For example, a strike is indicated on the chart paper as north-east.

4.18.2.2 Subsurface Structural Contour Map

Subsurface structural features are mapped similarly to a topographic contour map. The map gives the depth (elevation) from the bottom to the top of the strata. The depth is reported below sea level. Stratigraphic contour lines of equal value (usually depth of the layer) are selected and plotted, to yield a contour map of the subsurface. Contour maps are employed to show the depth of subsurface dome anticline and faults, etc. An 'isopach map' is used to show the variation of thickness in strata, by using contour lines of equal thickness over a subsurface area.

Apart from the contour map showing lines of equal depth/elevation or thickness of the subsurface rock layers, there are different kinds of contour maps showing other characteristics as described below:

- **A litho-facies map** is a contour map showing the proportion of sandstone, shale, carbonate and other mineral content of the underground rock formation.
- **A bio-facies map** is a contour map showing the fossil content of the formation.
- **An isopach map** shows the thickness variation of underground formation; it connects the points of equal true thickness measured perpendicular to the layer.
- **An isochore map** shows the thickness of the layers from top to bottom along a vertical line; it connects points of equal vertical thickness.

4.19 OCEAN BOTTOM TOPOGRAPHY

General features of the ocean bottom and relative distribution of ocean crust according to the water depth are shown in Figure 4.7.

The part of the earth's crust comprising the ocean bottom is different from the land crust. The land crust is mostly covered by thicker sedimentary rock, whereas

Continental margin

| 1 Continental | 2 Continental shelf 7.5% | 3 Continental slope 8.0% | 4 Continental rise 5.0% Submarine delta fan | 5 Deep sea plain 77.5% trench 2% |

0-200 m deep   200-400 m deep   400-5000 m deep   >50000 m deep

0 m
Sea surface
200 m

Depth

Earth Crust

Sedimentary rock

Metamorphic rock

Gravity flow, turbidity current

Igneous rock

Sea

Continental margin

400 m

Trench (2%)

5000 m
1C000m

Asthenosphere
Earth mantle

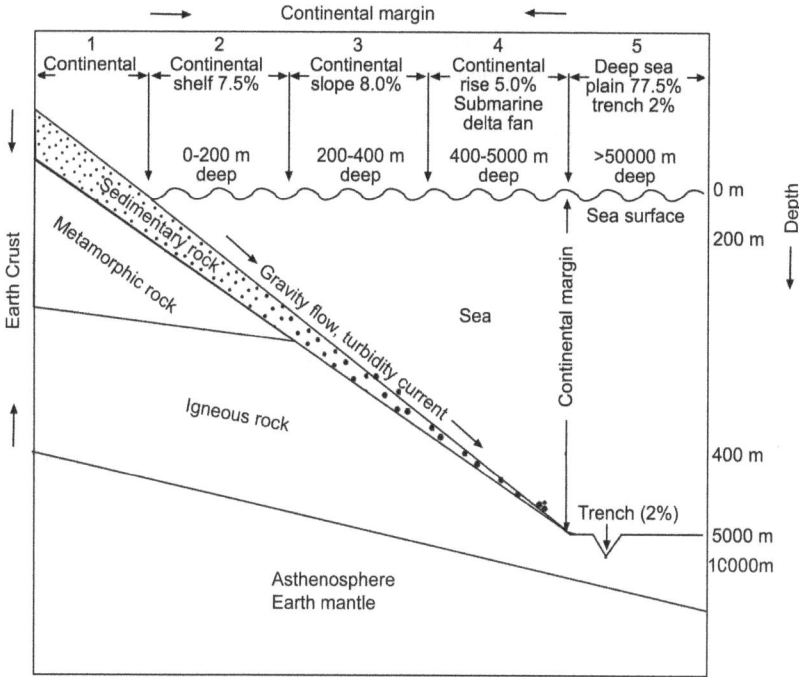

FIGURE 4.7 General features of the ocean bottom and earth crust are explained: relative distribution of ocean bottom according to the depth. Stratification by gravity flow and turbidity current on the inclined sea bottom of the continental margin. The position of sedimentary rock in the earth's crust along with igneous and metamorphic rocks.

the thickness of the sedimentary layer in the sea bottom is less than that of the land crust (<1 km on average). The sedimentary rock extends from the coastal line into the sea gradually. The sedimentary rock becomes thinner and thinner in the advancing sea bottom and finally disappears near the middle ocean ridge range. The sedimentary rock of the sea bottom is younger than land sedimentary rock. The ocean bottom surface is defined and characterized by three distinct zones:

- Continental margin
- Deep sea plane
- Mid-ocean ridges

4.19.1 CONTINENTAL MARGIN

A continent is one of seven main division of the earth's land. Here continental means land. The continental margin is the transition zone between coastal land and the deep sea bottom. The continental margin is defined "the zone of shallower ocean bottom that separates the deep ocean crust (bottom) from continental (land) crust". The bottom of the continental margin is the earth's surface submerged in ocean water. The continental margin is further subdivided into three distinguishable bottom layers, the continental shelf, continental slope and continental rise, according to their features; beyond the continental margin there is the deep sea plain containing trenches and ridges. Their share of the total sea bottom is as follows:

Continental shelf = 7.5%
Continental slope = 8%
Continental rise = 5%
Deep sea plane plus trench and ridge = 77.5%

4.19.1.1   Continental Shelf

The continental shelf is considered the border between the land and the ocean. The continental shelf is regarded as an extension of geographical land into the water. It is the shallowest part of the continental margin. The continental shelf starts from the coast and has a gently sloping surface of 1° till the end of the continental shelf and the beginning of the continental slope. After the end of the shelf, the water bottom slope increases abruptly.

The average depth of the shelf is 100–200 meters. The width of the shelf varies considerably from 1 km to 500 km according to geographical conditions in different locations of the earth. The average width of the shelf is reported as 78 km. The continental shelf is of very short width or may be non-existent in a mountain range. On the other hand, a broader continental shelf is found in plain areas of the earth, and is particularly associated with desert plains.

The geological features of the shelf and adjacent land show some similarity as well as differences as expected in all border or transition zones. The characteristics of coastal beach sand sediment are similar to the sediment of the continental shelf. In fact, the continental shelf sediments are an extension of the coastal beach sediments. The beach surface sediments are exposed to erosion and weathering, whereas the continental shelf bottom is a place for the deposition, preservation and accumulation of sediments. This sediment is transported from the continental land surface. The drained sediments overlay the already consolidated bottom rock. The sediments submerged under the continental shelf water and coastal beach sand vary constantly due to high and low tides. The high tide and high level of sea water result in the submerging of more coastal land under sea water. The water recedes at low tide and, at a low level of sea water, the coastal land is exposed to atmospheric conditions.

The area under the continental shelf is very important. Large human activities are witnessed involving commercial, economic, defense and recreational functions. One of the major activities is the exploration and production of oil and gas due to the presence of many promising sedimentary structures in the area. Anticline structures that partly exist in water and partly exist on land form good reservoir rock. Sedimentary fault and fold structures continue to protrude from the sea bottom to the beach land. The continental shelf is a potential area for oil and gas production. The whole petroleum system, namely the source rock, reservoir rock and trap rocks that appear on land continue into the adjacent continental shelf. Living organisms, plants and animals sustain and thrive in large numbers in the continental shelf. One of the many factors that determines the sovereignty of a nation over its territorial water boundary is the geographical features of the continental shelf.

4.19.1.2 Continental Slope

The continental slope begins at the end of continental shelf and goes deeper into the sea. The continental slope bottom is steeper than the continental shelf with a depth between 200 and 400 meters. The inclination of the continental slope bottom is about 5°. The width of the continental slope is reported as 12.5 km. The continental rise starts immediately after the continental slope.

4.19.1.3 Continental Rise

At the end of continental slope there is a rapid transportation and deposition of sediments that ultimately forms the 'continental rise' between sea depths from 400 to 4000 meters. The continent rise witnesses the rise of consolidated sediments and rock structures. The rock structures are formed by the heavily loaded sediments transported from the continental shelf and the continental slope by gravity, underground turbidity current and flowing debris. The drained turbid water from the continental shelf and slope contains mostly sandstone and shale sediments.

At depths below 4000 m, the sea bottom is a flat plain surface with calm water. The turbidity water current offloads its sediments, according to the law of the 'superimposition of sediments'. This forms a graded bed according to the size of the particles on the abyssal (greater depth) plain of the sea bottom.

4.19.1.4 Submarine Canyon and Fan

An important feature of the continental slope/rise is the rapid transportation, settling and deposition of sediments carried by water current along the inclined bottom surface. The heavily loaded sediments first form the continental rise, followed by the formation of 'submarine canyons' under different condition. The submarine canyons are narrow deeper water channels and trenches (valley-like) in the floor of the continental slope. A canyon can be very deep, up to a few km. The land canyon acts as a channel for the flow of rain water. In the same way, a submarine canyon provides a means for the flow of heavier turbid water in the sea floor. The turbid water is saturated with suspended sediments, originating from river discharge to the sea. The turbid water also originates from the turbulent and agitated sea bottom water. Turbid water creates a fast-moving current from the continental shelf/slope, eroding or creating a continental rise. The turbidity current flow continues as long as sea bottom slope exists.

The flowing sediment debris with turbidity current, in addition to continental rise, forms an underground canyon and delta. The underground delta is known as a 'submarine fan'. A submarine fan is the splitting of a canyon into smaller distributary channels, similar to a land river delta. The deposits in the submarine fan area by turbidity current are known as 'turbidity accumulation'. The water in the submarine area is in a dynamic state. Most of the finer shale sediments drain onward, leaving course-grained sandstone sediment. The submarine fan slope area forms good

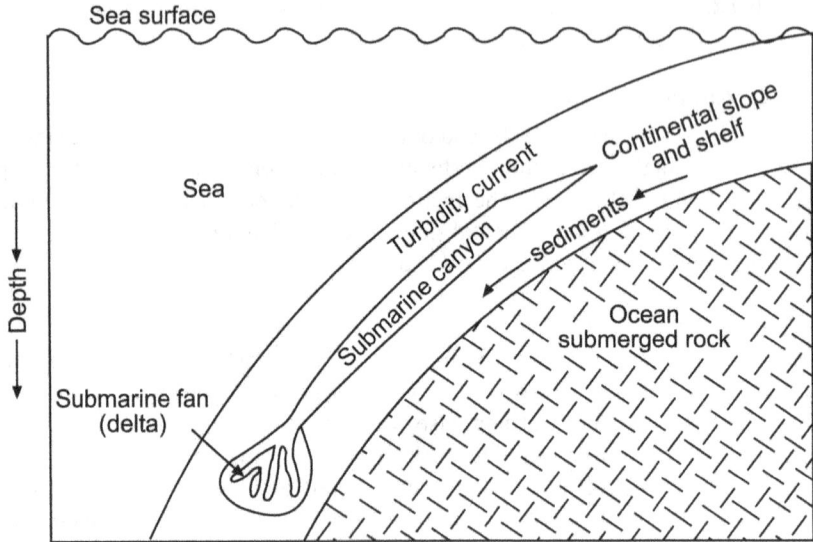

FIGURE 4.8 Formation of submarine canyon and fan by turbidity current in continental slope. Source: Modified from J. Hyne Ph.D., *Non-Technical Guide to Petroleum Geology, Exploration, Drilling and Production, Penn Well Book* (e-book), Oklahoma, USA, 1995 (Figure 10.2, page 146).

sandstone reservoir rock, capped by recent shale sediments. The formation of a submarine canyon and fan is shown in Figure 4.8.

4.19.2 Deep Sea Plane

Beyond 4000 meters deep, after the continental rise there is the vast open sea, constituting about 77.5% of the total ocean of the earth. Its depth is between 5000 and 10,000 meters. The under ocean crust and continental (land) crust basement rock are different. The ocean bottom crust rocks are formed by upwelling basalt lava from the mantle below. The average thickness of the under-ocean crust is 8 km. The continental crust, in addition to basalt, also contains granite minerals, making an average thickness of about 17 km. The molten lava contains both granite and basalt minerals. Granite, being lighter, floats up and becomes part of the land crust. The deep floor contains some mountains and trenches, but otherwise has a flat bottom throughout.

4.19.3 Mid-Ocean Ridges

Long underwater mountain (ridge) ranges are located in the middle of the bottom floor of the Atlantic Ocean. The ridge range is extended in both directions toward the Pacific Ocean in the west and the Indian Ocean toward the east and also from a few locations the ridges extends toward the north and south. The ridges are formed by volcanic activity and the spreading and cooling of the basalt minerals from adjacent magma. In a few locations the ridges appears at the surface.

4.20   CONTINENTAL AND OCEAN CRUSTS

The earth's crust is a thin layer of igneous, metamorphic and sedimentary rock covering the earth's surface (already discussed). The earth's crust can be divided into the continental (land) crust and the ocean crust. The oceanic crust consists of the sea bottom. Both crusts exhibit similarities as well as clear distinctions. The continental crust is lighter, lower-density and thicker than the ocean crust. The thickness of the continental crust varies from 20 to 70 km, whereas the thickness of the ocean crust is almost uniform and constant (8–10 km). The continental crust is made up of lighter minerals containing lighter elements such as silicon, aluminum and calcium; the ocean crust is predominantly made up of the heavier minerals containing heavier elements magnesium, iron and nickel. The ocean bottom crust is more active geologically. Underground geological phenomena are more pronounced in the ocean bottom than on land. Both types of crust are constantly being made and broken. However the continental crust is comparatively stable.

4.20.1   CONTINENTAL DRIFTING AND OCEAN FLOOR SPREADING

Ocean and continental crusts are not in static conditions, both are dynamic (very slow movement). The movement of oceanic crust is termed 'ocean floor spreading' and the motion of land crust is known as 'continental drifting'. The ocean bottom crust is spreading from the mid-ocean ridges. Similarly the continental crust is drifting from its position slowly. Continental drifting and ocean floor spreading are due to underground geological processes. All geological processes originate from the movement of hot magma in the earth's mantle. The upwelling molten magma affects the ocean and continental crusts. These geological processes are also responsible for sea floor spreading (formation of new crust), and continental drifting for the creation of mid-ocean ridges, oceanic trenches, the collision of crusts, rock subduction, etc. Important geological processes are volcanic activities, earthquakes and plate tectonic.

The concept of continental drifting is extended to explain the formation of today's seven continents. The seven continents of the earth were formed by their relative drifting from each other in the ocean. Earlier all seven continents were joined together into one supercontinent called Pangaea. Super Pangaea was formed about 350 million years ago and the Pangaea was broken and split into several parts about 200 million years ago, due to underground tectonic activities. The parts of super Pangaea drifted slowly across the earth's surface and ultimately attained their present geographical positions. The drifting of continents is a continuous geological process. Every continent is drifting in one direction or the other. The rate of drifting varies from continent to continent. The drifting rate may vary from un-noticeable to 1 mm to 100 mm per year.

4.20.2 Generation of Convection Current and Ocean Floor Spreading

The process of ocean floor spreading is defined as 'the formation of new oceanic crust in the mid-ocean ridge region by volcanic activity'. The newly formed crust gradually moves away from the mid-ocean region in geological time. The driving force for the ocean floor spreading and creation of new crust is the 'convection current' generated in the hot molten magma. Molten magma is found in the upper part of the earth's mantle (asthenosphere) below the earth's crust (lithosphere). The earth's mantle is heated from the radiation and conduction of heat from the earth's hot core below. The mantle materials melt, forming hot molten magma (liquid). The density of hot magma reduces. The lighter hot magma is slowly uplifted (upwelling) by buoyancy pressure. The upwelling magma pushes cooler mid-ocean crust. The pushing thrust results in the displacement of the mid-ocean crust, formation of new crust, formation of the mid-ocean ridge, trenches and the subduction of rock. Meanwhile striking magma cools down due to heat transfer. The density of the magma increases. The heavier cooler magma drops down (sinks) in the earth's mantle from where it was originated. Magma in the mantle is again heated by the conduction and radiation from the earth's hot core. Therefore a constant circulation, the uplifting (upwelling) of lighter magma and dropping down (sink) of cooler heavier magma, generates a convection current. Convection current is generated in liquid where a sufficient temperature variation exists. Convection is the transfer of heat by the actual movement of molecules of matter, as is observed in the circulation of magma in the earth's mantle.

4.20.3 Formation of Ridge, Trench and New Crust

The mid-ocean ridges are under sea mountain ranges, linked together in a chain. The ridges are formed along a divergent tectonic plate boundary. A trench is a long, narrow pit in the ocean bottom typically found parallel to the plate boundary. Crust spreading is the fresh addition or formation of new ocean crustal rock.

The uplifting (upwelling) magma from the mantle (asthenosphere) pushes the mid-ocean crust. The ocean crust is pushed upward. The upward thrust is most effective and active beneath the ocean crust where two divergent tectonic plates meet. The joints between rocks are favorable spots for any change or modification or alteration. The upward thrust brings about many changes. First the formation of a ridge/hill at the ocean crust takes place. Simultaneously in the area, trenches also appear, separating the ridge into two hillocks. In this way a series of ridges is formed in the area. With the further thrust of incoming molten magma from the mantle, the ridges rises and the trench widens, and at certain stage of pressure the magma erupts out from the crest of the ridge or mouth of the trench. The erupted magma (lava) spreads out on all sides of the ridge. On cooling and re-crystallization, the lava is solidified as basalt rock. The formation of basalt rock is the beginning of the formation of new crust. The eruption and spreading of lava/basalt by additional upwelling magma continue to add to the formation of the new crust. A similar situation is witnessed in volcano eruption. From the mouth (crest) of the volcano, the lava continues to spread all around during the eruption period with the formation of mid-ocean ridges, trenches and new crust.

4.20.4   Collision and Subduction of Ocean/Continental Crust

The ocean bottom crust spreads from the mid-ocean ridges. Similarly the continental crust is drifting from its position slowly. Both the ocean and land crusts may collide or may not collide depending on their directions of movement. One can expect a collision between these two crusts in motion, if they are moving in a convergent direction. Also a collision between two continental crusts or two ocean crusts is possible. Before describing the collision of two crusts or plates, it is necessary to define a new term, 'subduction'. Subduction is a subsurface process involving the sliding of one crust beneath another. Subduction is a pushing phenomenon of ocean bottom crust/plate beneath another continental crust/plate. The subduction (sliding) of the heavier crust/plate always occurs beneath the other lighter crust.

- The upper crust is always lighter than the crust/plate below.
- The heavier oceanic bottom crust/plate is always beneath the lighter continental crust/plate.

The subsurface subduction zone is the oceanic bottom. The zone is marked with frequent volcanic and earthquake activities, the emergence and formation of ridge ranges and trenches and the sliding of tectonic plates. The tectonic plate (portion of earth lithosphere) may be a part of either the continental crust or the ocean bottom crust. The boundary of two crusts (continental–oceanic or oceanic–oceanic), may be a boundary between two tectonic plates. At the boundary a collision between the two crusts occurs. As a result of collision, the crusts undergo alteration. Examples of different types of crust collision along with subduction are described below:

4.20.4.1   Collision of Two Ocean Bottom Crusts

The subduction of the heavier ocean crust below the lighter ocean crust, trench formation and volcano eruption are the salient features of the collision of two ocean crusts. In this case two crusts from different ocean bottoms meet and collide. Under the sliding force provided by upwelling hot magma, the lighter ocean crust pushes up, making room for the heavier ocean crust. The heavier oceanic crust penetrates below the lighter crust. A long narrow trench is created between the lighter and heavier subducted crusts. The subducted heavy ocean crust may go deeper into the hot earth mantle. The subducted rock materials are converted into lighter molten magmatic material. The lighter materials mix with upwelling hot magma from the earth's mantle. This provides further upward thrust, facilitating the formation of ridge ranges and volcano eruption spreading volatile lava. Volcano activities are seen adjacent to the oceanic trench.

4.20.4.2   Collision between Ocean Crust and Continental Crust

The basement of a continental crust and ocean bottom crust mostly consist of igneous basalt rock. The surface of the ocean bottom crust does not differ from the bulk of the crust material. Ocean crust is almost uniform. The surface of the land crust is marked with wide variations in sediments, soft and hard structure, topography, plain terrain, mountains, hills, etc. The collision of an ocean bottom crust with a lighter continental crust results similar to the collision of two ocean crusts. The subduction of the heavier ocean bottom beneath the lighter continental crust takes place. Simultaneously the creation of an oceanic trench adjacent to the subduction zone is observed.

4.20.4.3 Collision between Continental Crusts

The collision of two continental crusts is an encounter between similar objects. The formation of land mountain ranges and crust downward intrusion are the main effects of collision between two continental crusts.. Both crusts are made of similar minerals and have the same density. Neither is subducted. Instead the faces of crust that collide are compressed and crushed. The compressed faces may be uplifted to form a terrestrial mountain range or may go down as intrusion rock, making a depression at the surface. It depends on the extent of collision thrust and direction.

4.21 PLATE TECTONICS AND UNDERGROUND GEOLOGICAL PROCESS

A plate tectonic is a massive irregular-shaped block of solid earth lithosphere comprising land and ocean earth crust. It is also called a lithospheric plate. A tectonic plate is of various sizes, from a few hundred to thousands of kilometers across. The earth's crust is made up of eight major and eight to ten minor tectonic plates. Tectonic describes the motion and movements of these plates, and it is related to the underground geological forces that bring about changes in the earth's crust. The plate tectonics theory explains how the plates move across the earth's surface and how the crust is being destroyed and created. The word tectonics means a 'builder'. The place where two plates meet is known as a plate boundary.

The earth's crust, lithosphere, is the outer shell of the earth. Below the crust is the asthenosphere. Asthenosphere is the upper part of the earth's mantle and is molten hot rock. Lithosphere is cooler, rigid and solid. Asthenosphere is a hot and mobile viscous liquid. The solid lithosphere/crust floats above the molten asthenosphere. The crust is divided into various tectonic plates. It means all the tectonic plates are floating on the asthenosphere. Every place and object, including water, of the earth's crust is placed on one of the tectonic plates. Plate tectonics are interrelated with every geological event and process. The relative motion of two tectonic plates is related to continental drift, ocean floor spreading, earthquakes, volcano activity, subduction of rock and the formation of terrestrial mountains, depressions, underwater ridges and trenches. The characteristics of continental crust and the ocean crust differ considerably. In the same way, the tectonic plates belonging to continental and ocean crusts differ considerably.

- Continental plates are part continental crust. Continental plates are lighter, lower density and thicker than the ocean plates. They are made from lighter elements, for example silicon, aluminum and oxygen.

- Oceanic plates consist of the ocean crust. The plates are thinner but denser, harder and more rigid than the continental plates. The plates are made of heavy elements such as magnesium, iron and nickel.

The differences in the features of the continental and ocean crusts are reflected in their continental and ocean tectonic plates. Tectonic activities are related to magma movement geological processes.

The plates are in a dynamic state. The plates move, slide, glide and float over the hot magma of the mantle. Under normal conditions the movements of the plates are not observable. During earthquakes, the movement is noticeable and is recorded. The movement of the plates can be horizontal or vertical. The phenomenon of 'continental drifting' may be interpreted as the movement of the continental tectonic plates. In the same way the spreading of the ocean floor is the spreading of oceanic tectonic plates. The size and the shape of tectonic plates change with geological time. The sizes may be decreased or increased with the drifting and spreading of the plates. Major changes occur at the outer margin and edges of the plates. Plate edges interact with different plates under geological forces. The plate rock is broken and crushed. The broken rock material is either cemented or attached to the other plate, or the broken material sinks down into the asthenosphere. In the asthenosphere the broken material becomes part of the molten magma. The tectonic plates move in different directions and with different force. Modes of movement of the tectonic plates determine the type of effect.

4.21.1 Divergent Plate Movement (←→)

The divergent movement of two plates exerts a pulling force on both; sea floor spreading is best explained by this theory. The formation of new crust and the appearance of islands at the ocean surface are the result of divergent plate movement. It is the movement of two oceanic plates away from each other in opposite directions. The movement starts from the mid-ocean ridge. Incoming magma from the mantle zone below results in the formation of new ocean crust bottom. Divergent plates moving apart create space for the molten magma rock from the mantle to rise above. After cooling, the molten materials form new crust that appears as an island. The area also witnesses volcanoes and earthquakes.

4.21.2 Convergent Plates Movement (→←)

The convergent movement of two plates results in a head-on collision. It can be destructive or constructive. Convergent plate movement takes the following forms:

4.21.2.1 Ocean and Continental Plate Collision

This is the most prominent geological phenomena. The site is the junction (boundary) between the moving thicker/dense ocean tectonic plate and the relatively light/thinner continental plate. A continental plate moves toward the west from the east (from right to left). The submerged oceanic plate moves north-east. A crushing collision between the two plates occurs. The ocean plate is forced to bend and slide under the continental tectonic plate. The collision results in the formation of land mountains, trenches, rock subduction zones and mid-ocean ridges. The bent or curved ocean bottom plates continue to move under the continental plate. They may end in the asthenosphere, where the ocean plate melts and becomes part of the molten magma in the earth's mantle.

4.21.2.2   Collision of Two Ocean Plates

Collision between two oceanic tectonic plates results in a pushing movement. One plate may be pushed under the other. The downward plate forces the magma to rise upward. The molten magma appears as a volcano at the site.

4.21.2.3   Collision of Two Convergent Continental Plates

The collision of two continental plates creates mountains at the earth's surface. The colliding surface boundary of the plate is crushed, compressed and pushed upward in the form of a mountain range.

4.21.2.4   Slipping Plate Movement

Two tectonic plates move side by side against each other, creating a slipping plate movement. The force required for the slipping motion of the plates is provided by a flowing mantle lava convection current. While sliding and rubbing against each other, a lot of frictional heat is produced between the plates. The frictional force between the plates tries to stop sliding and causing sticking of the plates. Upwelling lava tries to slide the plates and frictional force tries to stop the movement. The slipping movement and stickiness depend upon the respective forces of convection lava current and material frictional resistance force acting between the two plates.

4.22   GEO-PHOTOGRAPHY

Geo-photography is an aerial/satellite survey of the earth's land, sea and atmosphere by using airplanes, helicopters, unmanned aerial vehicles (UAV), balloons, drones and satellites to record pictures/images of the targeted object of geological interest. Aerial photography can reach and record locations and processes that are invisible from the ground or located in difficult and dangerous places. For example, it gives information regarding glaciers melting and sliding, ice caps, ocean drift, volcano eruption (live and dormant), river flow/track changes and elevation of terrain along with sea bottom topography. The photographs are arranged according to location altitude, longitude and time and purpose of taking the image. Geo-photography in the laboratory is used to magnify, enhance and improve the resolution of small objects such as rock detrital pieces, mineral constituents, grain structure, geo-chemical markers and microfossils. Analytical tools coupled with cameras are used to study an object at a very small scale (micro or nano). Such tools, for example, are scanning electron microscopes (SEM), X-ray diffraction and absorption, ultraviolet (UV), infrared (IR), atomic absorption and emission, chromatographic spectrographs, etc.

4.23 CONCLUSION

A petroleum geological survey is the initial step for determining the potential rock where there could possibly be the presence of petroleum accumulation. The survey involves detailed study of surface and subsurface geology. The earth's crust is made of igneous, metamorphic and sedimentary rocks. The three types of rocks have different characteristics and mineralogy. The main focus of the survey is on the characterization of surface and underground sedimentary rock and the factors which affect its formation and alterations. The study of physical features, surface structure, structural deformation and stratigraphy of sedimentary rock gives useful information. In the same way, study of the formation of the petroleum basin, by erosion, weathering and underground geological processes, provides useful information. Deformation of rock creates folds, faults and anticline structures which play vital roles in the petroleum system. Stratigraphy tells about the layer sequence of the rock; it helps to distinguish each layer on the basis of litho-logy, mode of layering and favorable location of formation. Bio-stratigraphy, paleontology and palynology identify the strata on the basis of fossil content and biological processes. The response of some minerals to the earth's magnetic field forms the basis of magneto-stratigraphy. Geo-chronological records of past events are compiled in terms of (i) chrono-geological time units/chrono-stratigraphic time units, (ii) bio-chronology, and (iii) magneto-chronology. Relative and absolute rock dating methods give geological information about an event and strata. Bio-stratigraphy, along with fossils, has greatly helped in relative age determination. Absolute age determined by the decay of radioactive element is more significant for the study of universe evolution. The mapping of underground rock features and surface facies is known as geological and topographic mapping. Ocean bottom crust geological survey includes the study of the continental margin (continental shelf, continental slope and continental rise) and the deep sea plain beyond 4000 meters deep and far into the ocean. The earth's crust, both on land and submerged in water, including sedimentary rock, is affected by underground geological processes and tectonic activities, originating from the upward thrust of hot molten lava from the earth's mantle. As a result of the lava thrust, the earth's crust is modified or destroyed or generated. Earthqukes and volcano eruptions are associated activities. The formation of undersea ridges, new crust, continental drift, ocean floor spreading, collision & subduction of the sea floor and land floors usually occur close to the coastal area. The earth's crust is structurally divided into 16–18 minor and major tectonic plates. A plate is a massive part of the earth's solid lithosphere. Every item/location in the sea or on the surface of the land is placed on one of the tectonic plates. Geological survey has been greatly aided by geo-photography using airborne gadgets, taking images/photographs of geological interest in harsh and difficult locations in different parts of the earth. Geological survey for petroleum has been greatly supported by Geo-physical techniques, describe is coming chapters.

5 Petroleum Geophysical Survey

5.1 INTRODUCTION

A geophysical survey is the practical application of the principles of physics to study the earth. The earth's crust is heterogeneous and anisotropic (different colors/properties in different areas). So variations in geology and in physical properties are expected. Geophysical data of the sedimentary rock down to about 10 km depth, both vertically and horizontally, are recorded and correlated in a geophysical survey. After obtaining enough experimental field data, the results are processed and interpreted with the help of rock geology. The possibility or absence of possibility of finding oil/gas in the surveyed area is worked out.

Geophysical measurements have assumed a high level of sophistication due to computerized processing and interpretation of the data, including the imaging and modeling of the subsurface features in two- or three-dimensional networks. Theoretical mathematical/algorithmic treatment of data, utilizing computers, is now routinely practiced. This book avoids such rigorous treatment and gives only the principles of physical measurement as applied to underground geological structures.

5.2 GEOPHYSICAL METHOD

A geophysical method studies a particular and sensitive physical property of the earth. The property is measured at the land/sea surface or by an airborne measuring station. The data are correlated with the internal rock geology. On the other hand, if the internal earth geology is known, then the physical properties of the structure can be worked out. The physical properties normally used for geophysical survey are divided into two groups: earth's natural or passive properties and artificial or active properties.

5.2.1 EARTH'S NATURAL OR PASSIVE PROPERTIES

The natural passive physical properties are natural characteristics of the earth's surface and underground structure. A passive property exists almost uniformly throughout the structure of the earth.

Earth properties such as the earth's gravitational force, magnetic force, electromagnetic, electrical field, self-potential, telluric current etc., are common examples of the earth's natural passive properties. The natural passive property method is inexpensive, quick and can cover large areas, both horizontally and vertically inside the ground.

5.2.2 Artificial or Active Properties

An active property is artificially induced in the earth. For example, artificial seismic waves are introduced into the earth. Active methods include seismology, induced electrical resistivity, induced electrical polarization and radar survey, whereas radioactivity surveys can be active or passive.

These geophysical methods are also applicable to ore prospecting, hydrology, earth science, archaeology and geology, but some methods are more specific than others to a particular application.

5.2.3 Comparison of Geophysical Methods

Among all of the geophysical methods, seismology finds greatest application, followed by gravity measurement for oil/gas prospecting. Practically the gravity method is followed by the seismological method. The gravity method is simple, cheap and requires little equipment and short training. The method does not give specific subsurface information about the presence of oil/gas. It gives just a feeling of the formation. On the other hand, the seismology method provides information about the presence of oil/gas with more certainty and confidence. The method gives detailed information about rock geology. However the seismological technique needs high-cost equipment and personnel with high expertise. It is a high-cost activity. The seismological method has a clear advantage over other geophysical surveys in prospecting for oil/gas.

Besides seismology and gravity methods, other methods have little application in the oil/gas industry. These methods are based on electric and magnetic measurements and applicable to underground conducting, magnetic and dipolar materials.

Though all the methods are geophysical experiments, they involve different physics laws. There is a clear distinction in the theories, accuracy and applications of each. Here the geophysical methods are divided into three groups and accordingly described in three chapters. Chapter 5 deals with gravity and magnetic measurements. Chapter 6 is exclusively written on seismological methods owing to their direct application in oil/gas findings. Chapter 7 describes all of the passive and active geo-magneto-electrical methods applicable to subsurface geology.

5.2.4 Geological and Physical Anomalies

Geological variation/discontinuity is a 'geological anomaly'; the corresponding physical variation in sedimentary rock is a 'physical anomaly'. Geological discontinuity is defined as the change of geological features at the rock surface or inside the earth's structure. The discontinuity is generated by geological, environmental, physicochemical and artificial activities acting on the rock. Underground disturbances due to geological phenomena, such as magma movement, tectonic processes, volcano activity, erosion and compactness, are all geological phenomena creating subsurface discontinuity. Over geological time new features appear and old features disappear. New structures of rock layering (stratification), foliation, cleavage, fracture, fissure, fold and fault are formed. That is to say, a discontinuity has occurred in the rock. A discontinuity is a well-marked and isolated feature against the vast continuous background of the earth's crustal rock. Identification and determination of the discontinuity are the key elements in geophysical survey.

The discontinuity is accompanied by a corresponding variation in the physical and chemical properties of the rock. Constant or normal physical properties of the rock are disturbed and changed due to geological discontinuity. The geological discontinuity and the accompanying variation of physical parameters are commonly expressed by the term 'anomaly'. As a geological discontinuity is a geological anomaly, the physical variation is a physical anomaly. A physical anomaly is the change or variation or deviation in normal physical measurement value. Physical anomaly is the key parameter of rock which is reported in the geophysical data. The determination and identification of physical anomaly is the objective of the geophysical survey.

An anomaly is measured as a relative value. It is not reported in absolute terms. However the physical parameters of the rock may be measured in absolute or relative terms. An absolute value is standard, normal, independent, and intrinsic, not relative to other values and is constant irrespective of time and location. The absolute physical parameter is rarely used and is of little value to the petroleum geologist. Relative value is of interest and significant. A relative value signifies the variation/deviation/anomaly with respect to the normal/standard/actual/background value. The relative value is obtained by comparison with standard or normal values. Relative value is a comparison between observed and absolute values. The relative value is easily measured, whereas the determination of absolute value is difficult. It is customary to call a relative property an 'anomaly' for any variation in any parameter. The geophysical methods record anomalies by noting the difference between the property of the targeted object relative to the prevailing (normal) property in a small area or region. The anomaly refers to a specific property value (number) of specified rock. The anomaly data are recorded, plotted (mapping), reported and interpreted in terms of subsurface earth geology. It is obvious that the presence of oil/gas in a rock structure is a geological discontinuity and can be predicted by geophysical methods.

5.2.5 Reliability of Geophysical Data

The primary aim of geophysical measurements is to know the internal geological structure of the earth. A survey identifies only a probable situation. The geophysical survey is associated with reliability as well as with uncertainty. Both reliability and uncertainty exist in all physical measurements to a certain extent. Imagining and constructing the model of an object far inside the earth's crust on the basis of some tests conducted at the earth's surface is bound to have some element of speculation. On the other hand the same geophysical data are interpreted differently by different petroleum geologists. The uncertain situation is minimized by comparison with geophysical data gathered from other geophysical methods. One geophysical method supplements the result of another method. In spite of this seeming short coming, the geophysical knowledge of the subsurface has greatly helped in taking decisions regarding borehole drilling. It minimizes the risk and failure of finding oil. The most certain procedure for exact knowledge of the subsurface is to drill a borehole and see what exists there. Drilling is an expensive business and cannot be undertaken over a very large area. The drilling is carried out only in specific targeted, identified, promising and well-demarcated areas, on the basis of geophysical survey.

A.   PETROLEUM GRAVITATIONAL SURVEY

5.3   INTRODUCTION

The earth's gravity is a natural force around the earth. Gravity is the downward pulling force that attracts a body towards the center of the earth. The force creates a field around itself. An object placed on the surface of the earth experiences an inward pulling force due to gravity. Gravity generates acceleration on the falling object. 'Acceleration due to gravity' is the speed gained by an object because of gravitational force. Acceleration defines the speed of a falling body. Acceleration is a vector quantity. It describes the magnitude and direction of the speed. The average value of the acceleration due to gravity is 9.81 m/s^2 (32.2 ft/s^2). It means that the speed of a falling object increases by 9.81 meters per second for every second near the earth's surface. Acceleration due to gravity is not a constant. It varies slightly from place to place on the earth. There are many causes for gravity variation. Among the many causes, one of interest is underground oil-/gas-bearing reservoir rock. The presence of underground oil/gas affects the gravity of the rock; correspondingly an anomaly in gravity is recorded at the measurement station on the surface of the earth. The underground targeted rock which is affecting gravity measurement at the surface is called an 'anomaly rock'. An oil/gas reservoir is an anomaly rock. Effected variation in gravity is called an 'anomaly in gravity value'. The knowledge of gravity anomaly is important for the prospecting of oil/gas.

The scope of gravity measurement (anomaly) survey is limited, and it is only indicative or speculative in nature. For the confirmation of the gravity findings, simultaneously other methods such as seismology, well logging and exploratory drilling are required.

5.4   THEORY OF EARTH'S GRAVITY MEASUREMENT

The earth's gravity is governed by Newton's law of universal gravitation, popularly known as the 'inverse squared law'. The force of attraction (F) between two bodies of different masses is given by the formula:

$$F = \frac{Gm_1m_2}{r^2}$$

Where
  G = gravitational constant ($6.67 \times 10^{-11} m^2/kg/s^2$) in the MKS (SI) system
  m_1 and m_2 = masses of the two bodies
  r = distance between masses m_1 and m_2

Likewise, the force of attraction (F) between the earth and a small mass at the surface is given by

$$F = \frac{GMm}{R^2}$$

Where
  G = gravitational constant
  M = mass of the earth (constant)
  R = radius of the earth, 6371 km (constant)
  m = mass of the small body at the earth's surface

The earth and the small body are supposed to be spherical, non-rotating and uniform; therefore, attraction between them is uniform and constant.

The above equation can be written as:

$$F = \frac{GM}{R^2} \times m$$

GM/R^2 = constant (g), therefore

$$F = g \times m$$

The 'g' is acceleration due to the earth's gravitational force that is acting on the mass body (m). The "mg" is the weight (W) of the body.

$$W = m \times g$$

The weight (mg) of an object is the mass (m) times the acceleration due to gravity (g). The weight of an object is the inward pulling force and depends on the acceleration due to gravity (g). The mass of any object is constant. Therefore, the weight depends on the acceleration due to earth's gravity (gravity inward pulling force). There may be some difficulty in distinguishing between mass and weight. The mass of an object is the amount of matter in the body, regardless of any force acting on it. Mass is independent of the volume. Weight is defined as the 'earth inward pulling gravitational force (gravity) acting on the mass of the body'. The mass of an object is the same on the earth and moon. The weight differs by an amount of 1/6 of the earth's weight to the moon's weight. The moon's gravitational inward pull is 1/6 of the earth's.

5.5 GRAVITY DETERMINATION

Absolute gravity is not required in geological survey. Its determination is not only difficult but also time consuming. The determination of 'relative gravity' or the 'gravity anomaly' is needed for the interpretation and correlation of the subsurface targeted small area in terms of rock features including oil/gas. Relative anomaly is the difference in the gravity of the region (background) from that of the targeted survey location area. If the regional gravity (normal/standard gravity) is higher than that at the targeted area, the difference is referred to as a positive anomaly, and if the regional gravity is smaller it is a negative anomaly.

A variety of commercial gravimeters are available. The principle of working of the gravity determination apparatus is either of a 'spring elongation' or 'pendulum oscillation time period' under the influence of gravitational force. Again, spring elongations (extension) or the time period of pendulum oscillation is not measured in absolute terms. Only the initial and final meter readings (number) are recorded on an already calibrated instrument. Gravity variation or anomaly is estimated with the help of elongation or time period recorded by the instrument. Two types of gravimeters, based on the above-mentioned principle, are the spring elongation balance and the pendulum torsion instrument. A third type of gravimeter is based on the combination of these two features.

5.5.1   Spring Balance

The downward pulling gravitational force is measured by a spring. A spring is made by applying a twisting force to a thin, soft, flexible and long metallic bar. The mechanical force needed to twist the metallic bar is resisted by the opposing resistance of the bar. The resistance force is equal to the applied mechanical force but in the opposite direction. Thus a spring is a store of mechanical energy. A vertically suspended spring will resist the downward gravitational force, at the cost of the elongation of the spring. Hooke's law states that 'the force needed to elongate the spring by some distance, is proportional to the change in length (Δr) of the spring'. The elongation force is provided by the earth's gravitational downward force (F) acting on the mass (m) of the spring. Therefore

$$F = mg = K\Delta r$$

K = spring constant; g = acceleration due to gravity.

The elongation force is provided by the earth's gravitational downward force F (mg) acting on the mass (m) of the spring. Therefore elongation (Δr) is directly proportional to gravity's downward pull or acceleration due to gravity ($\Delta r \propto g$), that may be computed. The sensitivity (elongation) of the instrument is increased by attaching a mass ball at the lower end of the spring.

The working of the balance with three equal length (r_0) springs each attached with balls of increasing mass (m_0, m_1, m_2) is shown in Figure 5.1. The initial spring lengths have been elongated by an amount of (Δr_1) and (Δr_2) by masses m_1 and m_2, respectively. Due to the different masses, the gravitational force (mg) acting on each ball is different. Different weights pull the spring differently. By increasing the mass, the corresponding elongation is also increased, which is measured by the optical scale attached to the instrument. A change in spring elongation (Δr) is related to the gravity anomaly or change in acceleration due to gravity.

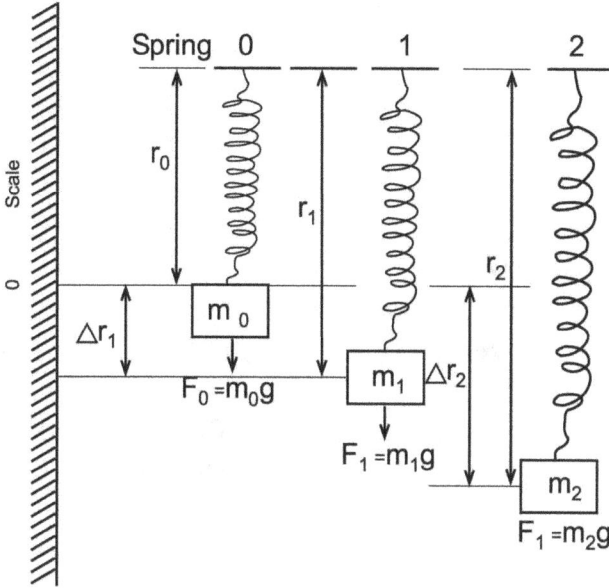

FIGURE 5.1   Working principle of spring balance.

5.5.2 PENDULUM TORSION BALANCE

Torsion is a twisting force. A torsion balance measures a small torque (rotating push) applied to a mass so as to set the pendulum in oscillation (to-and-fro motion). The working principle of pendulum torsion balance is illustrated in Figure 5.2.

A pendulum balance consists of a small heavy mass metallic ball suspended from a fixed frictionless pivot at the top with a fine wire. A pushing torque is applied to the mass body to set the pendulum in swing motion. The motion will continue indefinitely provided there is no air and pivot resistance. Oscillatory motion is characterized by amplitude and time period. The time period (T) for one complete cycle of the swing is related to gravity (g) as follows:

$$T = 2\pi\sqrt{L/g}$$

$$g = 4\pi^2 \frac{L}{T}$$

Where

  L = length of pendulum wire
  g = acceleration due to earth gravitational force which can be computed

In addition to the time period, gravity (g) is also proportional to the swing amplitude (displacement of the ball in to-and-fro motion.). The amplitude of swing is related to the twisting angle 'θ' as shown in Figure 5.2.

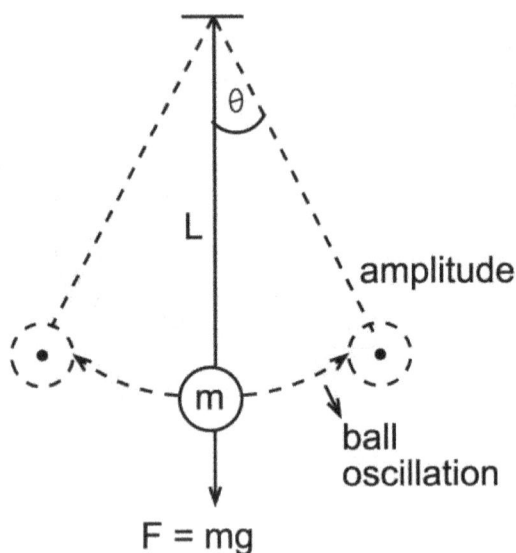

FIGURE 5.2   Working of pendulum torsion balance.

5.5.3   LA COSTE ROMBERG GRAVIMETER

Several commercial gravity determination instruments are available, including the common La Coste Romberg gravimeter. The La Coste Romberg gravimeter assimilates the features of both spring and pendulum torsion balances. A beam is hung on a wall vertical plane surface. The spring is fixed above the beam at a suitable distance away from the beam on the wall. The free end of the beam is fixed with a body of mass (m). The mass is also supported by the above-hanging spring as shown in Figure 5.3. Gravity is measured by recording the tilted angle or by noting the change in elongation of the spring. Working of the balance is explained as under:

- The gravitational force (mg) acting on the mass set the beam in motion, creating tilted angles Θ_1 and Θ_2. The angle Θ_0 signifies the original or null position. The gravitational force (g) is proportional to the tilted angles of the beam at the respective positions of the mass body. The lower and higher gravitational forces correspond to the angles Θ_1 and Θ_2 respectively.
- Likewise the elongation or contraction of the spring (Δr) at different positions of the mass is proportional to the value of the earth's gravitational force (g). The change in length of the spring is measured by adjusting and bringing back the beam and mass to their original or null position (angle Θ_0).

5.6   ANOMALY EXPRESSION

A gravity anomaly is a variation in gravity. Gravity variations occur due to different locations of measurements or one object influencing another. Variation in gravity is

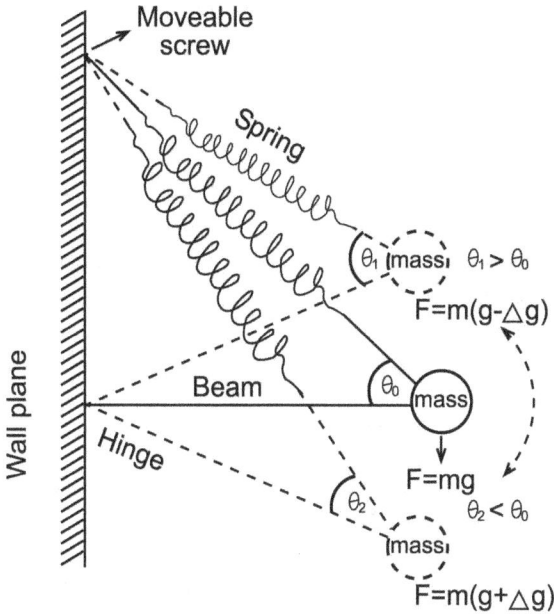

FIGURE 5.3 Principle of La Coste Romberg beam and spring balance.

also expressed as 'change in acceleration due to gravity'. Gravity and acceleration due to gravity can be expressed by single word 'g'.

5.6.1 RELATIVE GRAVITY

A gravity anomaly expressed as relative gravity is the difference between two gravities at two locations. One location with standard/normal gravity value gravity value (ρ_2) is referred to as the 'reference location' and the other station with gravity (ρ_1) is known as the 'observation location'. The reference location is chosen among the pole, equator and sea level. The reference value is also known as the regional or background value.

$$\Delta\rho = (\rho_2) - (\rho_1)$$

$\Delta\rho$ is called 'relative gravity' or the 'gravity anomaly'.

5.6.2 GRAVITY ANOMALY

A gravity anomaly is the difference in gravity between the regional/standard/normal value (ρ) in an area and the observed value (ρ_1) at the same location. The observed value is the sum of the normal value in a region and the gravity influenced by other object (ρ_1). Other object is called as anomaly object. For an object having gravity

(ρ_1), influencing other object of gravity (ρ_2), the resultant 'gravity anomaly ($\Delta\rho$)' is given by their difference.

5.6.3 Acceleration due to Gravity Anomaly

A gravity anomaly is also expressed by noting the change in 'acceleration due to gravitational ', instead of gravity.

The change in acceleration due to gravity (g) either from one location to another or one object influencing another is a 'gravity anomaly' and expressed as 'Δg'.

The anomaly may be positive or negative, depending upon the gravity contrast. A higher reference value gives a positive anomaly, and a lower reference value gives a negative anomaly. To find out the cause of a gravity anomaly either in terms of change in acceleration in gravitational force or as the change in gravity is the purpose of geological survey. The units of gravity and acceleration due to gravitational force are g/cc and cm/s^2 respectively. The gravity anomaly is interpreted to work out the subsurface geology. A typical gravity anomaly due to geological factors is around 100 micro-gals (0.1 milli-gal).

Gal is also a unit of acceleration due to gravitational force. One gal is equal to a change in rate of motion of 1 centimeter per second per second. Earth gravity at the surface is 976 to 981 gal.

$$\text{One gal} = 1 \text{ cm/s}^2$$

The maximum difference in earth's gravity from the pole to the equator is 5 gal or 5000 milli-gal or 5,000,000 micro-gal.

5.6.4 Gravity Anomaly due to Earth Shape and Rotation

The gravitational field originates from the center of the sphere. It is expected that a non-rotating uniform sphere would generate a uniform gravitational field around it. However the earth's shape is not perfectly spherical but elliptical spheroidal, and it is not stationery but rotating. Therefore the downward pulling force (g) is not uniform; it changes from place to place on the earth. The gravity and the weight of the body change from location to location accordingly. The resultant gravity of the earth is affected by the following factors:

- A major factor in resutant gravity of the earth is its huge mass along with the uneven mass distributions inside the earth's body.
- Deviation of the earth's shape from spherical to elliptical spheroidal
- Earth's rotation or spin.
- Tilted axis of earth rotation or spin (tilted to 23°).
- Earth's centrifugal force generated by the earth's rotation.

At the two poles the earth's surface is flat and uniform. It is expanded out (bulged) at the equator. The flat pole surface means there is a smaller distance to the center of the earth. The bulging equator means there is a larger distance to the center of the earth. The gravity force acting on an object at the earth's surface is inversely proportional to the distance from the center of the earth. The earth's rotation and tilted axis are accompanied by different centrifugal forces at different locations of the earth. The centrifugal force is acting in the outward direction, counteracting the inward pulling gravitational force. These factors contribute to different gravity or acceleration (g) values at the poles and the equator or any other location on the earth. The earth's gravity values expressed as acceleration due to gravity at different reference locations are given below:

g = 983 cm/s^2 (maximum at earth's poles)
g = 976 cm/s^2 (minimum at earth's equator)
g = 981 cm/s^2 (average value)
g = 32.2 ft/s^2 (average in FPS units)

5.7   GRAVITY CORRECTION

The terms gravity (earth's downward pulling force), gravitational force and acceleration due to gravitational force are used here with the same meaning.

The gravity anomaly is either useful or interfering. The former type of anomaly is targeted which is used in geological surveys for interpretation. The other kind of anomaly is interfering noise and must be segregated to obtain a 'corrected gravity anomaly'. The causes of anomalies need to be identified in order to separate them from the anomaly of the targeted object of interest. Besides the geological factors for the gravity anomaly, the other causes are due to the geographical position of the targeted rock along with interfering environmental objects. The gravity anomaly due to the targeted rock gives geological information of interest. All the other causes of gravity anomalies are useless and are identified and segregated. The observed gravity reading is the result of all the contributory geological and non-geological factors. A corrected gravity anomaly is obtained by computation with a normal value of gravity 'ρ' or acceleration due to gravity 'g' at sea level. Interfering factors and correction required are described below.

5.7.1   ALTITUDE (ELEVATION) AND LATITUDE (HORIZONTAL) CORRECTION

A change of gravity in vertical direction is significant whereas the change is negligible horizontally. Gravity decreases with elevation above the earth's surface and increases with depth inside the earth. Elevation means a greater distance from the earth's center, and depth means a shorter distance from the center of the earth. Observation stations at greater heights give lower gravity and the corrected gravity anomaly positive. The gravity measurement in presence of air or in a vacuum at the same station gives different result. The air upward buoyancy decreases the weight/gravity of an object. The rate of change of gravity with horizontal direction is small compared to the change with altitude (elevation), which is formidable.

5.7.2   Topographic Correction

The topographic features like hills (more elevation) and valleys (less elevation) exert different gravitation pulling forces on the body. Hills behave like elevation so the gravity (g) is reduced. Valleys effect a small downward pull, increasing the gravity (g) of the object. These two corrections are applied on the observed gravity value at the station, so as to give a reading as on a flat surface at a certain elevation from topographic features.

5.7.3   Bouguer Correction

The Bouguer gravity correction is the combined correction for elevation, weight of material (excess mass) and air between the field measuring station and reference location (usually sea level). It is the gravity correction for height, attraction of the excess material slab and air buoyancy. Field topography, plateau, hill, valley and other structure are excess material slab between measuring station and reference sea level. The effect of elevation, attractive force of the material slab and air buoyancy on the gravity measurements are summed and subtracted from the observed value to get a corrected gravity.

5.7.4   Miscellaneous Corrections

Planet orbital correction. The force of attraction among the moon, sun and earth changes with their orbital positions. The earth's gravity changes with different orbital positions of the rotating sun, moon and earth with time.

Environmental correction. The artificial processes at the surface of the earth, such as industrial, agriculture, commercial and other human activities, disturb the earth's gravity (g) and need to be filtered in order to obtain a corrected gravity.

Seismic event correction. A seismic event disturbs the gravity measurement. The experiment is stopped during the event.

Instrument correction. Fluctuation in the instrument reading is known as instrument drift. The drift is computed by taking a number of readings at regular intervals. The repeatability of the data is evaluated and due correction is made in the observed gravity. The drift of a gravimeter is usually due to the elastic fatigue of the sensitive spring with time and usage. The instrument drift continues to increase with time and usage.

Motion correction. A survey is often required to cover a large area with minimum time, so different types of transportation are used. A gravimeter placed in an aircraft or vehicle experiences attractive or repulsive forces depending on the direction and speed of the mobile unit. The generated positive or negative effect on the gravity anomaly.

5.7.5   Net Gravity Anomaly

The net gravity anomaly is computed by the subtraction of all of the above anomalies from the observed anomaly. The purpose of all the correction is to get the residual net anomaly specifically caused by the underground 'anomaly rock' observed at the surface measuring station.

5.8 GRAVITY SURVEY METHOD

On the basis of available geological information, a subsurface zone is chosen whose geology is to be determined. Several gravimeter instruments are placed on the surface at regular intervals in the longitudinal and horizontal positions, making a grid network above the entire subsurface targeted zone (anomaly rock). The separation distance between measuring gravimeters depends on the area of survey. Normally a large area is covered by placing gravimeters at large distances from each other. A small area is surveyed with small distances between stations. Large spacing between gravimeters covers a large underground survey area, but sensitivity is reduced. The distance between gravity gravimeters at the surface varies from a few meters to a few kilometers. Gravimeters are calibrated and a reading is taken covering the demarcated area. Taking account of regional gravity and other corrections, the net gravity anomaly is estimated.

Gravity anomaly data are used to draw a contour map and two-dimensional graphs (distance/depth versus gravity anomaly) which is correlated with the geology of the underground targeted zone. Also gravity anomaly data are subjected to processing by simple/complex mathematics and computer techniques.

5.8.1 CONTOUR MAP

A contour map is drawn by joining the point of 'equal gravity anomaly values' to obtain a contour of each set of gravity value. Contour maps are used to find out the structure of the subsurface rock. The contour interval, the separation between two gravity readings, is usually 1 milli-gal to 100 μgal. A contour map also specifics the elevation. From the center of the contour map, elevation decreases outward.

5.8.2 TWO-DIMENSIONAL GRAPHS (DISTANCE/ DEPTH VERSUS GRAVITY ANOMALY)

A two-dimensional graph is a plot of x-axis versus y-axis. On the x-axis, subsurface depth is recorded or the distance between gravity meters at the surface and on the y-axis gravity anomaly (Δg) is recorded. If a significant gravity anomaly rock is present, a well-resolved peak (x versus Δg) is obtained. A peak is a part of the curve where the maximum change in gravitational acceleration (Δg) occurs due to the pull of the targeted anomaly object. Quantitatively the change in gravity can be measured by noting the peak height or area under peak. The maximum change in gravity (Δg) occurs at the maximum peak height that is also called the maximum amplitude. Peak height has been utilized to determine the depth location of the anomaly in the rock. Peak height decreases with depth. There can be two depth measurements. One is from the center of the rock to the base measuring station at the surface. Another depth measurement is taken from the top of the targeted rock to the surface. The former has more distance (depth) from measuring point at the surface, therefore the gravity signals are weak. The latter depth is of less distance therefore the signal is strong. The thickness of anomaly rock can be computed by the knowledge of the difference in gravity anomalies in terms of peak height and area under peak.

5.9   GRAVITY DATA INTERPRETATION

The subsurface geological features generate specific and significant gravity anomalies and lead to the finding of oil/gas. The gravity anomaly is related to the geology, material distribution, depth and shape of the subsurface target rock.

A dense and compact rock generates more pulling force, resulting in a greater gravity anomaly than soft and porous rock. A feeble gravity anomaly is generated due to unconsolidated soft sedimentary rock. A basement rock that has been pushed up into sedimentary rock gives rise to an appreciable anomaly, due to reduced distance from the surface and the denser character of the rock. Different underground structures (denser rock, anticlines, synclines, salt domes, folds, faults, porosity and fluids) generate gravity anomalies differently so they are identifiable. Salt rock is less dense and softer and has little impact on gravity reading, so the presence of a salt dome can be established. Fault-, fold- and fluid-bearing rock gives different anomalies.

The gravity anomaly depends on the density contrast of the subsurface layers. Density distribution is not uniform in sedimentary rock. It varies from 1.6 to 3.2 g/cc for different mineral compositions. The materials (minerals, oil, gas, water and void space) in sedimentary rock are not distributed uniformly.

The material distribution and heterogeneity in subsurface rock affect the gravity measurement at the surface.

The attractive gravitational force between the gravimeter at the surface and an underground object decreases with depth. The gravity anomaly decreases with increasing depth (increasing distance) as observed at the surface. At a certain depth the 'anomaly rock' generates a minimum noticeable anomaly at the surface; the corresponding position (depth) of the rock is known as the 'minimum or limiting depth'.

Below the limiting depth, the rock body does not exert any attractive force at the surface, so no anomaly effect is observed at the surface. Shallow rock exerts an effective attraction force at the surface, compared to the deeper rock, thus creating a distinct and distinct anomaly (Δg). The depth of the underground rock can be computed as a clear anomaly contrast exists between a shallow and deeper rock. The shape of the anomaly rock affects the gravity readings and peak shape. From the gravity anomaly data, the shape of the anomaly rock is predicted. For a spherical mass object, the maximum gravity anomaly (peak point for Δg) is observed on a vertical line passing from the surface through the center of the underground spherical body. The anomaly gradually and symmetrically decreases on either side of maximum peak point corresponding to the spherical surface. The shape of the peak and area under the peak determine the size and shape of the underground object.

B. PETROLEUM GEOMAGNETIC SURVEY

5.10 INTRODUCTION

The earth's magnetic field (geomagnetic field) is a natural force around the earth. The field originates from the earth's interior and appears at the surface and atmosphere. A geomagnetic survey is the study of the earth's subsurface rock, through measuring anomalies in the earth's magnetic field at surface measuring stations. It is one of the geophysical methods used for finding oil/gas, ores and the remains of buried structures (archaeology) in sedimentary structures. Geological features such as geothermal, tectonic, litho-logy, stratigraphy and the presence of magnetic materials are correlated with magnetic fields. The magnetic data are also utilized for the identification of underground utility piping, underwater submarines, and all other metallic materials.

A magnetic field is present around the earth, just like a small magnet produces a magnetic field around itself which cannot be seen. A freely placed magnet aligns itself toward the north-pole because the earth acts like a big magnet. A magnetic bar has equal amounts of opposite polarity; a north-pole and a south pole or positive and negative poles. The poles of a small magnetic do not align to the geographic poles; it aligns to the magnetic north pole. The locations of the earth's magnetic poles are not fixed but variable. The magnetic north pole drifts to the west about 20 km per year. The magnetic field originates from the earth's core. The earth's core contains hot moving magma. Magma is mainly composed of iron and nickel magnetic conducting minerals along with other magnetic materials. The moving charged conducting materials generate a convection current and magnetic field. The associated magnetic field from the earth's core passes through the earth's mantle & crust and penetrates the atmosphere. This advancing magnetic field counters the incoming charged particles from the sun. The harmful charged particles, called the 'charged wind storm', are prevented from reaching the earth by the magnetic field, thereby saving the earth. The earth's magnetic field behaves similarly to the earth's gravitation field in many ways, but differs in theory and application, and geomagnetic surveys follow a different path from the gravity geological survey. The findings of geomagnetic surveys reinforce the results of gravity surveys.

5.11 THEORY OF MAGNETISM

Magnetic material produces around itself an invisible magnetic field. The field attracts other magnetic materials. A magnetic element has a north and a south pole. Two poles with similar polarity exert a force of repulsion, and opposite poles exert a force of attraction of equal strength but opposite direction.

Similarly to Newton's law of universal gravitation, Charles Augustin de Coulomb discovered that the force of attraction or repulsion (F) between two magnetic poles and two electrically charged bodies, placed at a distance 'ɤ' apart, obeys the 'inverse squared law' as given as follows:

$$F = \frac{1}{\mu} \times \frac{p^1 p^2}{r}$$

where

    p^1 and p^2 = pole strength
    $\mathbf{r}$ = distance between the poles
    μ = magnetic permeability constant

Permeability is the ability of a material to acquire magnetism in response to the applied magnetic field.

The magnetic permeability constant (μ) is given by the ratio of magnetic flux to the magnetic field strength (B/H). The permeability depends on the magnetic properties of the medium.

5.11.1 Magnetic Field Strength (H) and Magnetic Flux (B)

A magnetic field is the area where magnetic lines are effectively observed. The magnetic field strength is one of the physical measurements that expresses the intensity of the magnetic field. The magnetic field strength, denoted by H, is the density of the field and is defined as the force exerted on a pole of unit strength placed at that point. The magnetic field strength is similar to the gravitational acceleration (g).

Flux indicates the presence of a force field in a particular area.

Magnetic field and magnetic flux are related terms. The area around the magnet where the pole and the moving charge experience the force of attraction or repulsion is called the magnetic field, whereas the magnetic flux represents the quantities of magnetic lines of force passing through the area. The magnetic flux density, denoted by B, is the magnetic flux per unit area perpendicular to the direction of magnetic field. It is a vector quantity measuring the strength and direction of the magnetizing field.

The units of magnetism are complicated. The units are expressed in term of force, flow of electric current and distance. Weber, tesla and gauss are common units (1 tesla = 10^4 gauss). The tesla is a large unit and in earth's magnetic field only small variations are encountered, so the units are subdivided as follows:

    1 nano tesla = 10^{-9} tesla (nT) and 1 micro tesla = 10^{-6} tesla (μT)

5.11.2 Susceptibility of Magnetization

Magnetic susceptibility indicates the degree of magnetization (I) of a material in response to the strength of the applied magnetic field (H):

$$I \propto H$$

$$I = kH$$

Where 'k' is a dimensionless proportionality constant and is termed 'susceptibility'. For a non-magnetic material (diamagnetic), 'k' is zero, including air and vacuum, that is they cannot be magnetized.

5.11.3 MAGNETIC INDUCTION

Magnetic induction is a process in which a magnetic material is magnetized through an externally applied magnetic field. Induction is a magnetism acquired by the material. An external applied field (H) introduces magnetic properties in magnetic permeable material, say an iron bar. Iron itself does not have a magnetic field, because the atoms are randomly arranged. In an external field, the atoms of the bar align themselves in a north–south direction, producing magnetic dipoles. The bar acquires an induced magnetism till a 'saturation' point is reached. The magnetized material produces its own induced magnetic field and the material as a whole behaves like a dipole magnet. The induced magnetic field (H*) is directly proportional to the applied magnetic field strength (H).

5.11.4 MAGNETIC DIPOLES MOMENT

The law governing magnetic attraction or repulsion force or the inverse squared law is applicable to isolated monopoles placed at a certain distance apart. In practice a magnetic material is composed of two monopoles and is a dipole. A magnetic element is composed of north and south poles and is called a magnetic dipole. The positive and negative poles cannot be separated.

A dipole (magnet) has 'magnetic moment' since it tends to rotate in a magnetic field. Just as an electric field originates from positive and negative charges, similarly a magnetic field originates from the positive and negative poles of a magnet. Two poles of equal strength, but of opposite charge and separated by a distance 'ɤ' from each other, will produce a dipole moment. If the strength of the pole is 'F', then the magnetic moment of the dipole will be the product of 'F' and 'ɤ' (Fɤ). Magnetic dipole moment is a vector quantity associated with the magnetic properties of the magnet. The direction of the moment is along the line of the north pole.

5.11.5 ELECTROMAGNETISM

Electromagnetism is the study of electric/magnetic fields associated with a positive or negative charge and magnetic poles. Electromagnetism describes how the flowing current produces an invisible magnetic field around it and a moving magnet creates an electric field around itself. The magnetic field is perpendicular to the direction of the flowing current. The magnetic strength produced by the current flowing in a loop of wire of radius 'ɤ' is directly proportional to the current intensity. The dual nature of electromagnetism is utilized for the prospecting of underground oil/gas and magnetic materials. Electromagnetism has an enormous effect on industrialization, and its detail is given in Chapter 7.

5.12 EARTH'S MAGNETIC FIELD

The earth's magnetic field is similar to the magnetic field around a bar magnet. Both are invisible magnetic lines of force around the earth and bar magnet. The field around a bar magnet can be visualized and mapped from the direction of a compass needle placed around the bar magnet. A bar's magnetic field flows from the south pole and converges at the north pole. In the same way, a freely suspended bar magnet aligns in the direction of the earth's magnetic field. The north pole of the magnet aligns itself toward the earth's north pole, which is the north-seeking or positive pole. The other pole of the magnet is the south-seeking or negative pole.

The origin of the earth's magnetic field is the ferromagnetic substances (Fe/Ni/Co minerals) in the hot outer part of the earth's core. Earth's rotation and the motion of the core materials generate a convection current in the molten magma. The convection current induces magnetism in the ferromagnetic minerals. This is similar to an iron bar that is magnetized by moving a magnet at the surface of the bar. The shape of the earth's magnetic field is similar to the field around the bar magnet but tilted to 11° from the spin axis of the earth. The earth's magnetic field originates from the south pole and ends at the north pole; it has both vertical and horizontal components as follows:

- Vertical upward. The magnetic field originates from the south pole. The direction of the field is vertically upward from the south pole.
- Horizontal. After that the field proceeds to the earth's equator horizontally.
- Vertical downward. After crossing the equator horizontally the magnetic field takes a downward dip toward the north pole. The magnetic field reaches the north pole almost vertically downward.

The direction of the earth's magnetic field, originating from the south pole vertically upward and ending in the north pole vertically downward, is shown in Figure 5.4.

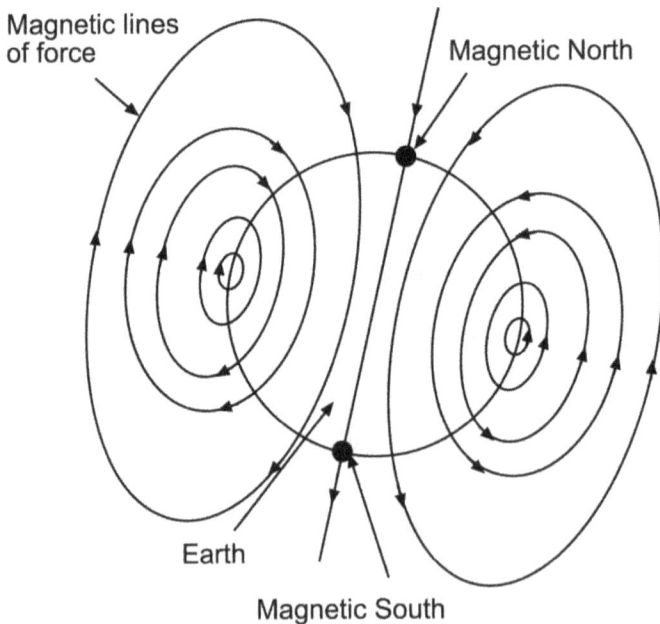

FIGURE 5.4   Earth magnetic field from south pole to north pole.

5.12.1 VARIATION OF THE EARTH'S MAGNETIC FIELD

The variation of the direction of the geomagnetic field around the earth can be understood by considering a freely moving compass needle in space. The compass needle aligns itself according to the earth's magnetic field. At the north hemisphere, the north-seeking compass needle points in an upward direction; at the magnetic north pole the compass needle points vertically upward. In the south hemisphere the other end of the needle, that is the south-seeking needle, points in an upward direction and at the magnetic south pole the compass needle points upward vertically. In between these two locations there is a place where the needle is horizontal and no dip witnessed. This is the location where the magnetic equator is passing east–west. The magnetic equator is not the same as the geographic equator, but passes closely to it in an east–west direction.

The direction of the earth's magnetic field changes from location to location in space. It is defined by three-dimensional x, y and z coordinates. All the three coordinate components are mutually perpendicular to each other (see Figure 5.5). The 'x' and 'y' coordinates define (2-D graph) a horizontal plane. Together all three coordinate axes determine a magnetic field position in space. Since the total magnetic field F, is a vector quantity, it can be resolved into the components F_x, F_y and F_z along the x, y and z coordinates (3-D). Applying mathematics:

$$F^2 = F_x^2 + F_y^2 + F_z^2$$

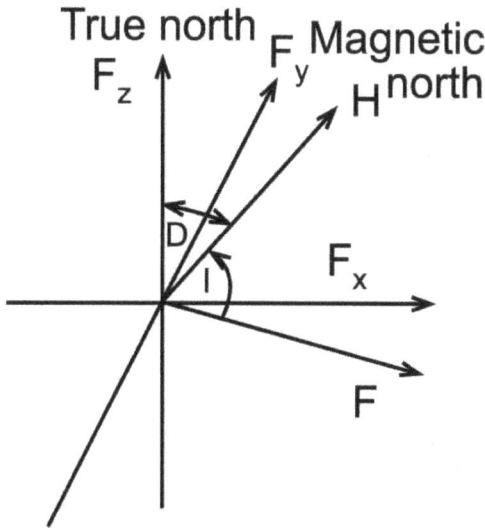

FIGURE 5.5 Resolution of total of total magnetic force (F) into components; $F^2 = F_x^2 + F_y^2 + F_z^2$ and horizontal components $H^2 = F_x^2 + F_y^2$. I = inclination varies from the equator (zero) to pole (90°) and D = declination angle between true north and magnetic north.

The magnetic field (H) acting on the horizontal plane (x, y) is the sum of the x-y components. Therefore

$$H^2 = F_x^2 + F_y^2$$

The total magnetic field (F) at the two poles of the earth is almost resolved to vertical field components (F$_z$). The horizontal component is almost nil at the poles. On the other hand, at the equator the horizontal component (H) represents the total magnetic field (F). The total magnetic field between the equator and poles is the combination of vertical and horizontal vector components (F$_x$, F$_y$ and F$_z$).

Two important magnetic terms derived from the above discussion are:

- **Inclination (I)** is the angle between magnetic field on horizontal plane (H) and earth's total magnetic field (F). It is denoted as 'I' in the figure. The angle varies at different points on the earth's surface. This is the angle that the needle makes with the horizontal plane at any location. The magnetic inclination at the equator is 0° and 90° at the poles. An inclination angle greater than 60° indicates that the dominant factor is the vertical component (F$_z$) in the total magnetic field. An inclination angle less than 40° indicates a place near to the equator, so that the dominant factor is the horizontal component (H). The place in between inclinations 40° and 60° represents a combination of vertical and horizontal magnetic components.
- **Declination (D)** is the angle between magnetic north (magnetic field pointing north) and true geographic north.

All together F, H, F$_x$, F$_y$, F$_z$, I and D are known as the elements of magnetism since they characterize the field at any point.

5.12.2 EARTH'S MAGNETIC MATERIAL

The earth's magnetic materials, on the basis of their magnetic properties, are classified into three types:

- Diamagnetic material
- Para-magnetic material
- Ferromagnetic material

Almost all the non-metal elements are diamagnetic, that is to say they do not respond to a magnetic field. The non-metal cannot be magnetized. Para-magnetic elements are called 'metalloids'; they response to magnetic fields and are magnetized temporarily. As soon as the external magnetic field is withdrawn, the induced magnetic field vanishes. Examples of this class are aluminum (Al) and antimony (Sb) and their minerals. Ferromagnetic materials are those elements that are readily magnetized by an external magnetic field. They retain the induced magnetic field after withdrawal of the external field. Examples of ferro-magnetic materials are minerals containing iron (Fe), nickel (Ni) and cobalt (Co).

5.12.3 MAGNETIC RESPONSE AND RESIDUAL/REMNANT MAGNETISM

The response of rock to the external magnetic field not only depends on the mineral constituents but also on its geological structure. Rock compactness, density, faults and litho-logy all affect magnetization. A compact, hard and dense rock (igneous) is expected to be more susceptible (sensitive) than soft, coarse-grained and rarefied rock. A sedimentary rock may have some small magnetic property, but its intensity is too low to be used for geological information. Some kinds of earth rock can be magnetized appreciably and other types of rock may not or rarely.

A magnetic material, for example iron, initially with zero magnetic property can be magnetized by an external field. How much will the material be magnetized or how long will the material remain magnetized after withdrawal of the external magnetic field? The residual magnetism depends on the magnetic susceptibility of the material. 'Residual (remnant) magnetism' is defined as 'the magnetization left behind in the acquired ferromagnetic material, after the withdrawal of the external magnetic field'. Residual magnetism observed in a rock is of the following kinds.

5.12.3.1 Natural Residual Magnetism (NRM)

'Natural residual magnetism (NRM)' observed in rock is of two kinds:

- Natural remnant magnetism is the induced magnetism in the original rock strata by the earth's magnetic field during the geological history of sediment deposition, consolidation, stratification and lithification. The magnetism is original and is known as 'primary natural remnant magnetism' (PNRM).
- Secondary natural remnant magnetic (SNRM) is induced magnetism in the rock after the geological processes of lithification and formation of the rock. It is due to further geological processes of strata alternations, chemical changes in ferromagnetic minerals, chemical transformation of minerals, thunder and lightening, fluctuation of earth's magnetic field and temperature. The total 'natural remnant magnetism' (TNRM) is the sum of primary remnant magnetism (PNRM) and secondary remnant magnetism (SNRM).

5.12.3.2 Thermal Residual Magnetization

The rock has retained the original magnetic field acquired during early geological history when the earth's molten state was cooling. Specifically it is the residual magnetism, observed in cooling from high temperatures (above the Curie temperature) of hot magnetic material. The Curie temperature (T_c) is the temperature at which certain material loses permanent magnetism and acquires induced magnetism. The magnetism left upon cooling the earth's hot (700°C) core magnetized minerals to a lower temperature (300°C) is the thermal residual magnetism.

5.12.3.3 Depositional Residual Magnetization

During the deposition of different sediments, some magnetic sediment particles (iron) are incorporated into the deposits. The incorporation of magnetic minerals facilitates the induced magnetism in the structure. During the initial stage of deposition, when the deposit is unconsolidated, the iron particles are able to align themselves along the natural earth field, thus creating dipole magnetism. It is known as depositional remnant magnetism.

5.12.3.4 Chemical Residual Magnetism

Chemical remnant magnetism is exhibited by ferromagnetic materials due to chemical processes below the Curie temperature (T_c). In the earth's core ferromagnetic materials are formed as a result of the chemical transformation of minerals.

5.12.3.5 Demagnetization and Curie Temperature (T_c)

The 'Curie temperature (T_c)' may also be explained as follows, 'at the higher temperature the thermal vibration and motion of the material particles are so great that alignment (mechanical coupling) of ferromagnetic materials is not possible'. At higher temperatures the alignment of iron particles is disturbed due to the acquired kinetic motion of the particles. Therefore magnetization is not possible at higher temperatures due to the kinetic energy. A varying and fluctuating magnetization field also renders demagnetization of material. A control thermal treatment of the magnetic body reduces or eliminates the secondary natural remnant magnetism but retains the primary natural remnant magnetism. A record of original magnetic field (PNMR) is utilized for correlation of the rock formation environment.

5.13 GEOMAGNETIC ANOMALY AND CORRECTION

Variation in the geomagnetic field due to a targeted underground source is a geomagnetic anomaly. The earth's magnetic field varies in three dimensions (spatial). The geomagnetic field is not constant, neither in direction nor in quantity (magnitude). It is a vector quantity.

A geomagnetic field anomaly (ΔF) is defined as 'the difference in earth geomagnetic field (F) and magnetic field (F^*) created by other factors':

$$\Delta F = (F - F^*)$$

The total magnetic anomaly (ΔF) caused by a subsurface object along with other factors is combined with the earth's geomagnetic field causing a geomagnetic field value of ($F + \Delta F$). Accordingly, the horizontal component (H) of the geomagnetic field changes to new value of ($H + \Delta H$) and the vertical component has changed to ($F_z + \Delta F_z$). A geomagnetic anomaly due to a targeted underground object is related to the object and gives useful information about the object.

Not only does the targeted rock generate an anomaly but there are a number of other factors that contribute to a geomagnetic field anomaly. These anomalies are of non-geological interest and are considered as disturbing or interfering anomalies. The subtraction of interfering geomagnetic anomalies from the observed geomagnetic anomaly gives the net (corrected) anomaly caused by the subsurface target source.

The factors that contribute to anomalies of non-geological interest (interfering) are given below:

- Secular anomaly correction. A secular geomagnetic anomaly is observed at a particular location and time. The magnetic field keeps on changing with time. The variation of the magnetic field from the north pole to south pole is from 30,000 nT to 65,000 nT. The average value of the earth's field is about 50,000 nT.
- Diurnal (daily) variation correction. Diurnal variation is the hourly and daily change in the magnetic field. It is due to the magnetic radiation reaching the earth in the daytime from the sun. The normal diurnal field variation in a day is 20–80 nT.
- Magnetic storm anomaly. A magnetic storm is defined as a sudden, severe and short-duration variation in the earth's magnetic field. Magnetic disturbance is due to the storming of the earth's atmosphere by the charged particles from the sun. The magnetic storm causes severe interference in the magnetic field. During a magnetic storm the survey is stopped to avoid fluctuation in the magnetic field. The variation due to a magnetic storm may be as high as 1000 nT.

5.14   GEOMAGNETIC FIELD DETERMINATION

A magnetometer is an instrument for determining the magnetic field. It may be noted that variation due to instrument drift is minimal and not significant as compared to other geophysical measuring instruments. The instruments are robust and sturdy. Descriptions of four different magnetic instruments are given below.

5.14.1   FLUXGATE MAGNETOMETER

The fluxgate magnetometer is a practical application of the theory of electromagnetism. The working of a fluxgate magnetometer is described stepwise as follows:

- I. Two cores (bars). A fluxgate magnetometer consists of two cores (bars) made of highly permeable ferromagnetic material such as iron or ferrite.
- II. Primary and secondary windings. Around each bar, identical primary and secondary electrical windings (coils) are wound in clockwise and anticlockwise directions.
- III. External input primary current. An external electrical current (DC), passing through the primary coil wound around a bar, induces magnetism in the bar. 'Magnetic saturation' point is reached after some time and no more magnetization of the bar is possible. On the other hand, if a strong alternating current (AC) (input primary current) is passed through the primary coil wound around the bar, then the bar witness spontaneously the maximum and minimum of magnetic polarity during each cycle of AC current. The

AC current magnetizes the bar to saturation and unsaturation that is magnetizing and demagnetizing the bars alternatively.

IV. Output secondary current. The alternating magnetic field from the bar induces an electrical current in the secondary winding coil. The current 'output' from the secondary coil is measured by an ampere meter. In a uniform magnetic field, the primary input current is equal to the secondary output current. If the uniform magnetic field is disturbed by underground magnetic materials or other sources, then the induced output AC secondary current oscillates between maximum and minimum and out of phase (different) with the input primary current. The difference between input and output current is related to magnetic anomaly. The magnitude of magnetic anomaly depends upon the type and magnetic strength of the anomaly source.

5.14.2 PROTON RESONANCE MAGNETOMETER

A proton resonance magnetometer is commonly used for measuring the magnetic field. The instrument is also known as a nucleus precession magnetometer. The nucleus of an atom contains protons and neutrons and outer-shell orbiting electrons. However the nucleus of a hydrogen atom contains only one proton and an outer-shell electron. The proton magnetometer instrument uses only the lone nucleus proton of a hydrogen atom. The working of the instrument is described stepwise as follows:

I. Normal magnetic field → protons at ground state (uniformly aligned). In a normal geomagnetic (ambient) field, protons are aligned in the direction of the field and said to be at ground level.

II. Applied strong direct current (DC) → protons are in an excited state (oriented). In the presence of an external strong direct current, the hydrogen nuclei (protons) absorb the electric field and are said to be in an excited state (polarization/orientation). Hydrogen nuclei are chosen from a liquid rich in hydrogen content, for example kerosene filled in a glass cell. The cell is surrounded by a DC coil.

III. Withdrawal of applied DC field → protons are in a transition state (spinning, precession, and excitation). On withdrawal of the applied field, the proton releases the electromagnetic radiation of particular frequency along with precession of the proton. This is a transition stage of spinning/precession and excitation also called relaxation. Precession of the proton is similar to the changing motion of a spinning top that wobbles to form a rotating cone. The duration of the precession is 1–2 seconds. After that the protons realign themselves and return to original ground state I. The precession frequency cycle (f) of the rotating proton is proportional to the normal geomagnetic field (F). The frequency cycles are converted into electrical signals (voltage). The voltage is sent to an electronic detector and printer to obtain a computed ambient magnetic field.

The experiment is repeated, completing the cycles of excitation and precession of protons. It gives a series of frequency readings at regular short intervals.

The proton magnetometer is a reliable instrument for geomagnetic field measurement.

5.14.3 ALKALI METAL VAPOR MAGNETOMETER

The alkali metal magnetometer is also known as the 'optical or electron pumping magnetometer'. The theoretical basis of the instrument is derived from quantum physics where classical physical laws do not apply. Alkali metals sodium (Na), potassium (K), rubidium (Rb) and cesium (Cs) are soft, can be vaporized under vacuum and have only one electron in the outer shell. Cesium is almost a liquid at ambient conditions.

A single electron has electric charge, spin and an associated magnetic field. The electron freely exchanges energy levels from one orbital level to another, under influence of other forces. Absorption of energy occurs when an electron is polarized (excited) to a higher level, and energy is depolarized (emitted) when the electron returns to a lower level.

In practice an energy lamp, mostly cesium lamps or any other alkali metal lamp, is used to generate incidence/excitation/energizing light. The incident light from the cesium lamp goes through a slit, lens, filter and then to the light absorption cell containing vapors of cesium metal. Some of the incidence light is absorbed and the remainder is transmitted through the cesium vapor glass cell. The incident light, called resonating light, is of a particular frequency that matches the frequency of the vibration (changing energy level) of the charged cesium metal electrons in the light-absorbing cell. A photocell converts the transmitted light from the cesium cell into electric signals which are recorded. Briefly the working of magnetometer is stated below:

* The partly evacuated light absorption glass cell contains the Cs metal vapor in discreet atomic forms.
* The outer electron or the valence electron of the Cs atom occupies the orbit designated by level 1 and level 2 at the ground state.
* Incident light/excitation light from the cesium lamp excites the electrons from level 2 to 3, in the absorption cell containing Cs atoms, a process known as polarization. The electron in level 1 remains unmoved. Level 1 is not affected.
* If the energy of the incident light from the Cs lamp is more than the requirement, the outer electron will be ejected out from the orbit and lost. The cesium atom will not remain neutral; the cesium atom will be converted to a positive ion due to the ejection of the negative electron. This situation is avoided in the experiment.
* The electron at level 3 is unstable and quickly falls back to either level 1 or level 2, with the release or 'desorption' of absorbed energy.
* Repetition of 'absorption' and 'desorption' of energy leads to the saturation of level 1 at the expense of level 2 which becomes deficient in or unoccupied by electrons.

- The transferring of electrons to different levels, that is from level 2 to level 3 and then from level 3 to levels 1 and 2, is known as 'optical or electron pumping'. With the continuous electron pumping the level 2 becomes empty and level 1 is fully occupied by the electrons. So the cesium atoms in the cell stop the absorption of incident light energy because no more level 2 electrons are available to be exited. The total incident light is transmitted through the absorption cell since hardly any level 2 electrons remain to absorb incident light. The cell atoms are said to be at saturation stage or fully polarized.
- Cesium lamp light cannot excite the electron from level 1 to level 2.

Now the role of radio frequency energy is as follows:

- The electron of the Cs atom in level 1 is excited to higher level 2 by the application of external radio frequency (RF) energy (radio frequency electromagnetic radiation). The wavelength of the RF radiation corresponds to the energy difference between levels 1 and 2 which excite the electron and is a measure of magnetic field strength.
- The energy difference between level 1 and level 2 is proportional to the earth's magnetic field strength. The difference in energy between level 1 and level 2 is measured.
- Measurement is enacted by the application of proper wavelength radio frequency (RF) energy around the cell. A balance is achieved in absorption and transmission which is monitored through photocell detection. A radio frequency wavelength that matches the absorption and transmission is a measure of ambient magnetic field strength.

5.14.4 MAGNETIC GRADIOMETER

A magnetic gradiometer is also known as a 'differential magnetometer'. The gradiometer is employed for the determination of the magnetic field and also used for the measurement of the magnetic gradient (rate of change of the magnetic field with distance).

A gradiometer consists of two magnetometers. It may be two alkali metal vapor or proton resonance magnetometers. One instrument is placed vertically above the other, say at a height of 30 meters. It is known as 'axial gradiometer arrangement' and measures the vertical component of the magnetic field. Another combination of gradiometer is horizontally in which two magnetometers are placed side by side at a distance say 30 meters apart. It is called 'planer gradiometer arrangement' and measures the horizontal component of the field.

The advantages of the magnetic gradiometer are as follows:

- The method is useful in regions with a lot of magnetic field variation. Gradiometer readings are not affected by regional, diurnal or any other environmental object.

- The method has improved sensitivity, better resolution and clearly differentiates between mirror magnetic anomalies.
- As usual the method is applicable for the determination of shape and depth of anomaly rock.

5.15 GEOMAGNETIC SURVEY METHOD

Normally a geomagnetic survey is carried out by aircraft and ships. It is known as a motion/mobile survey and covers both land and sea areas. Exclusive land survey is also practiced. The basics of all surveys are the same. There are some obvious differences between mobile and land surveys.

5.15.1 LAND GEOMAGNETIC SURVEY

The scope of land magnetic survey is limited. The survey is mostly related to the measurement of the horizontal component of the magnetic field and covers a small area. Spacing between magnetic measuring stations in a land survey is much smaller (10–500 m) compared to the parallel line spacing of motion surveys (5 km). The geometrical arrangement of the observation stations in the land survey is similar to the gravity measurement. The selected location for measuring station should be free from metal objects or structures.

The interfering magnetic anomalies from all sources, including due to diurnal, region and location, must be taken into consideration and subtracted from the observed field data. The magnetic field readings are taken from the chosen magnetometer at regular intervals.

5.15.2 MOTION GEOMAGNETIC SURVEY

Motion survey can be an aero-magnetic survey or a marine survey. Aero-magnetic surveys are rapid and cost effective compared to marine mobile and land survey and cover large areas with different topography. The position and flying route at different heights is planned with respect to the targeted underground object by considering the initial geological survey data.

An aerial survey measures the total magnetic field. The survey covers the horizontal ($H = F_x + F_y$) and vertical (F_z) components of the field. Aerial and marine surveys are carried out by towing the instrument, usually fluxgate or proton resonance magnetometer, behind the aircraft, helicopter or marine ship with a cable. A magnetic survey is usually carried out in conjunction with another survey such as a gravity and seismological survey from the same ship. Flight height is determined on the basis of the purpose of the survey. An elevated height covers a greater underground survey area, compared to the lower height flight that covers much less or limited surface area. On the other hand, in the lower height flight the anomaly readings are much more resolved, well defined and distinct and sharp, having higher amplitude (more response) but covering a smaller survey area. In case of higher elevation, the anomalies recorded are broad and diffuse and have lower amplitude. Likewise a shallower magnetic rock produces a well-resolved, well-defined, distinct and sharp anomaly having higher amplitude (more response) than the deeper anomaly rock for a given flight height survey.

5.15.3 CONTOUR MAP

After gathering geomagnetic anomaly data through a survey, they are further processed. More information is obtained by drawing a contour map of the magnetic anomaly data. The contour from sedimentary rocks is smooth and simple. The contours drawn from magnetic anomalies of igneous and volcanic rocks are complicated. A contour may be either closed or open. Closed contours are of interest since they correspond to a ferromagnetic underground source.

5.16 INTERPRETATION OF MAGNETIC ANOMALY

The magnetic anomaly is not related to rock, in fact the anomaly is related to the presence of magnetic material in the rock. Some specific examples of magnetic anomaly are as follows:

The ferromagnetic materials progressively decrease from igneous to metamorphic and sedimentary rocks. The highest magnetic anomaly (higher amplitude) is observed for igneous rock, moderate for metamorphic rock and is almost absent in sedimentary rock (feeble magnetic field). The presence of igneous rock and intrusion of igneous rock and absence of sedimentary rock and oil/gas are predicted with confidence.

The magnetic anomaly created by an underground ferromagnetic rock may be either positive or negative depending on the dipole nature of the rock constituents. Different orientations of the dipole particles of the rock matrix give rise to varied magnetic fields. Each pole of the material particle contributes to the anomaly.

Variation of the magnetic field occurs throughout the earth. Different locations give different magnetic anomalies. Hence magnetic anomaly data give different shapes of a single anomaly object. The determination of the depth/distance of an underground magnetic object is based on the principal that with the distance the strength of the magnetic anomaly decreases. At shallower depth (less distance) the anomaly response is strong (high amplitude). The depth (z) is inversely proportional to the magnetic field strength, horizontal component (H) as well as magnetic vertical component (f_z). The magnetic survey is not applicable directly for oil and gas prospecting because oil/gas are not ferromagnetic materials and do not respond to magnetic fields.

5.16.1 COMPARISON OF GRAVITY AND MAGNETIC ANOMALY

The earth's magnetic field closely resembles the earth's gravitation field in many ways. At the same time, it has clear distinctions in theory and application and follows a different path from the gravity geological survey as evident from Table 5.1. The theory of magnetism is difficult, whereas theory is simple for gravitational

TABLE 5.1
Comparison of Earth Magnetic and Gravitational Fields

	Earth Magnetic Field	Earth Gravitational Field
1	Invisible field	Invisible field
2	Natural potential field	Natural potential field
3	Magnetic anomaly is not sharp and distinct	Magnetic anomaly is sharp and distinct
4	Technique is simple and comparatively costly	Technique is simple and cheap
5	Sedimentary rock exerts feeble magnetic field	Sedimentary rock exerts little gravity anomaly
6	Mostly airborne and ship survey	Land, sea and airborne survey
7	The magnetic anomaly of the field is caused by subsurface magnetic materials	The gravity anomaly caused by density contrast of the subsurface materials
8	Minerals containing magnetic elements greatly influence the field	Minerals containing heavy elements greatly influence the field
9	Magnetic field exerts attractive and repulsive force	Gravity field exerts only attractive field
10	Magnetic field may originate from little material	Gravity field originates from bulk materials
11	Igneous rock exerts a significant magnetic field	Igneous rock exerts feeble gravity field
12	The magnetic field is three-dimensional. It is a vector quantity	Gravity field is almost vertical. It is scalar
13	Exploratory reconnaissance technique	Exploratory reconnaissance technique
14	It gives subsurface litho-logical information	It gives subsurface litho-logical formation
15	High-precision instruments are used	Moderate precision instruments are used
16	A magnetic field anomaly is positive/negative, arising from finite magnetic dipole material	Gravity anomaly is either positive or negative due to denser/softer source (gravity contrast)
17	Utilizes natural earth potential force and obeys law of inverse attraction	Utilizes natural earth potential force and obeys law of inverse attraction
18	Theory is difficult	Theory is simple
19	Survey result is indicative	Survey result is indicative
20	The magnetic field is generated in subsurface rock by the presence of ferromagnetic material	The gravitation field originates from the pull of a denser anomalous bulk subsurface rock

force. The gravitational field anomaly is either positive or negative depending on the density contrast (dense or soft) of the subsurface anomaly rock, whereas the magnetic anomaly is both positive and negative due to the dipole particles of the buried anomaly object.

The earth's gravity field is uni-directional and acts vertically downward, whereas the magnetic field is three-dimensional. A magnetic anomaly is a vector, related to the magnitude of the field and also to the direction of the field. The direction as well as the strength of the magnetic field vary from place to place.

The interpretation of magnetic anomaly data becomes similar to that of gravity anomaly data, when only the horizontal component of the earth's magnetic field is taken into consideration. It is only encountered in land surface surveys. The gravitational anomaly clearly shows the geometrical shape of the underground object. The magnetic anomaly is not as sharp and distinct compared to the gravity anomaly. The magnetic anomaly data do not give a specific shape of the anomaly rock, as compared to the gravitational anomaly data. Magnetic data has limited scope in oil/gas prospecting.

5.17 CONCLUSION

The gravity survey method is simple, cheap and covers large areas. The method is based on the natural inward pulling force of the earth, exerted on the surface materials. The inward force is uniform but has slight variations (anomaly), due to major subsurface heterogeneity observed at the surface. The anomaly is related to the geological structure. The gravity downward pulling force is not measured in absolute terms but as relative gravity or as a change in gravity or gravity anomaly. The principle of the measuring instrument is based on the (l) elongation (Δr) of a spring or time period (T) of pendulum or twisting angle (Θ) of a metallic bar, under gravitational force. The elongation of a spring, time period and twisting angle are related to the gravitational force. The anomaly data are recorded in a two-dimensional graph; the horizontal distance (km) between measuring stations or downward depth versus vertical gravity anomaly (Δg) is expressed as milli-gal or μgal. The shape and size of the anomaly curve correspond to the structure underneath. Through varied analytical, mathematical and computer techniques, details of the understructure are worked out. The gravity method is an initial and preliminary survey. The findings by other survey methods are needed to confirm the gravity results.

A geomagnetic survey is utilized for the studies of underground rock containing ferromagnetic materials. The magnetic survey is similar to the geo-gravitational survey along with some obvious dissimilarity. Gravity determination is carried out by simple apparatus whereas magnetic measurement requires a sophisticated instrument. The gravitational field is a vertically inward pulling force to the center of the earth. The geomagnetic field (F) is three-dimensional (x, y, z axes) is therefore more complicated. The x-y components are combined into the horizontal component (H). The gravitational anomaly originates from the bulk characteristics of rock. A magnetic anomaly is due the presence of dipole particles (magnetic material) in the rock. The magnetic survey is not applicable directly for oil and gas prospecting because they are paramagnetic materials, and do not respond to magnetic fields. Igneous rock contains a detectable amount of magnetic minerals, metamorphic rock contains some and sedimentary rock has negligible amounts of ferromagnetic materials The presence of underground shallow igneous rock is inferred as an absence of sedimentary rock. This leads to the conclusion that there is no possibility for finding oil/gas, thus avoiding costly exploration and drilling operations. The depth of subsurface magnetic rock is estimated by noting the decreasing magnetic field strength with distance. Magnetic surveys are gainfully applied in mineralogy and finding ore. The findings of gravity and magnetic survey is confirmed by seismological survey discussed in the next chapter.

6 Petroleum Seismological Survey

6.1 INTRODUCTION

Seismology is the study of earthquakes. Seismic waves are naturally generated during earthquakes and accompanied by movement or vibration of the earth's crust. The crust vibration is due to the traveling wave energy through the underground structure. Petroleum seismology is the study of the geological characteristics of underground sedimentary rock, through artificially created seismic waves. The seismic wave or energy produced at the surface travels into the ground and results in a variety of reflected, refracted, head refracted and transmitted waves. The reflected and head refracted waves appear back at the surface, whereas the refracted and transmitted waves travel deep into the ground and are lost. The study of reflected and head refracted waves lead to information on the path traveled through the subsurface rock strata in terms of change in the characteristics of the seismic wave. The nature of the strata affects the traveling seismic wave. Change in the characteristics of wave is related to the subsurface rock structure. Petroleum seismology for oil/gas prospecting gives more certain and reliable information than the other geophysical survey methods. The survey is close to the geological reality of the subsurface structure. Besides petroleum prospecting, other areas where seismology is employed are ground engineering, environmental, coal, minerals, hydrology and geothermal studies. The study ranges from shallow to deep earth crust.

The theory of seismic wave propagation is simple and easily understandable. But in practice it needs a great deal of involvement from various branches of science and technology. With the introduction of sophisticated computerized processing, recording, interpretation and modeling, it has assumed a new dimension in oil/gas prospecting with improved certainty. The whole exercise includes a long list of equipment and personnel at the surveying site. This chapter deals only the basic information of the subject. Seismological study is employed in two different ways:

- Study of earthquakes
- Study of subsurface geological rock structures for petroleum prospecting

The basic seismological principles for the two applications are the same, but they differ considerably when practically carried out. Former earthquakes are related to natural phenomena originating from tectonics. The latter application requires artificially created seismic waves.

The propagation of seismic wave is similar to other types of waves such as acoustic, sound, water, optic and electromagnetic waves. All exhibit reflection, refraction

High frequency (f)=10 Hz, low wave length (λ)=2cm, velocity=$f \times \lambda$=20 cm/sec

Low frequency (f)= 5 Hz, high wave length (λ)=4cm, velocity=$f \times \lambda$=20 cm/sec

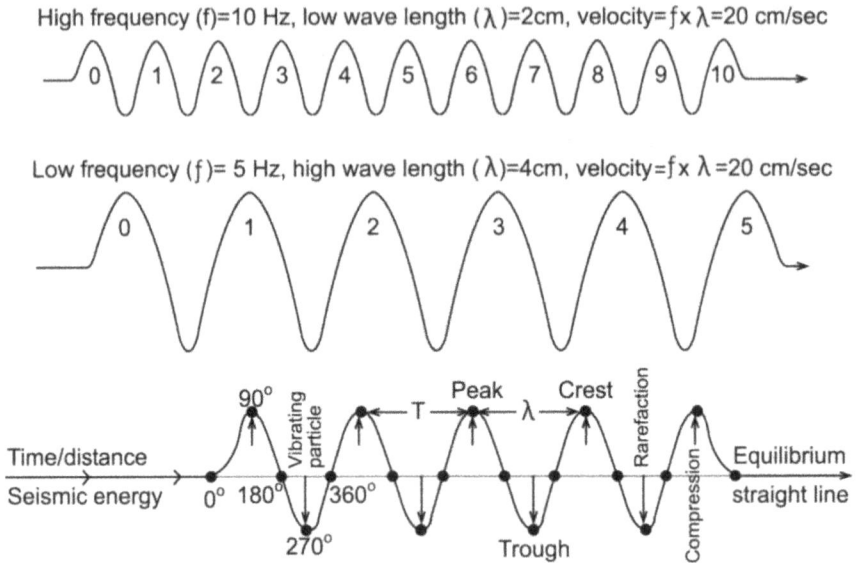

FIGURE 6.1 Characteristics of a propagating wave: Frequency (f) = number of waves per second (Hz). Wavelength (λ) = distance from crest to crest or trough to trough or one wave distance or phase 0° to 360°. Amplitude = peak height or phase 0° to 90°, maximum particle displacement. Cycle = number of waves that pass a given point on the equilibrium line in one second. Period (T) = time taken for one cycle. Phase = particle location during oscillation expressed as an angle (degree). Wave number = reciprocal of wavelength. Wave motion is due to particle motion about the equilibrium line. Source: Modified from Bhal B.S., Arun Bhal and Tuli G.D., *Essentials of Physical Chemistry*, Rajendra Ravindra Printers, Ram Nager, New Delhi, 2006. Page 13.

and transmission of waves. Electromagnetic waves originate from the sun, cosmos and artificial means. It may be mentioned that electromagnetic waves are transverse waves, whereas sound and seismic waves can be both transverse and longitudinal.

The characteristics of propagating waves are explained by two-dimensional graphical (time/amplitude plot) representations in Figure 6.1.

6.2   SEISMIC WAVE PROPAGATION

A wave is defined as 'the transfer of energy without the transfer of matter', or as the 'the repeated chain of interconnected particle motion' (up-down or to-and-fro). A wave is a disturbance or vibration that travels through a medium (rock) and transfers tiny amounts of energy from one point to another location. Vibration energy carries the wave through interconnected particles of the medium. The medium is the matter (solid, liquid, gas) in which the wave propagates. The medium is not a wave and it does not create waves. The medium is the means to carry the wave. The seismic wave is produced due to the oscillation in different orientations of the medium (rock) particles, along the line of wave propagation. The artificial generated seismic waves by explosion or pressure are of low frequency and high wavelength. For geological investigation, seismic waves of frequency up to 300 Hz are employed.

The artificially created seismic wave at the earth's surface travels inside the earth and is reflected and refracted back at the surface. During underground propagation, whenever the wave encounters a geological discontinuity (layers contrast), it is affected accordingly; the wave characteristics change. The recording of reflected and refracted waves, with the time elapsed, provides information about the traveled path, that can be related to the geological characteristics of the rock.

6.2.1 Basic Principles

The laws of optic and sound wave propagation are equally applied to seismic waves. The underground reflection and refraction of seismic waves obey Snell's law of wave propagation as follows:

$$\text{Sine } \theta_i/V_1 = \text{Sine } \theta_1/V_1 = \text{Sine } \theta_2/V_2$$

V_1 and V_2 are the wave velocities in the upper and lower strata (layers I, II) of the medium rock. And θ_i, θ_1 and θ_2 are the incidence, reflected and refracted angles. The reflected angle (θ_1) is equal to the incident angle (θ_i), whereas the refracted angle (θ_2) depends on the angle of incidence, properties of layers I and II (layers contrast) and the medium velocity in the two layers. The incidence seismic ray enters from a less dense (soft) layer I of rock to a denser (hard) layer II; the wave strikes the boundary between the two layers I and II. At the boundary interface, the incidence seismic wave energy is broadly split into direct, reflection, refraction, head and transmission waves. The sum of wave energies from all these waves is equal to the wave energy of the incidence ray. The 'direct wave' travels along the surface line.

The velocity of the 'reflected wave', back to the surface, has the same velocity as that of the incidence ray (V_1), since both waves travel in layer I homogeneous medium. The refraction of the wave takes place due to definite contrast between layers. The incidence ray after striking the interface is deviated to an angle away from the normal. The transmitted ray deviating from the straight line is called the 'refracted ray'. The angle of deviation is known as the refraction angle (θ_2), between the normal and refracted rays. The refracted waves continue to travel deep into the earth in the absence of definite geological contrast. The 'transmitted wave' through the interface continues to travel downward in a straight line and is lost in deep into the earth, not of use in seismological survey.

Reflection and refraction seismological methods are employed for underground geological survey. Underground reflected and refracted seismic waves propagate through different paths and distances. The waves bear different geological histories and undergo different degrees of attenuation (variation) during propagation. Reflection seismology is employed for recording the amplitude and shape of seismic waves and ground motion (wave velocity) as a result of reflection from the underground interface. Reflection seismology is used to determine subsurface geological structure. On the other hand, refraction seismology is based on the time of first arrival of ground motion, generated by the propagating seismic wave, from different subsurface distances. The first arrival time is interpreted in terms of depth (thickness) and change of ground motion in different layers.

6.2.2 Wave Field (Wave Sphere)

An expanding and penetrating downward seismic wave sphere (field) is shown in Figure 6.2 created by the contraction (compression) and expansion (rarefaction) of the waves.

In the above discussion, it is assumed that wave travels as a single ray in a straight line. The fact is that from the seismic wave-generating shot point, the waves travel all around in the form of an expanding wave field constituting a sphere. The expanding and advancing wave sphere is known as the 'wave field'. The sphere is the wave field where all the seismic waves are interconnected and overlapping each other.

6.2.3 Wave Front

The furthest ahead section of the wave field is known as the 'wave front' which is the vibratory particles of the medium and advancing wave.

6.2.4 Ray

A ray may be defined as a thin pencil of seismic energy. The ray represents the travel path of the seismic wave with the support of medium particles' oscillation in different orientations.

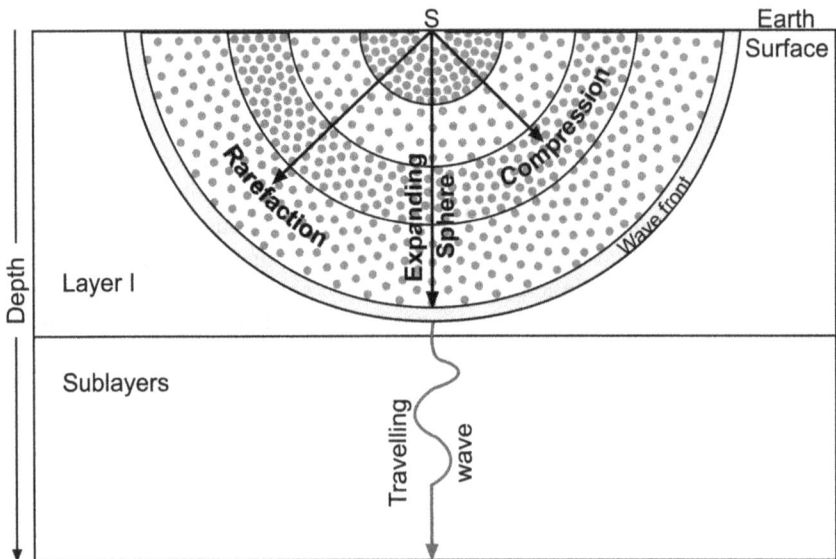

FIGURE 6.2 Expanding seismic wave sphere down into the ground.

6.2.5 PULSE

A pulse is a unit of a wave. A pulse is the single particle oscillation in the medium in one cycle or through a distance of one wave.

6.2.6 WAVE FORM

Normal practice for the interpretation of seismic reflected/refracted data is by using a seismic trace. The seismic trace is a plot of wave amplitude versus time elapsed/distance traveled (x-y trace plot). Due to the development of computer technology coupled with mathematics, the seismic 'wave form' is given more emphasis as it gives more details about the subsurface geological structure. The seismic wave form is a mathematical and computerized representation of different characteristics of waves such as velocity, amplitude, phase angle, frequency or wavelength (y-axis). All these characteristics of waves are affected by the underground geological character during propagation. The constructed wave is known as a 'sine wave'. The term sine wave is derived from trigonometry, on which the graph is based. In a 'sine wave' the amplitude of displacement at each point is proportional to the 'sine' of the phase angle of the displacement.

With increasing radius of the sphere and advancing wave front, the seismic energy decreases. The advancing and penetrating downward seismic wave front encounters different subsurface geological discontinuities. As a result, the spherical shape is disturbed. It is no longer a sphere with a shot point as its center. The expanding sphere may either be contracted or expanded according to the geology of the penetrating path. In case seismic energy is traveling from a soft to a denser layer medium, contraction of the sphere occurs. The expansion of the sphere results by traveling from a denser to a softer layer. The contraction and expansion of wave continue alternately with diminishing intensity till the wave is lost in deep ground.

6.3 ACOUSTIC IMPEDANCE AND SEISMIC WAVE

Acoustic impedance is defined as the resistance offered by the medium (rock or water) to the seismic wave propagation. This is similar to the resistance offered by a medium to the flow of electrical current. For a particular incidence seismic wave, the extent of the reflected and transmitted waves from the subsurface geological heterogeneity depends on the acoustic impedance of the rock. Due to the differences in the acoustic impedance of layers, the total seismic energy of the incidence wave is divided into reflected, refracted and transmitted waves; the wave energy and its velocity change accordingly. Acoustic impedance (I) is used to define the effect of layer contrast on seismic wave propagation in terms of density and velocity, that is:

$$I = \rho V$$

ρ = density of the medium (rock)
V = velocity of ground (wave) motion

When the acoustic impedance difference between two layers is nil, there will be total transmission, no reflection. In that case the densities and properties of the two layers are similar; incidence seismic wave is transmitted through the rock layers. Acoustic impedance is related to electrical resistance. Matching electrical resistances between two materials ensure smooth current flow between the two. Similarly maximum seismic energy is transmitted when the difference in the acoustic impedance between two layers is minimum. Impedance determines the contrast and heterogeneity between layers. Higher acoustic impedance is expected from hard (consolidated) rock compared to soft (unconsolidated) rock. A seismic wave travels faster in hard compact rock than in unconsolidated soft rock. The wave's velocity also depends on the types of minerals. The acoustic impedances of common materials like limestone = 16,500, shale = 15,500, sandstone = 10,500, water = 1500, oil = 1000 and gas = nil are reported in the literature.

6.4   ELASTIC PROPERTY OF ROCK, STRESS/STRAIN

A seismic wave travels through the earth's subsurface due to its energy and pressure. The seismic wave is actually elastic energy, generated as a result of artificial explosion/pressure or natural geological pressure. Elastic energy is exhibited when rock particles are compressed or pulled or deformed by a seismic wave. Due to the elastic energy the seismic wave propagates through the earth in subsurface and surface conditions.

Elasticity is the natural property of a material to regain its original position after the withdrawal of an applied force (stress). The elastic property of a material is governed by the stress (deforming applied force) and strain, damage produced in the material. The applied stress force tends to deform or break the material. On the other hand, the natural internal elastic force of the material resists the deforming force, which is an internal counterbalancing force exists in the elastic material. The deforming applied force (stress) is proportional to the deforming or damage (strain) done within the elastic limit.

All matter that has been subjected to stress are deformed and undergoe changes in shape or size or both. The change in shape or size of the matter is known as the elastic strain or simply strain. Up to a certain limit of stress (elastic limit), the elastic strain is reversible and the matter recovers its original position, after withdrawal of the stress force. If the stress elastic limit is exceeded, the strain (deformation) produced is non-reversible and permanent. There is a definite relationship between stress and strain.

Applied force/deformation = Stress/strain = Constant (modulus)

The constant is termed the 'modulus' and depends on the type of stress force applied and the types of resultant strain (deformation) produced. The modulus determines the elasticity of a solid and is a measure of the stiffness of the material. Four types of modulus are identified on which the elastic character of the solid, and a wave's traveling time through the solid, depend.

6.4.1 Longitudinal Young's Modulus (Y)

Longitudinal Young's modulus quantifies the extent of change in length of an elastic material upon the application of a stress force. A rod of length (L) and cross section area (A) is stretched by the application of a stress force (F) and increases in length (dL). Longitudinal Young's Modulus (Y) is given by:

Y = Applied force/change in length
Y = Longitudinal stress/longitudinal strain
Y = (F/A)/(dL/L)

While stretching the rod, in addition to longitudinal elongation, some lateral reduction of the rod also takes place.

6.4.2 Bulk Modulus (K)

The bulk modulus is related to the volume change (strain) in an elastic matter, due to an applied pressure force (stress). It represents the compressibility property of the matter. A pressure force (P) is applied to a cubic-shaped matter of volume (V) resulting in a change of volume (dV). The bulk modulus (K) is defined by:

K = Bulk pressure force/change in volume
K = Bulk stress/volume change
K = P/(dV/V)

Solids and liquids are less compressible (less change of volume) than air and gases. A soft rock compresses more (low K) than a compact consolidated rock (high K). The 'K' of the rock depends on the litho-logy and compactness of the rock.

6.4.3 Shear Modulus (M)

The shear modulus (μ) indicates the change of shape or deformation of matter due to applied stress pressure force (S). The change in shape (deformation) is expressed as tilted angle (θ):

μ = Stress force/shear strain
μ = S/(tan θ)

The gases can be deformed by a slight force. Liquids need more force, and solids require the highest force to deform. That is to say that air and liquid have a lower shear modulus and solids have higher 'μ' values. However solids differ widely in shear modulus (stability) depending on their properties. Due to their very low shear modulus, air and liquid do not transmit S-waves. The P-seismic waves travel in all media, solid, liquid and gas.

The above three types of modulus are the mechanical properties of the matter and are responsible for the elastic behavior of the respective matter. The wave energy, including all the other characteristics of propagating waves, is related to the elastic properties of the rock.

6.5   TYPES OF SEISMIC WAVE

The different modes of rock particle orientation (vibration) give rise to different types of seismic waves. The type of seismic wave depends upon the kind of ground motion brought about by the seismic energy. Important types of seismic waves are as follows.

6.5.1   BODY WAVE

Body waves are those seismic waves that propagate deep into the earth from a source shot. These waves are the main interest in geophysical survey, because they interact significantly with the rock body and undergo changes according to the geology of the rock. The body wave is non-dispersive, that is it travels without deformation. It propagates through a medium at a speed that depends on the elastic modulus and density of the medium. Body waves are of two types as follows.

6.5.1.1   Primary/Longitudinal Wave (P-Wave)

Primary waves are also called 'pressure waves', 'compressional waves', 'longi-tudinal waves' or 'P-waves'. The wave propagates through the medium (solid and fluids) by the pressure exerted by seismic wave energy. The pressure creates alternate compression (pushing inside) and expansion/rarefaction (pulling apart) strain in the medium in the longitudinal direction of wave propagation. The medium (rock and liquid) particle motion is an oscillation back and forth, along the direction of wave propagation. The P-waves are the most important and used for underground geological seismic studies. The seismic P-waves are faster than any other types of wave, appear earlier at the surface recording station and are easily identifiable.

6.5.1.2   Secondary/Transverse Wave (S-Wave)

Secondary waves are also body waves. Other terms used for secondary wave are 'shear wave', 'transverse wave' and 'S-wave'. The S-wave is generated by the shear strain produced in the direction perpendicular to the direction of wave propagation. The motion of the rock particles is an oscillation about a fixed point in the plane at a right angle (transverse) to the direction of wave propagation. The S-wave does not travel in fluid. The S-wave finds greater applications in engineering than oil/gas.

6.5.2   SURFACE WAVE/NOISE WAVE

The surface waves travel along the surface of the earth or at a shallow depth (1–2 m). The surface waves are of longer amplitude and slower in propagation speed. Surface waves are generated by uncontrolled ground motion and natural/artificial phenomena around the area. The surface waves are termed as 'disturbing waves'.

The disturbing waves are known in spectral terminology as 'noise' waves. Different types of noise waves/surface waves are as follows.

6.5.2.1 Rayleigh Wave

Rayleigh waves are also known as 'ground roll'. Rayleigh waves are generated by shear strain of the solid media, and are not possible in fluid media. The Rayleigh surface waves have variable shape and velocity. They are scattering and dispersive in nature. The Rayleigh wave travels along the boundary between two layers of different density. It can propagate along the earth's surface between solid crust and the atmosphere.

6.5.2.2 Love Wave

The propagation speed of the Love wave lies between the shear wave velocity (S-wave) of upper and adjacent lower layer. Similar to the Rayleigh wave, the Love wave is dispersive.

6.5.3 MISCELLANEOUS NOISE WAVES

Waves are generated at the sea bottom. These waves are reflected from the bottom of the ocean and appear at the surface. Atmospheric noise waves are generated at the point of source shot or explosion. These waves propagate along the surface line and directly reach the detector (geophone) located at the surface as direct signals. Environmental noise waves originate from enormous natural and man-made activities. Sensing and recording systems may be a source noise signal. The topography of the survey area may affect and produce noise signal. Refraction waves in reflection seismology and reflection waves in refraction seismology create noise signals. Diffraction phenomena produce noise signal.

6.6 SEISMIC WAVE VELOCITY IN ROCK

The propagation or traveling of seismic energy, ground motion and velocity of seismic wave are equivalent terms. Seismic wave velocity is the transferring of wave energy from one location to another (from particle to other particle) in the form of wave propagation through a medium. The propagation velocity of a seismic wave is the velocity with which the seismic energy travels. A seismic wave is drastically altered during underground propagation. A change in wave properties is related to the geology of the traveled path. The wave velocity plays an important role in the determination of subsurface geological structure and rock characteristics. The seismic wave's velocity behaves differently in different rocks and depths. A more homogenous, deeper and compact rock (igneous/metamorphic) transmit a faster seismic wave velocity than a soft shallow and unconsolidated sedimentary rock. Rock with more porosity results in reduced wave speed. And rock with less porosity produces more wave speed. With depth and compaction, the velocity of the rock increases. However the wave velocity attains a steady state value after a certain depth, overburden pressure due to absence of significant heterogeneity. At greater than 5000 meters deep the velocity starts to attain a saturation value of about 6000–7000 (m/s).

Different types of rocks transmit seismic wave speed differently. The velocity of a seismic wave is low at shallow depths compared to deep rock. The differences in velocities at shallow depths among different types of rock are appreciable. The differences in velocities and densities decrease with depth. In different rocks at about 5000 meters the difference in velocity is eliminated. The velocity attains a single value, which is about 6000 m/s.

6.6.1 P- AND S-WAVE VELOCITY

In terms of physics, seismic P- and S-wave velocity is related to the elastic properties of the media (rock and fluid). P-waves (V_p) are related to both shear modulus and bulk modulus. On the other hand, S-waves (V_s) depend only on the shear modulus, that is S-wave velocity depends only on rock mineral matrix. The wave does not travel in fluid. The S-wave is generated by the shear strain produced in the direction perpendicular to the direction of wave propagation. Fluids (oil/gas/water) do not withstand shear strain. For gas and oil, the bulk modulus (pressure stress) is insignificant and negligible, whereas it is important and substantial for solids. This creates a clear distinction between V_p and V_s traveling through oil/gas-filled rock. The P-wave velocity is susceptible to pore saturation and porosity of rock.

Differentiation between these two types (P- and S-waves) is made by considering their subsurface velocities. The P-wave is faster than the S-wave, so they can be separated on the basis of arrival times.

The shale rock is compact and dense with small porosity. It transmits the propagation of the P-wave faster than the S-wave. On the other hand sand rock varies in consistency. It can be soft or hard depending on sediment's litho-logical and consolidation processes. Compact rock contacts P-waves faster than the soft rock. The waves travel faster in solids (rock) than in water/oil accumulation and slowest in air/gas-filled rock. The P-wave is more useful than the S-wave in reflection/refraction seismological survey. The speed of P-waves increases with depth and overburden pressure. It means that the P-wave velocity is proportional to the density of the rock.

The density and degree of consolidation of the rock can be predicted by velocity value. The reflection of P-wave data is used to gather information about source, reservoir, trap, cap rocks, salt dome and pore fluids (oil/gas). The values of V_p and V_s can be correlated for oil/gas prospecting in the reservoir rock.

The P and S seismic wave velocities in some selected media (rocks), along with density, are given in Table 6.1.

6.7 ATTENUATION (WEAKENING) OF SEISMIC WAVES

Seismic waves generated at the shot point at the earth's surface travel all around. The splitting/distribution of the seismic energy takes place as follow:

- A majority of the waves go into the atmosphere and are lost.
- Some seismic waves travel as surface waves in a lateral direction.
- A small portion of the seismic waves penetrates into the ground. This is the main interest for subsurface geological study.

TABLE 6.1
Seismic Wave Velocity (km/s) in Different Media

Medium	Density	P-Wave	S-Wave
	g/cc	Velocity k m/s	Velocity km/s
Limestone rock	2.6–2.8	3.0–6.0	3.0
Shale rock	2.0–2.6	2.5–3.0	1.0
Clay	1.6–2.6	1.0–2.5	-
Sandstone rock	2.6–2.7	4.5	2.8
Sand	2.1–2.4	0.5–2.0	–
Anhydride	2.1–2.4	4.5–6.0	–
Oil reservoir rock	2.3	2.5	–
Gas reservoir rock	2.0	1.9	–
Water-bearing rock	2.3	2.0	
Water	1.0	1.5	–
Oil	0.8	1.2	–
Air	–	0.33	–
Igneous/metamorphic rock	3.3–4.0	5.5–8.0	–
Earth crust (4–6 meters)		6.0–7.0	3.0–4.0
Aluminum	2.7	6.1	–

Earth surface seismic wave (0–2 m/depth) velocity = 0.5–1.0 km/s.

Figures quoted are indicative as wide variation in P- and S-wave velocities are expected because of the anisotropic nature of the medium.

The interaction of the seismic wave with medium materials (rock) brings about definite changes in the wave. The changes first lead to the attenuation (thinning) of the wave and ultimately, with further propagation in the medium, the wave is lost. The major underground changes in seismic wave propagation are as follows:

- Absorption and attenuation of seismic wave energy
- Reflection of seismic waves
- Refraction wave, head wave and critical angle
- Direct surface waves
- Diffraction of seismic waves

With the increase of spherical wave field from the source point, the energy content per unit area decreases. The seismic energy (E) is related to the wave amplitude (A). Both seismic energy and wave amplitude decrease with time and distance (radius) traveled by the expanding wave sphere. Not only is the amplitude reduced but other characteristics of the wave are also affected. Energy, velocity and frequency are shortened and wavelength increased during wave propagation.

The main causes of the absorption and attenuation of seismic energy during wave propagation in the medium are given are as follows:

- Increasing distance of wave from source. The intensity of the wave form decreases with the increasing of the radius of the expanding spherical wave form. The distance attenuates the amplitude, frequency and energy and increases the wavelength of the traveling wave.
- The total wave energy of the incidence wave is distributed among all types of generated waves and also energy is dissipated as heat. The former energy attenuation is known as 'geometric dispersion' and the latter as 'non-elastic dispersion'.
- In non-elastic dispersion, the seismic energy is absorbed by the frictional forces of the medium, due to the imperfect elastic properties of the rock. The absorption of seismic energy is dissipated as heat energy due to the imperfect elastic properties of the rock medium. This phenomenon is similar to the dissipation of frictional heat, due to the resistance offered by the moving parts of a machine. There will be no loss of seismic energy during propagation in a perfect elastic medium.
- Flat rock layers reflect/refract/transmit the seismic wave according to their acoustic impedance.
- The geometrical structure of the subsurface affects the reflection. The reflections from an upward convex layer are scattered upward. On the other hand, a downward concave layer reflects a concentrated wave form on a limited area. In both forms the waves are modified.

Each complete wave particle oscillation (one cycle), or each wavelength distance traveled, absorbs a definite portion of the total seismic energy during wave propagation. The 'absorbed energy' during one cycle is known as the 'absorption coefficient', which is defined as the 'absorption/loss of seismic energy during one complete traveling wave cycle or wavelength'. The absorption coefficient is a measure of the rate of decrease of seismic energy as it travels through a medium. When the absorption coefficient is expressed as 'loss of speed' in one complete wave cycle during propagation, it is known as the 'attenuation coefficient'.

Additionally the 'attenuation coefficient' may be defined as a measure of the ratio of the decrease in the intensity of a wave during propagation to the incident energy. It is the fraction of incident wave energy absorbed per unit mass or thickness of the medium. Reflection coefficient is the ratio of amplitude of the incident wave to the amplitude of the reflected wave.

The absorption coefficient and attenuation coefficient depend on the characteristics of the medium materials. Both absorption and attenuation coefficients are responsible for progressive attenuation (thinning/reducing) of the wave frequency, velocity and seismic energy with increasing distance of wave propagation in the subsurface. A high-frequency (small wavelength) wave will lose more energy than a low-frequency (long wavelength) wave traveling the same distance. The speed of a seismic wave is drastically reduced in fluids, an indication of the presence of oil/gas.

6.8   DISTRIBUTION (SPLITTING) OF SEISMIC WAVES

Energy distribution and splitting of waves have the same meaning. The splitting of a seismic incidence wave into five rays, reflected, refracted, head, direct and transmitted, is known as the energy distribution of the wave, which depends on the angle of incidence in a given medium. The transmitted wave travels deep and is of no value in seismology. The relative division of the wave energy between two layers I and II depends on the rock properties related to acoustic impendence, absorption, attenuation and reflection coefficients.

The symbols used in the text and figures are:

The velocity of a seismic wave (ground motion) in upper layer I = V_1 and density = ρ_1.
The velocity of a seismic wave (ground motion) in lower layer II = V_2 and density = ρ_2.

The incident ray from the source point 'S' strikes the interface at depth 'h' on point 'O' or 'A' and the returning wave is detected by a receiver placed at the surface at point 'R'.

6.8.1   THE REFLECTION OF SEISMIC WAVES

An incidence seismic wave generated at the earth's surface travels down inside the ground. Let the speed of the seismic wave is (V_1) in layer I. The wave interacts with the boundary interface between layers I and II at a point (O). According to Huygens's principle, point 'O' is now a center for the generation of a new wave front. The extent and nature of the new waves depend upon the layers contrast and angle of incidence rays. The incidence ray may strike the boundary point between two layers either (a) normally (perpendicular) or (b) obliquely at an angle 'i' to the normal.

6.8.1.1   Normal Incidence and Reflection Rays

See Part A of Figure 6.3. The incidence ray is normal (almost) to the interface. The incidence ray generates the following new waves:

- Reflection wave (shown with a very small angle of reflection). Almost coinciding with the incidence ray.
- Transmitted wave going down in a straight line and lost in deep earth. It is not of much interest in seismological studies.

6.8.1.2   Oblique Incidence and Reflection Rays

See Part B of Figure 6.3. The reflections of two oblique, SO and SA (slanting), incidence rays strike the interface at points O and A. The oblique incidence ray strikes the layers' interface at an angle. The reflected wave is split into a P-wave and S-wave with different angles of reflection. Both rays are reflected back and travel upward into layer I and appear on the surface. The angle of reflection of the P-wave (θ_{p1}) for

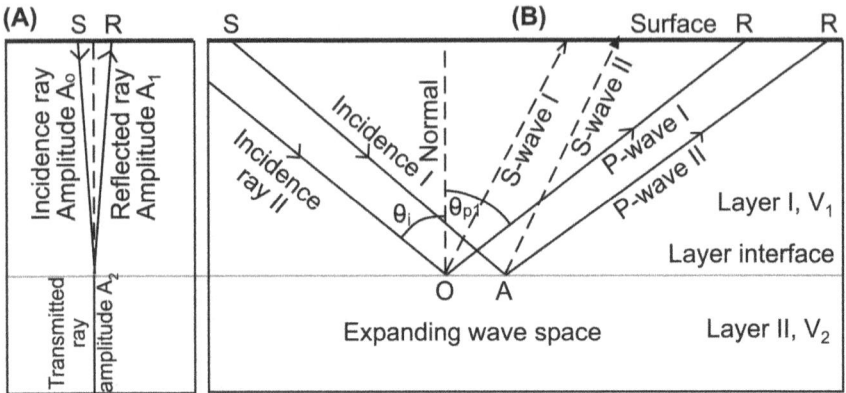

FIGURE 6.3   A = normal reflection. B = oblique reflection.

the OA ray is the same as that of the angle of incidence (θ_i), since the velocity of the reflected P-wave is the same as that of the incidence ray. The angle of reflection (θ_{s1}) between the normal and the reflected rays for the S-wave depends upon the properties of the adjacent layers (boundary contrast) and on the angle of incident ray. If the incidence ray is normal (perpendicular) to the boundary, the angle of incidence is zero, and then the ray reflected back is also normal.

6.8.2   REFRACTION OF SEISMIC WAVES

The incidence seismic ray, after striking the interface of the two layers, is transmitted into layer II, but deviates from straight line with an angle called the 'refraction angle' (θ_2). A deviated transmitted wave is a refracted wave. After refraction the wave is split into a P-wave and an S-wave, similar to the splitting of a reflected wave, creating separate angles of refraction θ_{p2} and θ_{s2}, respectively (Figure 6.4). The refraction angle is tilted away from normal; the size of the refracted angle depends on the angle of incidence and the geological properties of the rock layers. Refraction of the seismic wave occurs when the lower layer II is of greater density and velocity than the upper layer I (V_2 is more than V_1). High-velocity rock (dense) over low-velocity rock (less dense) does not generate a refracted wave since V_2 is less than V_1. Rock layer velocity means the velocity of a seismic wave in the layer. In the upper layers, both reflected and refracted waves travel with the same velocity; there is no practical separation of these waves. They are distinguished by the shape of traces (graph) and arrival time. The refracted rays can be of two kinds, a diving refracted wave and a head refracted wave.

6.8.2.1   Diving Refracted Wave

See Figure 6.4. The diving (dyeing) waves are those refracted waves that do not return back to the surface. The incidence ray after striking the interface at point 'O' continues to travel in layer II but with due refraction (deviation). After refraction at

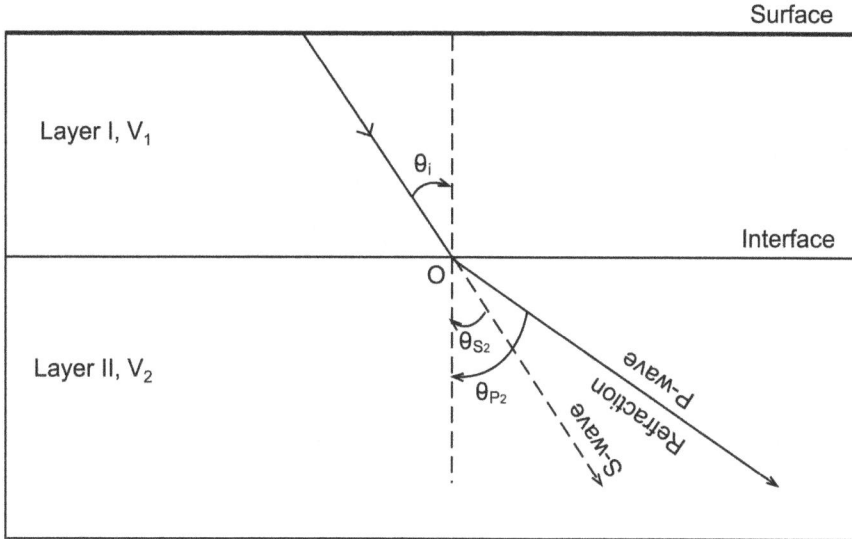

FIGURE 6.4   Refraction of a single incidence ray.

the boundary, the wave continues to propagate deep into the earth, similar to the transmitted wave. Ultimately the wave dies out in the deep zone and is lost. The angle of refraction θ_2 is not equal to the angle of incidence (θi). The angles of incidence and reflection are equal but not the angle of refraction.

6.8.2.2   Head Refracted Wave

See Figure 6.5. The figure shows two incidence rays; SO having incidence angle (θ_1) and critical incidence ray SO with critical incidence angle (θ_{ic}). A critically refracted wave is shown as OH and four head refracted rays (OR) along with the expanding sphere of a head refracted wave and a diving refracted wave front as a broken line circle.

The critically refracted wave is generated in the lower layer II (high V_2 than V_1) of the rock at a certain incidence angle called the 'critical angle'. Even if the lower layer has more velocity than the upper soft layer ($V_2 > V_1$), the head refraction is not generated if a proper incidence angle is not achieved. The proper incidence angle is called the 'critical incidence angle' (θ_{ic}) generated an angle refraction of 90° in the lower layer. The critical incidence angle is the largest incidence angle at which refraction takes place; beyond this incidence angle there will be no refraction but total reflection.

The critically refracted wave continues to travel forward, parallel to the interface line (OH) along the top of the lower layer II, causing the excitation and vibration of particles. Vibrating particles generate a 'head refracted wave' along the interface line adjacent to the upper layer I. The head wave (OR) travels up obliquely in upper layer I. The head refracted wave attains the ground motion (V_1) of layer I and appears on the surface. The head refracted wave doesn't propagate into the lower layer II;

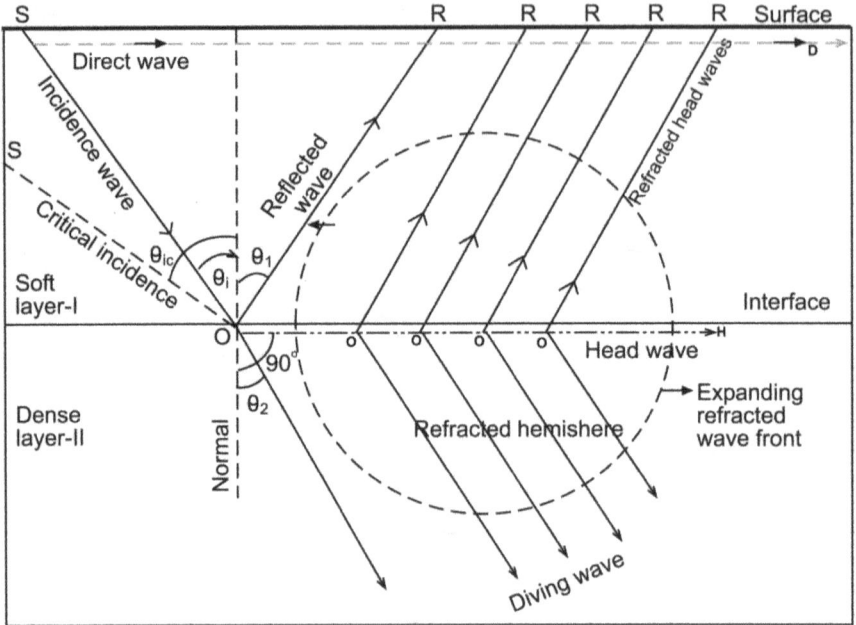

FIGURE 6.5 Critical incidence SO and direct SD waves (- - - -). Incidence SO, reflected OR and refracted head waves OR (—). Head wave OH (— • • —) and critical angle (θ_{ic}).

rather it travels along the horizontal boundary line OH. The head refracted wave in layer I makes an angle equal to the critical angle (θ_{ic}) with the horizontal interface. Simultaneously with the generation of the head refracted waves, downward refracted waves are also produced that propagate into the deep zone and are lost with the ground motion of layer II.

Head waves bear the travel history and part of the original seismic incidence energy by traveling first in layer I as critical incidence waves and then in layer II as critically refracted waves and again in layer I as head refracted waves.

6.8.3 DIRECT SEISMIC WAVE

See Figure 6.5. The direct wave, shown below the earth's surface as broken line (SD), is that wave which is generated at shot point and travels parallel to but slightly below (1–2 m) the earth's horizontal surface. It is directly detected by the geophone/ receiver. It travels along the surface with the velocity (V_1) of layer I. The wave is considered a disturbing noise wave.

6.8.4 DIFFRACTION OF SEISMIC WAVES

Diffraction is a phenomenon in which waves are bended, spread out and cause inter-ference between the wave forms. It occurs when a wave encounters the edges of an object or passes through an orifice. The main cause of diffraction of seismic waves is the non-homogeneity of the layers such as edges of faulted layers, isolated and discrete accumulation of minerals and other materials forming edges and small openings. A discontinuity in the layer structure, with a radius of curvature less than the wavelength of the seismic wave, does not obey the laws of reflection/refraction. Instead diffraction of the seismic wave takes place. Diffraction is the scattering of waves. The diffracted wave originates from underground, appears on the surface and is considered an interfering noise wave.

In addition to the above-stated splitting of incidence energy and generation of various kinds of waves, there are discreet points in the rock, which encounter the seismic wave creating disturbing rays. This kind of seismic wave encounter is best stated by Huygens's principle. The principle states that 'every distinct and discreet point encountered in the subsurface by a seismic wave front is the source of new waves'. The new waves originating from the point, spreading around as an expanding wave sphere, are not needed in seismological studies.

6.9   SEISMOLOGICAL SURVEY AND TECHNICAL ASPECT

The seismological survey is carried out in land and marine areas. Occasionally harsh coastal and marshy land is encountered. The planning of a land survey is carried out by considering the topography of the area.

Although the same principle of seismology is applied in all survey areas, the operational activities differ widely in each survey zone. The seismological survey is an involved exercise covering a broad range of activities on land or sea or harsh land in varied geographical conditions. Practically a high level of expertise and high cost are needed for carrying out the survey followed by complex processing, interpretation and modeling of the gathered data.

In general the seismic waves are artificially created at the earth's surface, introduced into the ground and the returning waves are monitored and recorded. Huge numbers of seismic waves, both genuine and interfering, are generated from the generating device as well as from the interaction of seismic waves with strata. Enormous wave signals are received at surface detector and recorded as 'traced curves' on a single sheet of paper or monitor screen. To get clear data and quality trace in the initial stage, the elimination of unwanted signals is carried out on the recording machine and seismograph. Distinct data and traces are obtained by proper automatic gain control (tuning), attenuation (reducing or elimination of spurious signals) and modulation (improving the clarity of the required signals). These electrical functions on the recording machine improve the resolution (making the trace clear and distinct), and enhance the signal/noise ratio. Final processing and improving of the data are done by computer personnel (discussed later). Here the discussion is concerned with explaining the methodology for surveying.

An explosion on the earth's surface, designated as source 'S' later on, is carried out with a suitable chemical/mechanical device to create seismic waves. The explosion or mechanical impact transfers an impulse force to the ground which propagates as seismic energy deep into the ground. However most of the explosive waves are scattered around. Only a small fraction of the waves travel underground and interact with the underground targeted area and are reflected/refracted back for monitoring at the ground surface.

Practically the methodology used for marine survey is quite different from a land survey. Logistic and operational equipment are different. The marine area is divided into shallow-depth and deep-ocean; accordingly different techniques are used. In shallow depth, one or two floating cables are tied to a surveying ship. The required numbers of watertight hydrophone detectors (receivers) are tied (hung) with the cable at regular spacing (spread). All the hydrophones are lowered into the water at shallow depth. The marine cable can run up to several km.

The hydrophones are kept at a safe distance from the ship so that noise from the ship or any other source may not interfere. The deep-sea survey is carried out by laying a series of hydrophones at the bottom of the sea bed according to the designed geometry. This is similar to creating offset geometry at the surface. A seismic source point, usually high-pressure compressed fluid energy in a steel cylinder, is mounted on the ship or on a towing ship or placed at the water bottom and triggered. The seismic source, hydrophone and recording system are interconnected so that all machines are triggered simultaneously. General information about the material, equipment and procedure employed in seismology is given below.

6.9.1 DEVICE FOR GENERATING SEISMIC WAVES

The device creates and suddenly releases artificial seismic wave energy and introduces it into the surrounding ground. Different seismic sources release different quantities of energy with varying frequencies (1.0–1000.0 Hz). A controlled seismic stress energy is created by chemical energy (explosion) or mechanical energy having different energy levels and frequencies. The amount of energy released and the delivery time should be repeatable at the location. The same amount of energy and the timing should be reproducible at different survey locations. The amount of energy should be enough to reach and propagate deep inside the underground targeted area. Any one of the following sources can be used for generating seismic waves.

6.9.1.1 Chemical Explosion Source (Chemical Energy)

Chemical explosion is the most appropriate method for producing and introducing seismic motion energy into the ground. Technically and economically the method of chemical explosion works well, but environmentalists have reservations about it. The location of chemical explosion is a small hole on the earth's surface providing a good contact of the chemical material with the ground. The hole is known as the 'shot point' or 'source point'.

The required quantity of the explosive – dynamite (TNT) or ammonium nitrate is commonly used – is placed in the bottom of the shot point hole. The explosion is carried out inside the hole by an electric cable. The explosion is triggered with the help of the electric cable extending to a far-off safe place so that the explosion may not cause harm. The trigger is also connected to a geophone (receivers) and recording system, so that at the moment of explosion the receivers and recorder are energized to give zero (t_0) time. Chemical explosives are dangerous material should be handle with care, precaution and according to the safety instructions.

6.9.1.2   Vibroseis Method (Mechanical Energy)

Vibrations on a steel plate are produced as a result of mechanical impact. Mechanical impact is the simplest way to produce seismic wave motion in the ground free from environmental pollution compared to the other devices. The method can be used in towns and other human dwellings. A pit hole is dug at the earth's surface. A heavy steel metal plate with suitable dimensions is placed in the pit hole. Impact energy is imparted to the plate by hitting it with heavy hammer. Mechanical vibrations produced in the steel plate are transferred underground resulting in the propagation of seismic waves. The hammer is electrically connected to a geophone detector recording system to start the recording ground motion from the initial (t_0) time. The method has limitations. The mechanical device generates a small amount of energy with a low frequency (10–80 Hz), and low amplitude of the seismic wave. The repeatability of the method is not good.

6.9.1.3   High-Pressure Fluid Compressed Energy

This method is good example of energy transformation from one form to another. A gas cylinder filled with a fluid at high pressure is used as storage or for release of compressed energy instantaneously. The 'fluid compressed energy' method is suitable where environmental laws are stringent for example marine areas. Two techniques are for generation of seismic wave by high pressure. The simplest way is to use a steel cylinder filled with water at high pressure. One end of the cylinder contains a number of steel balls. By opening the cylinder water rushes out and the kinetic motion forcefully ejects the steel balls to hit the ground, generating propagating seismic energy.

The other method for use of high pressure for generation of seismic wave is large amounts of compressed energy resulting from the combustion of highly flammable gases in a steel cylinder. The cylinder is fitted. with a moving steel ball, similar to the cylinder–piston arrangement in an automobile. The steel cylinder is filled with a flammable and explosive mixture of propane, butane, pentane and oxygen in proper proportions. A moving ball is placed inside the steel cylinder. The explosive mixture is detonated with the help of an electric spark. A spark cable is also connected to geophone/recorder for instant recording (t_0). High pressure is developed in the cylinder due to the combustion of the combustible mixture which moves the installed hard steal ball with force to hit the ground for generation of seismic waves. The system converts chemical energy (combustion) into kinetic energy (moving shell). By hitting the ground, kinetic energy is converted into seismic energy (ground motion). The dynamite explosions directly convert chemical energy into seismic energy. The method produces high-frequency seismic waves and minimizes and controls the generation of seismic surface noise waves. The method works better than the other using mechanical impact.

6.9.2   GROUND MOTION SENSING DEVICE

Seismic ground motion-sensing devices are known by many names, sensor, detector, geophone, hydrophone and receiver. A hydrophone is a watertight and well-insulated sensing device used in marine conditions.

Underground traveling seismic waves are also termed as ground motion. Geophone receiver equipment senses ground motion, receiving and detecting the wave signals. The equipment is placed on the land surface for sensing the propagating seismic waves. The incoming ground motion in the form of seismic waves from underground reaches the earth's surface and hit the geophone sensor.

The geophone works on the principle of electromagnetism. It consists of a metallic block surrounded by a coiled wire. It is hung by a spring in a magnetic field. The magnetic field is produced by the magnets placed around the coiled metallic block. Seismic ground motion creates a fluctuating magnetic field and causes the hanging metallic block to oscillate. The moving magnetic field and the metallic block induce an electrical voltage in the surrounded coiled wire. The induced voltage is proportional to the speed of ground motion and the intensity of the arriving seismic waves. Voltage output from the coiled wire is amplified and recorded on a seismogram. A hydrophone used in marine zones, and is a pressure sensor not an electromagnetic detector. It detects the water pressure arising out from the incoming seismic waves. The magnitude of the pressure is proportional to the velocity of the seismic wave traveling in water.

6.9.3   SEISMIC DATA RECORDING EQUIPMENT

The incoming seismic waves from underground are detected by a geophone which is connected to the seismic wave shot point and seismological recording system. All these machines are interconnected so that all instantaneously trigger the monitoring of the initiation of explosion, detection by geophone and recording of the seismic signal transmitted by the geophone. There are two kinds of recorders. One is analog continuous and the other is a digital discrete recorder.

Seismic data are recorded as a time/distance function versus the characteristics of a ground motion (amplitude, frequency, etc.) related to the seismic wave. The ground motion is represented as a traveling seismic wave and recorded as a trace. A wave is identified by amplitude, frequency, phase, etc. The recorded output is similar in nature but may be distinguished as follows.

6.9.3.1   Seismograph

The output of the seismograph machine is a graphical representation of ground motion, usually on an x-y axis. It is a time/distance axis versus some other characteristics of the wave associated with ground motion.

6.9.3.2   Seismogram

Both a seismograph and a seismogram are the same type of machine. But the latter gives three-dimensional graphical representations, one axis perpendicular to earth's surface and other two axes parallel to the surface. It shows the seismic record in detail, such as the intensity and duration of ground motion and amplitude width and height, etc. A synthetic seismogram has also been constructed on the basis of geological information.

6.9.3.3 Seismic Spectrum

'Seismic spectrum' is a generic term used for showing a seismic data record. The output of a seismograph or seismogram is a seismic spectrum. The enormous numbers of traces are recorded simultaneously on screen or on paper. The spectrum may be two-dimensional or three-dimensional.

6.9.4 Seismic Trace

A trace is the basic unit for obtaining geological information. Only a small portion of the incidence wave is returned back as reflected and refracted rays from the subsurface interface. These returning waves are received by the geophone. The proportion of the wave energy returning back is dependent on the properties (acoustic impendence) of the underground rock. The hydrophone receives the returning waves. These input signals are recorded and given a shape according to the seismic energy content. The incoming seismic waves, in the first instance, are recorded as an x-y graph on a sheet of paper. The traveling time (t) of the wave is plotted against the amplitude (A) of the wave to get a 'trace'. The trace is the basic recorded information, and the whole experiment is known as seismic trace acquisition. The trace reflects the geological characteristics of the traveled path of the wave. The seismic trace acquisition is comparatively a simple experiment which is followed by advanced computerized processing and interpretation systems. Each peak and amplitude of the trace is related to the earth (subsurface geology). A composite of seismic traces is a record of a series of reflected/refracted rays and noise signals from the subsurface. One trace (curve), printed on paper or shown on the screen, is gathered by one surface shot, one subsurface reflected/refracted point, along with noise signals. A number of traces are plotted side by side to produce an image known as a seismograph/seismogram.

6.9.5 Seismic Offset Array (Group) and Spread (Distance)

The seismic source (shot-point) and the receiver (geophone) are the main equipment for trace gathering. The relative location of source-receiver at the surface affects the quality of the gathered trace. To obtain a clear and workable reflection, the locations of the source-receiver array at the surface of the earth are carefully designed. The shot source and required number of receivers are placed in different holes, made into the earth's surface at appropriate distances. A seismic offset array (grid) is an arrangement of a series of sources (shot point) and receivers at the surface on a horizontal line. In a survey all are geometrically arranged in lateral (side by side) positions along a line on the surface of the earth. The numbers of receivers and shot points and their geometry of arrangement are utilized to get a clear reflection with minimum noise signals (surface waves). The distance between receivers is known as the 'spread'. The distance between the source and receiver is known as the 'offset spread'. The spread may vary from meters to several hundred meters. In a marine environment, an array of water tight hydrophones (receivers) are tied to a long cable at regular spacing (spread).

6.9.6 SEISMIC EVENT

An earthquake is a seismic event as a result of underground geological forces acting on the crust. The term 'event' is used to signify the presence of a subsurface geological feature. The interaction of seismic energy with subsurface geological discontinuity is a seismic event signifying the type of discontinuity. The seismic event generates multiple reflection/refraction and various noise waves and sets the particles of the ground in vibrating motion, transferring kinetic seismic energy from particle to particle.

6.9.7 SORTING AND ORDERING DATA

Sorting is a process that enables the arrangement of certain objects into certain categories or into a certain order. Huge numbers of traces are ordered (arranged) in sequence of increasing or decreasing offset, that is from the nearest (first) or from the farthest away (last) offset order.

6.9.8 WAVE AND PHASE ANGLE

The phase defines the 'position of particle in oscillatory motion in terms of angle from 0 to 360°'. In the same way, amplitude defines the particle oscillatory motion in terms of number from 0 to +1 (crest) or 0 to −1 (trough).The phase angle is zero at the starting point of the sine wave (0°) on the equilibrium wave propagating straight line (see Figure 6.1). A seismic wave is considered to be spherical and spread in all directions, as it is generated on the surface. The above hemisphere propagates upward into the atmosphere and is not of any interest to the geologist. The lower hemisphere penetrates and propagates inside the earth from the surface line. A wave consists of a series of peaks/crests (+) and troughs (−) alternatively joined by the center equilibrium straight line. The peak (amplitude, displacement) is defined by 0°, 90° and 180° angles corresponding to amplitude positions at 0, +1 and 0 heights. Similarly the lower trough is designated by 180°, 270° and 360° angles corresponding to amplitude positions at 0, −1 and 0 heights with respect to the equilibrium straight line.

6.9.9 SPECTRAL AND SEISMIC RESOLUTION

The resolution implies a clear and distinct separation of the wave spikes of interest in a trace from the unwanted noise signals. Two kind of resolution are practiced in seismology.

6.9.9.1 Spectral Resolution

Spectral resolution is the clarity and separation of overlapping waves in a seismic trace and spectrum.

Earlier methods for spectral resolution were based on pure electrical concepts such as amplification (enhance), modulation (adjust, regulate) and attenuation (thinning) of the incoming raw seismic signals on an electrical recording machine. These electrical terminologies together are known as 'gain control' or 'automatic gain control' (AGC). The characteristics of incoming seismic signal and consequently the trace recorded differ considerably due to the different distances traveled up to the several geophones at the surface. The amplitude of the seismic wave recorded at the last geophone may have been reduced too much compared to the wave amplitude from first geophone. Likewise the amplitude of the wave may be too large or too small to the extent that it cannot be plotted on the given chart paper. To overcome this difficulty the recorders are fitted with an automatic gain control (AGC) mechanism to fit the amplitude within the recoding chart. The AGC reduces or enhances the data to bring them under desired limit (resolution) for further processing.

6.9.9.2 Seismic Resolution

Seismic resolution is the process that is applied to improve the geological data and images. Better resolution gives the ability to read and differentiate each geological feature. Data obtained from shallow depth is not likely to produce a well-resolved reflection because of unconsolidated soft rock. The reflection will be highly contaminated with unwanted noise waves, giving poor resolution.

6.10 METHOD FOR THE ACQUISITION OF SEISMOGRAMS

The method for acquiring a seismic spectrum is illustrated in Figure 6.6. From the source point (S) the seismic wave travels into the ground and is reflected back to the surface by the boundary contrast of the rock layers. The reflected wave is accompanied by the particle vibratory ground motion. The reflected seismic ground motion energy is converted into an electrical pulse by the sensor receiver (R). The receivers are placed at several locations at the land surface in lateral positions (source-receiver arrangement). The amplified electrical signal from the detector passes through a filtering device; the signal is recorded as a trace and stored in analog or digital form. The shape of the traced curve reflects the geology of the travel path of the seismic wave and is related to the subsurface geological heterogeneity and structure. Each spike (peak/trough) of the traced curve has a defined amplitude related to the reflection coefficient of the subsurface layers and travel time. The travel time depends on the depth of the reflector/refractor interface and wave traveling speeds. With subsequent shot and recoding as time progresses, numbers of seismic traces are recorded on the chart. This gives a series of traces drawn parallel to each other, producing a seismic spectrum. The time between successive traces is very small (a fraction of a second). The traces are closely packed, separated by small time functions. The trace acquisition method is the first part of the seismological survey. The method of acquisition is more or less standardized and practiced accordingly.

```
┌─────────────────────┐  ┌─────────────┐          ┌───────────────────────────┐
│ Chemical or mechanical│  │ Electrical  │          │ Improving signal/noise ratio│
│   energy source     │  │   energy    │          │                           │
└─────────────────────┘  └─────────────┘          └───────────────────────────┘
          │                    │                            │
          ▼                    ▼                            ▼
┌─────────────────┐    ┌─────────────┐    ┌─────────────┐    ┌─────────────┐
│ Seismic energy  │───▶│  Geophone   │───▶│Amplification│───▶│  Filtering  │──*▶
│  ground motion  │    │  detector   │    │             │    │             │
└─────────────────┘    └─────────────┘    └─────────────┘    └─────────────┘

  *▶┌─────────────┐    ┌─────────────┐    ┌─────────────┐
    │Digital/analog│──▶│   Seismic   │──▶│ Seismogram /│
    │    data     │    │    trace    │    │ sesimograph │
    └─────────────┘    └─────────────┘    └─────────────┘
        ▲   │                │                  │
        │   ▼                ▼                  ▼
    ┌─────────────┐    ┌─────────────────────────┐
    │  Recording  │    │     Seismic spectrum    │
    └─────────────┘    └─────────────────────────┘
```

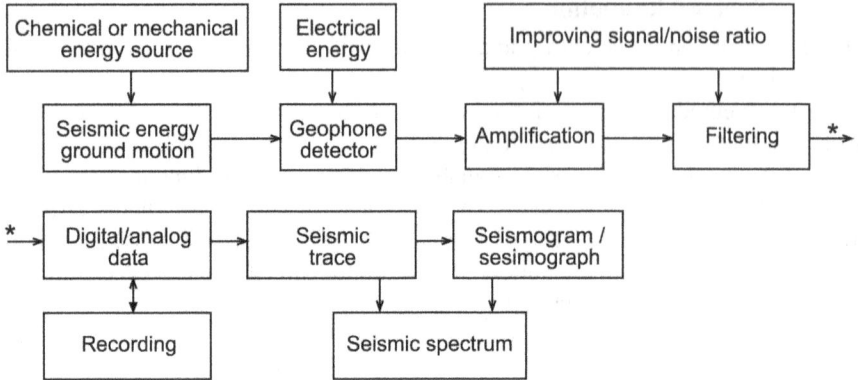

FIGURE 6.6   Acquisition and conversion of ground motion energy into seismic spectrum.

A seismograph spectrum is a complex arrangement of a number of traces representing reflected/refracted/noise signals. A trace is a unit of seismic spectrum. The next step is to read and analyze the seismic spectrum. Seismological data are obtained by identifying the genuine traces originating from the targeted underground object by the elimination of the unwanted traces due to noise signals. The shape, peak height and width of the trace are noted. The seismological data are subjected to processing and interpretation. Geological modeling of the subsurface is carried out. Processing, interpretation and modeling are seeing a great deal of innovation due to recent advancement in electronic/computer technologies.

A.   REFLECTION SEISMOLOGICAL SURVEY

6.11   INTRODUCTION

Among the two reflection and refraction waves, the reflection wave plays an important role in seismological surveying for oil/gas prospecting. A shallower sedimentary rock has more geological variation than a deeper one. The shallower rock is unconsolidated, soft, non-uniform and under dynamic conditions. The shallowest unconsolidated layer with low density contrast is unable to reflect the seismic wave or reflect the wave in a vertical direction. The incidence ray travels down almost in a straight line in the unconsolidated shallow layer until it encounters a significant geological discontinuity. The deeper rock layers are perfectly consolidated, compact and uniform. The interface between the layers reflects the incidence seismic wave in an oblique upward direction. The propagation history and overall shape of the reflected waves bear the characteristics of underground seismic events along with the geological structure.

6.12   TRACE GATHERING AND GEOMETRY OF SOURCE-RECEIVER

Trace gathering requires careful designing of source-receiver offset geometry at the earth's surface. Several geometrical arrangements are being practiced. The basic design is a lateral geometrical positioning of source (S) and receiver (R). The purpose of designing the source-receiver offset is to get a trace free from or with minimum noise signals. The desired seismic signals should be distinct and well resolved. The optimum design of a source-receiver offset is carried out by the variation of source/receiver positions. Several geometrical designs of source-receiver combination are available. Some may be of academic interest and others of practical use. A few designs and procedures for the gathering of seismic traces are given below.

6.12.1   SOURCE-RECEIVER AT SAME LOCATION AT SURFACE

The distance between the source point and receiver is almost zero. This is called zero offset arrangement. Zero offset is the simplest arrangement of source-receiver (SR). The incidence seismic wave travels almost vertically downward from the surface and strikes the interface at point 'O' and is reflected back. The reflected wave travels almost vertically upward. Wave is detected by the surface receiver (R) and recorded as trace. A trace recorded per seismic shot per subsurface reflection point on the interface is known as single fold trace. Zero offset reflection generates a lot of disturbing noise signals, giving a poor signal/noise ratio. Much interference in signals is witnessed. Only one trace is drawn from one subsurface interface reflected point per source shot. A single trace cannot differentiate between actual reflection and noise signal. It covers only one point at the subsurface interface at a time. It takes a long time and great effort to cover a sizable distance and area of the targeted interface.

6.12.2   MULTIPLE SOURCE-RECEIVER GATHERING

An offset arrangement consisting of multiple sources-receivers is called multiple channel. The sources-receivers are placed along the earth's horizontal surface line at regular distances of several km long. All the shot points and geophone receivers are interconnected to a recording system. The multiple arrangement works better because of the following reasons:

- Economical
- Efficient
- Covers larger targeted area
- Minimum time
- Proper trace

Seismic waves from the source points strike the interface at several points and are reflected accordingly. It gives more information over a longer length of the subsurface interface. Identification of the respective true reflection and noise signal becomes more possible in multiple traces than a single trace. Both reflected and noise waves travel along in the subsurface but with different characteristics and velocities. The difference facilitates the identification of the two waves and their arrival time at the receiver. The difference becomes more prominent with longer traveling distance and time to the receivers. Genuine reflections from the underground target of interest and are segregated from noise signals. Multiple-channel gathering consists of several geometrical source-receiver offset arrangements. The most common multiple gathering is described below.

6.12.2.1 Single Source Point and Multiple Reflector Points and Receivers

The procedure uses a single shot point at the surface, multiple receiver points on the horizontal surface and multiple reflector points on the horizontal subsurface interface between the upper and lower two layers. The seismic shot from the source strikes at a number of points on the subsurface horizontal interface line ('O-A)' as shown in Figure 6.3. The survey covers the interface distance between points (O-A). The reflected waves are monitored by multiple receivers at the surface.

6.12.2.2 Symmetrical Arrangement of Receivers on Either Sides of Shot Point

A single source point is located in the center and equal numbers of receivers are placed at equal distances at the surface, on either side of the shot point. There can be many reflections and traces between the first and last traces that cover the interface distance between points 'O to O' (Figure 6.5). The signal/noise ratio improves marginally compared to the two offset arrangements described earlier. If there are unequal numbers of receivers on either side of the shot point, the offset arrangement is called the 'asymmetric split spread'.

6.12.2.3 Common Midpoint (CMP) Gathering

Equal numbers of source points and receivers are placed on either side of a common point at the surface. It is an arrangement for multiple reflections from a common underground point on the interface. The method is known as 'common midpoint' (CMP) gathering. The reflecting point is vertically below between the source-receiver arrangements. The traces so obtained from the multiple shots are known as 'common midpoint' (CMP) gatherings and also as 'common depth point' (CDP) gatherings. The CMP method seemed to be difficult and time consuming. Only one point on the subsurface interface is surveyed through multiple shootings, receiving and recordings.

6.12.3 Reflection from a Dipping Layer Interface

A layer dipping to an angle (θ) from the horizontal line can be a tilted interface between the two layers. One type is an up-dipping layer and the other is a down-dipping layer. The travel time of the reflected ray from the tilted interface is dependent on the types of dip, down or up. Each wave reflection from each dipping point on the interface line travels from a different depth up to the receivers and with a different arrival time and velocity. A reflection from the dipping down line takes more time than from the dipping up reflection. A reflected wave from the dipping down interface line has to travel a longer distance. The total travel time of the incidence and reflected waves in a dipping layer can be computed by considering the geometry of the path of the waves. The subsurface dipping layer is detected by placing the geophones on either side of the seismic energy source and recording the reflection waves and noting the travel time.

6.12.4 Reflection from Multiple Layers

Rock with several layers is known as multiple rock. Incidence wave striking the multiple layers generates huge numbers of reflected and refracted waves and enormous numbers of traces are obtained making the spectrum complicated. A refracted wave from the upper layer becomes an incidence ray to the lower layer. The striking points of incidence rays become a source of upward reflection. Between layers there are huge numbers of reflections and refractions.

6.13 FIRST ARRIVAL TIME OF REFLECTED/DIRECT WAVE

The first arrival time of a seismic wave at the receiver, after triggering the shot, is an important parameter. It differentiates the type of wave and is also used for estimating the depth of the subsurface interface. The arrival time depends on the type of wave, geology of the rock and underground and surface distance traveled. Usually the first arrival at the receiver is a direct wave from the shot point that travels a straight short distance along the surface line up to receiver. The direct wave is followed by delayed reflected and head refracted waves.

6.13.1 First Arrival Time, Source–Receiver at Same Location at Surface

One shot source and one receiver are placed at the same location at the surface (zero offset). The first arrival at the receiver is the surface direct wave from the source and the later arrival is the reflected wave. From the knowledge of the first arrival time (t_0) of the reflected wave at zero offset ($x = 0$) and the seismic wave velocity (V_1) of layer I, the depth (d) of the reflected point at the underground interface can be estimated. A wave coming from a greater depth needs more time to reach the geophone at the surface.

6.13.2 Multiple Reflections/First Arrival Time

Multiple reflections take place when seismic waves strike the interface at more than one point, and subsequently reflections take place from each point on the interface. The waves return to the surface in multiple numbers. The arrival times of the propagating waves at different receivers progressively increase with the continuous offset spread distance (x-spread). It takes less time to arrive at the first receiver than at the other receiver on the horizontal spread line on the surface. The most time is taken by the last receiver located at the far end of the spread.

6.13.3 Direct Wave Normal Move-out (NMO)

The first arrival is the direct wave and the delayed arrival is the reflected wave. For the direct wave the first arrival time increases linearly with offset distance. A plot of 't' versus 'x' gives a straight line. The variation (increase) of arrival time with offset distance is known 'normal move-out' (NMO).

6.13.4   REFLECTED WAVE NORMAL MOVE-OUT

The reflected waves also show an increase of arrival time with increasing offset, but the increment is non-linear. A plot of 't' versus 'x' gives a curved path not a linear straight line. The different between the travel times, t_1 and t_2, of the reflected waves at two offset distances, x_1 and x_2, is known as the 'normal move-out' (NMO). The NMO is simply the increase of travel time with distance. NMO represents the extent of curvature in the t-x plot. It is noted that the NMO for a direct wave is a straight line; whereas a curve is obtained for reflected waves. This indicates the different propagating modes of direct and reflected waves. In the former case, the move out is linear and uniform, whereas for a reflected wave, the deviation is not linear and uniform. The NMO decreases with velocity, distance and depth. When the square of arrival time 't^2' versus the square of offset distance 'x^2' for the reflected waves is plotted, it gives a straight line for the same reflected data. From the slope ($1/V_1^2$) of the straight line, the velocity of ground motion in layer I can be calculated. The intercept of this straight line with the y-axis relates the first arrival time (t_0), the velocity (V_1) in layer I and depth (d).

6.13.5   DIPPING LAYER REFLECTED WAVE NORMAL MOVE-OUT

A dipping layer interface to an angle (θ) from the flat horizontal line is 'dipping down' on one end and 'dipping up' at the other end.

The reflected wave first arrival travel time for the down-dip layer (t_d) from the source to the detector is more than that of the up-dip layer (t_u). NMO for dipping down layer increases with offset distance.

6.13.6   MULTILAYERED REFLECTED WAVE NORMAL MOVE-OUT

Reflections from multilayered rock create a complex situation. Each layer of the subsurface rock produces multiple reflection/refraction rays that may be interfering with each other.

6.14   PROCESSING OF REFLECTED SEISMIC DATA

Processing is the analysis and manipulation of seismic data using a computer so that the gathered data can be interpreted. The processing converts the raw data, obtained during the initial gathering of primary reflection from the shot point to the receiver, into a machine-readable digital form. The processed data lead to information regarding the underground geology that may be useful for predicting the presence of oil/gas. Seismic data processing is a theoretical treatment to obtain an accurate image of the subsurface. The processing of seismic data requires advanced knowledge and high-level expertise. An introduction of the processing is given here.

The raw seismic data (raw spectrum) include both genuine reflection and interfering noise signals originating from the surface wave, diffraction pattern, multiple reflections/refraction, instrumental fluctuation, environmental disturbances and other miscellaneous noise signals. 'Natural and artificial filtering processes' also interfere with the seismic data. During the propagation of a seismic wave through rock, certain frequencies are eliminated or filtered out with increasing distance and time, by the rock matrix in a variable geological environment. Artificial filtering may not be able to eliminate some unwanted frequencies. The processing eliminates all the interfering noise signals and retains only genuine reflections. During the processing of the data, due attention is given to the shape, amplitude and spike (peak/trough) of the trace.

The processed seismic data are presented in different forms. The two-dimensional (2D) survey is carried out by recording the upward traveling wave form (vertical section) represented mostly by a plot of amplitude versus time/distance. Cartesian coordinates specify each point on the trace by a pair of wave functions, mostly in terms of time/distance (x) versus a property (y) of the wave such as wave amplitude (A), frequency (f), wavelength (λ), wave number (ν) or phase angle (degree).

A geological contour map draws an outline of a geological object. A contour line connects the lines of equal value of a particular property or elevation of the strata. A contour map gives the stratigraphic and structural features of the subsurface. A geological isopach map is used to study the thickness of strata. An isopach is a contour map representing lines of strata of equal thickness.

Computerized three-dimensional (x-y-z) cubic models of the subsurface rock are used for geological modeling and interpretation. The model is drawn on the theoretical consideration of processed seismic data. The model shows what a particular underground geological section would look like. The model is based on the seismic sampling on an area (volumetric column of the rock) of the subsurface interface rather than a point on a horizontal straight line. There can be more than one model depending on how the data are applied. A model gives qualitative as well as quantitative information. It may be noted that, in a 2D survey, the propagation of a seismic wave in a vertical or near-vertical direction is represented by an x-y plot. The fact is that the ground motion (wave velocity) is in all directions (x-y-z axis). To record the motion in all directions, to give three-dimensional models, two additional geophones (receivers) are installed. All three receivers, monitoring the signals in each direction, are placed at right angles to each other at the three receiver locations.

Briefly some aspects of processing are as follows.

6.14.1 FILTERING PROCESS

The removal of low/high unwanted frequencies and noise signals from the genuine frequency is known as the filtering process. Quantitatively the noise interferences are represented by the 'signal/noise' (S/N) ratio. Three scenarios may exist. First S/N < 1, second S/N = 1 and third S/N > 1. Obviously an S/N ratio greater than one is most desirable. The worst situation is when the S/N ratio is less than one. The signal to noise ratio is improved by filtering processes. A filter is a computerized machine; it receives the input raw contaminated seismic data, processes them and gives the output data in a clear and modified form indicating genuine reflection. The filtering process uses two terms, (a) convolution process/frequency filtering and (b) de-convolution process/inverse filtering. In seismology, convolution is an artificial mathematical operation of combining two wave forms to generate a third improved wave form. The convolution process is able to differentiate between seismic and noise frequency signals. The output signals from convolution filtering result in an improved S/N ratio. The convolution process is sensitive to the frequency; therefore, low and high unwanted frequencies are easily eliminated and the desirable frequency is retained. It cannot remove an unwanted frequency that has the same frequency as the desired one.

De-convolution is also called inverse filtering. The purpose of seismological de-convolution is to sharpen and enhance genuine reflection seismic waves. De-convolution inverse filtering is sensitive to other characteristics of the seismic wave such as wave number, phase angle, time period, etc., rather than the frequency filtering which is used in convolution filtering. Also de-convolution filtering can remove that noise signal which has the same frequency as that of the desirable seismic form. The end result of filtering is an improved S/N ratio with enhanced resolution of the seismic data spectrum.

6.14.2   SEISMIC MIGRATION

Seismic migration is a process by which an almost real image, in either space or time, of the underground structure is obtained from the observed seismic data recorded at the surface. It is the relocation of the seismic events to the real position rather than the location at which the event was observed at the surface. The purpose of seismic migration is to see the underground features in their proper locations both vertically and laterally. The concept of seismic migration is explained by citing a simple example of the geological non-conformity of dipping interface reflections. The recording of a seismic event (seismic reflection) from a dipping interface shows that the reflection point at the interface is displaced in an upward direction from its lower real position. The observed reflection point is not the real reflector.

6.14.3   STACKING OF TRACES

Stacking is a process by which huge numbers of different traces are stacked together to form a single trace. Stacking reduces and eliminates unwanted traces mostly comprising noise signals and improves the quality of data. The term 'fold' is used to indicate the number of traces that have been stacked together in a single bundle. The stacked bundle simulates the zero offset single trace reflection. The zero offset considerably improves the signal/noise ratio. In practice, huge numbers of traces are sorted out and stacked together by computerized processing.

6.14.4   VARIATION OF SEISMIC WAVE VELOCITY

A record of the variation of seismic wave velocity gives useful underground information. Geological heterogeneity affects the wave velocity; any change in velocity is attributed to a geological event (structure). Other factors affecting the change of velocity are surface offset source-receiver geometry, and distance traveled by the seismic wave. The distance traveled attenuates the amplitude, frequency and energy and increases the wavelength of the wave. Late-arriving seismic signals at the surface receivers are slow moving with high-wavelength and low-frequency events. There is a continuous increase of velocity with depth inside the earth.

6.14.5 CROSS CORRELATION

Cross correlation compares the different traces gathered at different times. It functions by determining the time delay between seismic signals.

6.15 INTERPRETATION OF SEISMIC REFLECTION DATA

After the processing and enhancing the S/N ratio, the data are correlated to and interpreted as subsurface geological features. The information so gathered is utilized for oil and gas prospecting. There are two important aspects of sedimentary rock, one is the stratigraphy and the other is structural geology. Reflection seismology is extensively applied in the study of structural geology, whereas it has limited use in stratigraphy studies. This is quite understandable. The geo-physical features of structural rock are more prominent, compact, distinct and resolved. The reflections from the structural rock are prominent and distinct. The reflection from the structural rock is characteristic of its features. The variations in stratigraphy layers are gradual. The sharp contrast among the rock layers does not exist, and as such the reflection response from the interface between the layers is not readable or distinct. Some examples of reflection from both categories of rocks are given.

With the help of traveling time and offset spread, the velocity of ground motion and depth could be calculated. The calculated underground velocity could be interpreted for characteristics of the rock section. Likewise, the attenuation of the amplitude, broadening of the wavelength and reduction of frequency with wave propagation are interpreted to reveal the nature of sedimentary rock. The reflections from the upward convex-shaped anticline dome rock structure are readable, clear, and distinguishable. A fault is a crack or boundary between tectonic plates. A faulted zone has several structural possibilities and so its reflection patterns vary. Most of the incidence seismic rays are transmitted or absorbed by the salt dome structure; it does not reflect the wave. Reflection from deep underground is enormously contaminated by multiple reflections, refraction and unwanted weak signals. However, with modern computerized trace gathering, recording and processing, it is possible to identify the surface structure of the basement rock.

Stratigraphic rock is formed by the gradual deposition of sediments, rock erosion and sometimes by the deformation of the strata. It forms rock layers with little litho-logical or density contrast. When the contrast between the layers is low, the incidence ray reads it as a uniform surface. Therefore reflection from such an interface is poor.

B. REFRACTION SEISMOLOGICAL SURVEY

6.16 INTRODUCTION

Basic laws and terms used in refraction seismology are already discussed in Section 6.8, including the concepts of refraction angles, critical incidence angles, transmission, head waves, head refracted waves and direct waves. Here only an application of refraction seismology is described. Numbers of refraction seismic traces are gathered and recorded to form a seismogram of refracted waves. The incidence seismic energy is distributed among different waves: direct, reflection, head, head refracted and transmission waves. The velocity of the seismic wave depends on the type of the wave and the characteristics of the medium that is on the heterogeneity/homogeneity of the rock. Direct, reflected, refracted and head waves travel different distances with different velocities. The longest distance is traveled by the head refracted wave: starting from the source point 'S', down to interface point 'O' with speed of V_1, parallel and just below the interface line 'OA' with faster speed V_2, finally upward in layer I again with speed V_1 and up to the receiver (R) at the surface (see Figure 6.5). The first arrival time and difference in velocities are utilized for the identification of different waves. The normal travel time or speed of the wave is expressed as the 'first arrival time'. The first arrival time is the instant at which the wave is detected by the receiver and recorded in a seismograph. It represents the total time elapsed between triggering the shot and detection by the receiver phone.

6.17 FIRST ARRIVAL TIME VERSUS OFFSET DISTANCE

The first arrival time at the detector is recorded for direct, reflected and head refracted waves.

6.17.1 Direct Wave First Arrival Time

The direct wave appears at the recorder instantly after the explosive shot from the source point. The plot of offset distance (x) versus first arrival time (t) gives a straight line that starts from the origin of the graph. With increasing offset distance (x-axis), the corresponding arrival time of the direct wave at subsequent geophones increases accordingly. The direct wave plot is almost a diagonal straight line with zero intercept and with a slope = $(1/V_1)$. The slope of the line is used to calculate the seismic speed (V_1) in layer I.

6.17.2 Reflected Wave First Arrival Time

The reflected wave appears on the seismograph after the direct wave, but before the head refracted wave. It travels with the same velocity (V_1) as layer I. The reflected wave signals not relevant to refraction seismology.

6.17.3 The Head Refracted Wave First Arrival Time

The head refracted wave does not appear in the detector/seismograph immediately as direct and reflected waves do. The wave is delayed until a critical offset distance

(x_c) and first arrival travel time (t_1) are achieved. The distance corresponds to an elapsed time (t_1) before the appearance of the wave on the trace. Critical distance (x_c) is defined as the 'minimum or critical offset distance' at which the head refracted wave first appears on the graph.

The following points are important and are relevant to the head refracted wave.

- The head refracted wave travels a longer distance and appears later than the direct wave. The direct wave travels a short distance. But the head wave travels with a faster speed. After covering more distance (crossover distance) the faster head refracted wave over takes the direct wave and appears first than the direct wave.
- The head wave travels with a higher velocity (V_2) in layer II along the horizontal interface for a larger distance. The higher speed compensates the time delayed due to the larger distance traversed. Both direct and head refracted waves travel with a slow speed (V_1) in the upper layer I for a shorter distance. That is to say, (V_2) >> (V_1) and OA > AR.
- The critical offset distance (x_c) is travelled by the head refracted wave before appearance at the receiver (geophone). The critical incidence angle is required to produce a head refracted wave. The refracted wave in layer II makes an angle of 90° with normal and propagates along the interface, just below the boundary with a speed of layer II (V_2).
- After the 'crossover' distance (x_d) the head refracted wave becomes the first arrival instead of the direct and reflected wave arrival. The offset distance at which the head refracted wave overtakes the direct wave is known as the crossover distance (x_d).
- A plot of various offset distances (x) versus first arrival time (t_1) at different receivers gives a straight line. The first arrival time (t_1) for the head refracted wave continue to increase with increasing offset distance (x).s
- The slope ($1/V_2$) of the straight line is used to calculate the seismic velocity of the refracted waves in the lower layer II. The intercept of the straight line (t_0) with y- axis is correlated with the thickness/depth (h) of layer I.

6.18   NORMAL MOVE-OUT (NMO)

Normal move-out (NMO) is defined as the difference in arrival time between two consecutive traces or the increment in the time of arrival with increasing offset distance. The increment in the arrival time is different for different types of waves. Thus the difference in arrival time (NMO) can be utilized for the differentiation and identification of different waves.

The important features of normal move-out are as follows:

- The normal move-out for a direct wave is linear and uniform and passes from the origin of an x-t plot (linear straight line) with zero intercept. The straight line is of comparatively higher slope than the straight (x-t) plot for a refracted wave.

- The normal move-out for a reflected wave is not uniform. It observes a substantial difference in each trace, represented by a curve. It was discussed earlier in reflection seismology.
- The normal move-out for a head refracted wave is similar to that for a direct wave, but with a smaller slope.

From the slope, intercept and move-out, the type of wave, seismic wave velocities in the lower and upper layers and the depth of the underground interface could be estimated.

6.19   CONCLUSION

Petroleum seismological survey is one of the most important geophysical studies of the earth including for oil/gas prospecting. The theory is based on the elastic disturbance (vibration) of the medium (rock) particles, by artificially generated and propagating seismic energy. The propagating wave interacts with strata, is affected accordingly and bears the subsurface geological history of the traveled path. The horizontal and vertical variation in the wave velocity, along with the modification in the characteristics of the seismic wave, forms the basis of seismological survey. The survey takes into consideration three kinds of waves, reflection, refraction and direct waves. A transmitted wave is of no value in seismology. The reflection and refraction propagating waves through a medium (rock) are mainly utilized for the characterizing of subsurface rock structure and to a small extent the rock stratigraphy. Surface, direct, ghost reflection and refraction, diffraction and environmental waves are unwanted noise signals and are eliminated through raw seismic data processing for improving the signal to noise ratio. Artificial seismic waves are generated by chemical explosion or mechanical thrust through laterally laid shot points at the earth's surface. The waves propagate all around including underground, which is of interest to the survey. The seismic waves are monitored through laterally placed receivers (detector/geophone) offset array at the ground surface. Seismic data are gathered in the form of a seismogram, a collection of huge numbers of seismic traces drawn side by side as a function of travel time/offset distance versus amplitude/frequency on two-dimensional seismographs (x-y plot). Reflection and refraction seismology bears the same theory but the application varies. Underground reflected and refracted seismic waves propagate through different paths and distances. The waves bear different degrees of attenuation (variation) during propagation. Reflection seismology is employed for recording the amplitude and shape of a seismic wave and ground motion (wave velocity) as a result of reflection from an underground interface. Reflection seismology is used to determine a subsurface geological structure. On the other hand, refraction seismology is based on the time of first arrival of ground motion, generated by the propagating seismic wave, from different subsurface distances. The first arrival time is computed and interpreted in terms of depth (thickness) and change of ground motion in different layers. The acquired data are improved, first by electrical manipulation automatic gain control (AGC) and then by computerized processing, and then interpreted. Three-dimensional (3D) seismic data modeling has introduced greater certainty in oil/gas findings. A three-dimensional survey is carried out by laying down three sets of sources-receivers offset (grid) at right angles to each other above the subsurface targeted volumetric area. The processed data are interpreted and related to the geological features. There can be different interpretations and different models depending upon the opinion and skill of the interpreter. The most probable geological model of the subsurface is constructed.

7 Petroleum Geo-Electrical Survey

7.1 INTRODUCTION

An electrical survey is the study of the interaction of an electric current and the associated magnetic field with materials of rock. Different materials respond differently to the applied electromagnetic current; some conduct electricity and others resist the flow of current. The difference in electric response is related to different types of subsurface geology. When there is geological discontinuity there is also electrical discontinuity. Geo-electrical surveys are gainfully utilized for the detection of underground buried conducting bodies at shallower depths. The survey offers little opportunity for oil/gas prospecting.

Electrical surveys employ two types of electric current. The first one is an electric current artificially introduced into the earth and see its effects on rock material (active method). The other is the study of the natural electric field present in the rock (passive method), similar to gravitational and magnetic fields. Distinct methodology is used in both categories.

Active electrical methods are as follows:

A. Geo-electro resistivity survey. This monitors the conductivity/resistivity offered by the subsurface material to an artificially induced electric current.
B. Electromagnetism survey. This monitors the conductivity/resistivity of the earth rock in presence of an induced electromagnetic field.
C. Induced polarization survey. This survey identifies the presence of conductive underground materials.
D. Ground penetration radar survey.

Natural underground current (passive method) survey is carried out by the following methods:

E. Self-potential field. The current generated by subsurface electro-chemical reactions.
F. Natural telluric field:
   • Natural telluric electric current field (frequency ranges from 10^{-1} to 10 Hz).
   • Natural telluric magnetic field (frequency ranges from 10^{-1} to 10 Hz).
   • Natural audio frequency magnetic field (AFMAG) (frequency ranges from 1 to 10^2 Hz).

A. GEO-ELECTRO RESISTIVITY SURVEY

7.2 INTRODUCTION

Electrical resistivity surveys measure the electrical resistance offered by an underground targeted buried body. A high-frequency DC or low frequency AC is introduced into the ground that penetrates deeply and interacts with the rock of interest. Reflected back to surface, electric signals (potential) are monitored by resistivity and potential meters. The purpose is to determine the geological features by the difference in resistance (anomaly) offered by different geological discontinuities. The principle of rock resistivity measurement is the same as that applied to measure the resistance of an electrical circuit. The rock resistivity is measured by an ohmmeter. Resistance to electrical current depends on the electrical properties of the rock materials. The resistance value is independent of the size or shape of the material body.

7.3 ELECTRIC RESISTIVITY OF ROCK

Different rock types react differently to the flowing current. Of the three major classes of rock, igneous rock contains conducting materials, offering little electro resistivity. Sedimentary rock, containing void pores, offers considerable resistivity to current. Metamorphic rock offers intermediate overlapping resistivity. Various factors affect the electrical response of the rock. An unconsolidated rock offers more resistance than fully consolidated compact rock. A young sedimentary rock offers greater resistivity than old consolidated rock. A rock contains mostly non-conducting material but it may also contain conducting matter. Therefore, resistivity differs from rock to rock and from mineral to mineral considerably (1–10^9 ohm m). The ranges of resistivity for different rocks are as follows: igneous (10–10^4), sandstone (1–10^9), shale (10–10^6) and clay rock (10–10^5). On the basis of resistivity study alone, the identification of rock is not possible as considerable overlapping exists. Pore fluid, especially water, plays an important role in resistivity measurement. Although sedimentary rocks are insulators, they become conductive in the presence of pore water and their resistivity drops considerably; therefore an electrical anomaly is observed. The conduction of electricity through rock takes place by two different path ways.

- Ohmic conduction. The current flows by an electron-transferring mechanism, as observed in metallic wire conductors.
- Electrolytic ions transferring mechanism. Minerals are dissociated into positive cations and negative anions in the presence of an electrolyte. An electrolyte is a conducting fluid, mostly pore water. The ions flow under the influence of a potential difference, and a current is generated.

7.3.1 CURRENT FLOW

Electrical resistivity (resistance) is the opposition to the electric current passing through a material. The opposite of electrical resistivity is electrical conductance (allowing current). Electrical resistance is similar to mechanical friction (moving machine) that opposes the motion. The resistance (R) of a material is related to the current (I) and potential difference (V) by Ohm's law (R = V/I). The resistivity of a material is defined as 'the resistance in ohms between the opposite faces of a unit cube of material'. For a conducting cylindrical material of resistance (R), length (L) and cross-sectional area (A), the resistivity (ρ) is given by ρ = RA/L. The SI unit of resistivity is the ohm. The reciprocal of resistivity is conductivity.

7.3.2 CURRENT FLOW IN EARTH

A single current electrode of uniform resistivity is inserted on the surface of the earth. The circuit is completed by a current sink electrode located far away. Current flows radially away from the electrode into the ground. The current distribution is uniform over a hemispherical downward path. The path of the circle represented by a radius 'r' from the electrode has a uniform voltage. The circular path with uniform voltage is known as the 'equipotential line' from the electrode. The current density and potential decay as the radius (r) increases away from the electrode. The decrease of potential with respect to radius of the circle is denoted by (–dV/dr). At a certain distance, the current is reduced to nil and the receiving potential signal at the surface is negligible.

7.4 FIELD SURVEY METHOD

Figure 7.1 shows the principle of resistivity measurement by placing four electrodes at the earth's surface. Practically, a single electrode is not used; it was described above to show the path way of the flowing of current in subsurface conditions. The practical device consists of two circuits; one is a 'current circuit' consisting of two electrodes and a battery, and the other a 'potential circuit' also with a pair of electrodes and a

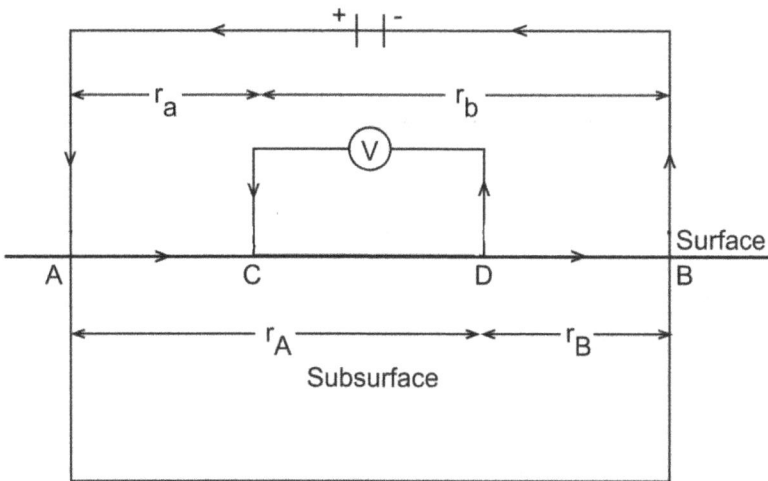

FIGURE 7.1 Electrode configuration for subsurface resistivity determination. Current electrodes = (A, B). Potential electrodes = (C, D).

potentiometer (voltmeter). Current is introduced in the ground by the current circuit. The returning electrical signal from the underground is monitored by the potential circuit as the potential difference (ΔV) developed by the flowing current in the ground. Either AC or DC current may be used. AC current has an advantage over DC current. AC avoids the accumulation of ions in the circuit and minimizes the regional interfering electrical anomalies.

The current electrodes are designated as A and B and the potential electrodes by C and D. The distance between current electrode A and potential electrode C is denoted by 'r_a.' The distance between current electrode B and potential electrode C is shown as r_b in the figure. Similarly the distances between potential electrode D and current electrodes A and B are given as r_A and r_B. It is supposed that current is spreading from the electrodes uniformly in a uniform-resistivity subsurface rock. The change in potential is measured at the surface by the potential electrodes (C and D) and a voltmeter, which leads to the identification of the resistivity of anomalous undersurface material. The potential difference between electrodes C and D is given by the following equation:

$$\Delta V = \frac{\rho I}{2\pi}\left[\left(\frac{1}{\gamma A} - \frac{1}{\gamma B}\right) - \left(\frac{1}{\gamma a} - \frac{1}{\gamma b}\right)\right]$$

All the parameters, potential difference (ΔV), electrode spreading distances (r_x) and current (I), are known. Therefore resistivity (ρ) can be calculated. It is also clear from the equation that the resistivity does not only depend on the nature of the underground but also on the relative distances of the electrodes at the surface. The resistivity thus calculated is known as 'apparent resistivity' (ρ_a) in underground heterogeneous rock. To cover a large underground survey area, the spacing between the current electrodes (A and B) is increased. With the increase of separation distances between current electrodes, the amounts of current, depth and area penetrated into the subsurface increase. The electrode separation distance is chosen so that sufficient energy (current) goes into the ground for creating a measurable potential difference.

Practically different electrode configurations consisting of four electrodes are used at the surface to conduct the subsurface electrical resistivity measurement. Two commonly used arrangements of the electrodes are shown in Figure 7.2 and described as follows.

7.4.1 WENNER CONFIGURATION

The Wenner electrode configuration is the simplest arrangement. All four electrodes are placed on the survey line, at the same constant distance (x) from each other. Current is introduced into the ground by the current electrodes (A, B). The potential (ΔV) developed in the ground is measured by the potential circuit with voltmeter. The apparent ground resistivity (ρ_a) is given by:

$$(\rho_a) = 2\pi x \frac{\Delta V}{I}$$

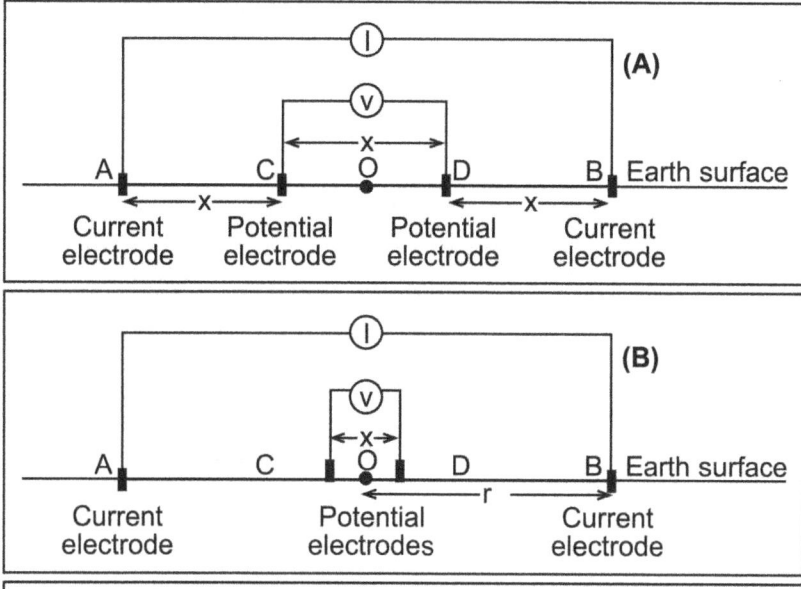

FIGURE 7.2 Practical electrode configuration at the surface for subsurface resistivity determination. A = Wenner configuration keeps constant distance (x) between electrodes. B = Schlumberger configuration keeps the small distance (x) between potential electrodes constant and current electrodes are placed at a distance (r) laterally.

This simplified Wenner equation is used for the determination of the resistivity, because all the parameters are known.

7.4.2 SCHLUMBERGER CONFIGURATION

The Schlumberger configuration uses different distances between the current and potential electrodes. The distance between the current and potential electrodes is increased (ɤ). The distance between the potential electrodes is a small distance (x). A simplified equation for the computation of the subsurface apparent resistivity (ρ_a) by the Schlumberger method is given below:

$$(\rho_a) = \frac{\pi \gamma^2}{\varkappa} \frac{\Delta V}{I}$$

Resistivity can be calculated, because all the parameters are known.

7.5   INTERPRETATION OF RESISTIVITY DATA

The resistivity method is applicable to the determination of shallower rock electrical characteristics and relates them to geology. The electric current does not penetrate into deeper layers or multilayered rock. The direction, depth and strength of electrical signal depend on the path way traveled as well as on the geology of the medium (rock). Weaker signals are received from deeper rock, and in the presence of multilayered or heterogeneous rock the returning voltage signals are complex. Different underground materials offer different resistivity to the flowing current. Geo-electrical heterogeneity in underground materials creates anomalous current. The anomaly or discontinuity in current/potential is correlated to the geological discontinuity as well as material type. The flowing current deflects in the rock as it encounters a high-resistance insulating material. The deflection in potential is a 'resistivity anomaly'. In high-conductivity rock the current maintains its original path and passes through the rock in a straight line. The electrical method works well when there is a clear electrical discontinuity (resistivity/conductivity contrast) in the rock. The method is not applicable for geological, litho-logical and structural and density contrasts. There is a clear resistivity contrast between basement igneous rock and overlying sedimentary rock. On the basis of resistivity data, the depth of basement rock and the thickness of sedimentary rock can be determined. The depth of a buried rock is estimated on the principle of the decrease of apparent resistivity (ρ_a) with depth. The depth penetrated by AC is dependent on the frequency employed. Normal practice is to use a low frequency for high-depth and a high frequency for low-depth current penetration. Faults, fissures, stratigraphy and oil/gas/water accumulation may be identified if a readable resistivity (voltage) anomaly is observed. The presence of ores can be established with confidence.

The flow of current in two-layered rock not only depends on the electro resistivity/ geology of the respective layers but also on the distance between electrodes. At the interface there will be variation of current flow and distortion of equipotential voltage. Therefore, the potential reading measured by the surface voltmeter also shows distortion (deflection) which is taken as an anomaly due to the resistivity contrast of the two layers.

Three scenarios of two-layered rock are described as follows:

- If the resistivity of the two layers is the same ($\rho_1 = \rho_2$), the current will not differentiate between the two layers. The current will flow uniformly in spherical form as in a single homogeneous rock.
- If the current electrodes are placed too close together, the current will not penetrate deeper, leaving the layer interface untouched.
- If the resistivity of the two subsurface layers are different (ρ_1, ρ_2) the current will not flow in a uniform spherical form. There will be a sudden variation (anomaly) in the flow of current at the interface of the two layers. The anomaly is due to the geological heterogeneity and different resistivity. If the resistivity of layer I is greater than that of layer II ($\rho_1 > \rho_2$), the current moves from a more resistive layer to more conductive (less resistivity) layer that facilitates smooth flow of the current. It means more current will flow in layer II compared to layer I for the same thickness of the layers. Similarly if the resistivity of layer II is more than that of layer I, ($\rho_1 < \rho_2$) more current will flow in layer I than in layer II.

B ELECTROMAGNETISM SURVEY

7.6 INTRODUCTION

Electromagnetism surveying is an example of an active method. Electromagnetism (EM) is a subject of physics. It describes the force and field associated with a charge, positive or negative. The sun's nuclear reactions are the main source of electromagnetic radiation reaching earth. Electromagnetism is also produced artificially. Electromagnetic surveying is used for the prospecting of subsurface conducting materials. The survey has limited less scope for oil/gas accumulation.

The full electromagnetic spectrum is shown in Figure 7.3 along with the nomenclature of each band of the spectrum. The whole electromagnetic spectrum ranges from very high frequency (10^{24} Hz) to very low frequency (1 Hz) radiation consisting of seven bands: (1) gamma (γ) rays, (2) X-rays, (3) ultraviolet waves, (4) visible waves, (5) infrared waves, (6) microwaves, (7) radio and broadcasting waves. Each band, from the total seven bands, is recognized by its specific wavelength, frequency and time period.

Precisely the wavelength of the radio wave portion of the electromagnetic spectrum is expressed in meters and kilometers, the wavelength of microwaves is expressed in a few centimeters, infrared wavelength is quoted in ten-thousandths of a centimeter and visible light has a wavelength range of 40–80-millionths of a centimeter. Further shorter wavelength (less than a billionth of a centimeter) are high-frequency spectrum bands (ultraviolet, X-rays and gamma rays).

Each band of the electromagnetic spectrum is associated with a specific energy and finds its specific application accordingly in diversified sciences and engineering fields. It is interesting to note that each band of the electromagnetism spectrum differs in wave properties, but the speed of all the bands is the same. The velocity of electromagnetic radiation (light) is 186,000 miles per second (300,000 km/s).

A very low-frequency (1^{-1} -10^8 Hz) and high-wavelength electromagnetic band is applicable to geo-electrical surveying which may be further classified as follows:

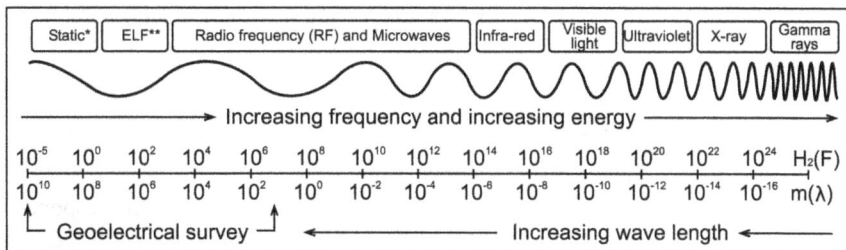

FIGURE 7.3 Full spectrum of electromagnetism wave.

**Extremely low frequency 3 to 30 Hz and long wave length.

*Static, very very low frequency with very low varying electrical field.

Survey Method	Applicable Frequency (Hz)
Ground penetration radar	10^2 to 10^8
Electromagnetic induction	10 to 10^3
Induced polarization	5 to 10^2
Self-potential	1 to 10
Telluric current and magnetism	10^{-1} to 10
Audio frequency magnetism (AFMAG)	1 to 10^2
Electro-magnetism survey	10–10^3

'Electromagnetic radiation' and 'electromagnetic wave' have same meaning. Electromagnetic radiation is the wave of electromagnetic fields, propagating in space, carrying electrical as well as magnetic radiation energy. The electromagnetic waves are generated by the up and down motion of the charged particles (electrons), in the direction of the wave progression. The vibrating charged electron particles create an oscillatory electric/magnetic field as a 'halo' around the propagating wave. The electromagnetic wave possesses dual properties of 'particles' and 'waves' simultaneously:

- Particle motion imparts mechanical properties in the propagating wave.
- Electrical and magnetic fields are generated along electromagnetic wave propagation line. It may be noted that a neutral field is created along the seismic wave propagation path.

Electric and magnetic fields were supposed to be different forces by earlier scientists, who treated them separately. The origin of both fields is common, that is from charged particles. The fact is that both fields are inter-convertible under specified conditions. A comparison of electric and magnetic fields is given in Table 7.1.

Electromagnetic waves do not travel deeper into rock. The survey is limited to a shallow depth.

Electromagnetism surveying is simple, quick and cost effective. It has many useful applications in mineralogy and engineering science, rather than in oil/gas prospecting. Oil and gas are basically non-conducting materials and do not respond significantly to electrical and magnetic fields.

7.7   THEORY OF ELECTROMAGNETISM (EM)

The theory describes how a magnetic field is generated by flowing current and how an electric current is created by a moving magnet. Different materials respond to electrical/magnetic field differently. On a theoretical basis, all elements/material should exhibit magnetism. The circular motion of the electrons of the atoms and their spin create a dipole (fundamental property) moment. The dipole moment produces an electrical current and magnetism. On the basis of materials' responses to magnetic and electrical fields, materials are divided in two types. One type is called

TABLE 7.1
Comparison of Electric and Magnetic Waves

	Electricity	Magnetism
1	Electric current is produced by a charged particle. The charge may be at rest or in motion.	A magnetic field is generated only by the motion of a charge. Its function depends on motion.
2	An electric field behaves like a magnetic field.	A magnetic field is generated by only by the motion of a charge. It behaves like an electric field.
3	A changing electric field produces a magnetic field (James Maxwell).	A changing magnetic field produces an electric field (Michael Faraday).
4	Electric and magnetic fields propagate through space simultaneously.	Magnetic and electric fields travel together as waves in space.
5	An electric field is aligned in the direction of the charged source.	A magnetic field acts perpendicularly to the moving charged source.
6	The strength of an electric field decreases with distance.	The strength of a magnetic field decreases with distance.
7	The initial and final points of an electric current are the positive and negative charges respectively.	A magnetic field acts in a closed or cyclic loop.
8	Electricity is the flow of electrons in a conductor or electrochemical reaction in electrolyte.	Magnetism depicts a force between a negative charge (electron) and a positive charge (proton).
9	Measurement at the surface is conducted by inserting electrodes in the ground.	Measurement is conducted by a wire loop laid down on the surface, or on an airplane.
10	The electrical signals are more strong and sensitive.	Electromagnetic induction signals are weak and less sensitive.

conductors and the other kind is insulators. Most materials belong to the insulator type. An insulator cannot conduct electricity because of the hindered motion of the electrons in the matter. The electrons are not aligned in a particular direction so as to show conductivity by free motion. The electrons are distributed in the matter in a non-uniform or random fashion. They nullify each other's fields. In conductors, mostly ferromagnetic materials (iron, cobalt, nickel, etc.), electrons are aligned in a particular direction and are free to move. Ferromagnetic materials are transition elements and contain free electrons in outer orbit. The movement of the free electrons generates electric and magnetic fields. Every charged material or particle generates an electric field around itself. The field is detected by another charged body. Like charges repel each other and opposite charges attract each other. The repulsive/attractive force between the two charged bodies is inversely proportional to the square of the distance between the two bodies. This is like gravitational force between two masses. How a magnetic field is created by a flowing current and how an electric field is generated by a moving magnet are explained below.

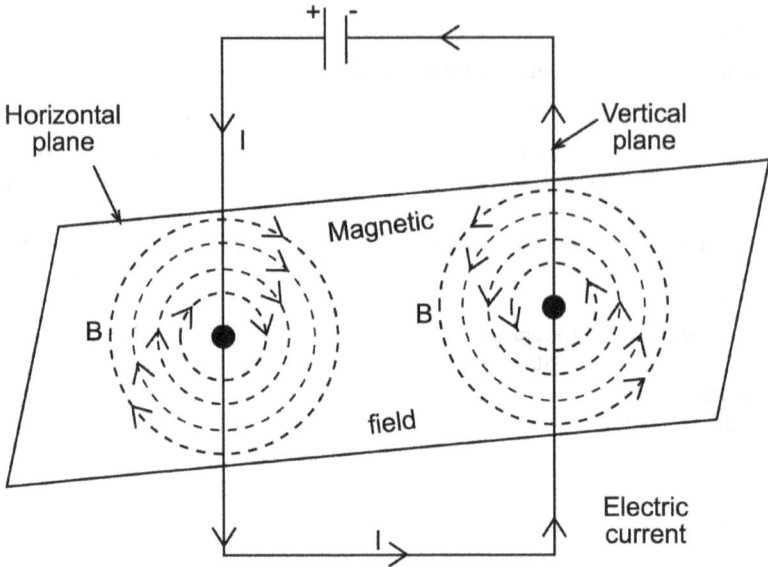

FIGURE 7.4 Maxwell's law of electromagnetism; flowing current (I) generates a magnetic field (B).

7.7.1 MAXWELL'S LAW OF ELECTROMAGNETIC INDUCTION

Maxwell's law states that 'an electric current flowing in a wire (loop) produces a circular magnetic field around the wire'. A time-varying electric field produces a time-varying magnetic field. The direction (clockwise or anti-clockwise) of the generated magnetic field depends on the direction of the flowing current in the wire loop. A magnetic field (B) generated by the flowing current (I) is shown as dotted circles in Figure 7.4.

7.7.2 FARADAY'S LAW OF ELECTROMAGNETISM

Faraday's law states that 'a varying magnetic field induces an electromotive force (emf) in a conducting loop', that is a moving magnet generates an electric field as shown in Figure 7.5. The motion of the magnet is towards the conducting wire loop (V). An induced electromotive force (emf) causes a current to flow in the wire. The direction of the induced current is such that the magnetic field inside the loop opposes the increase of magnet flux passing through the loop from the magnet. Flux is the measure of the amount of magnetic/electric field that passes through a specific area or loop. The magnitude of the emf (V) induced in the loop is equal to the flux passing (ϕ) through the loop in unit time, that is the rate of magnetic flux ($d\phi/dt$) passing is directly proportional to the induced emf in the loop. An induced emf (current) flows in the loop when the magnet moves toward the loop. The magnetic field inside the loop opposes the increase of magnetic flux through the loop from the approaching magnetic field. The direction of the induced current is opposite to the

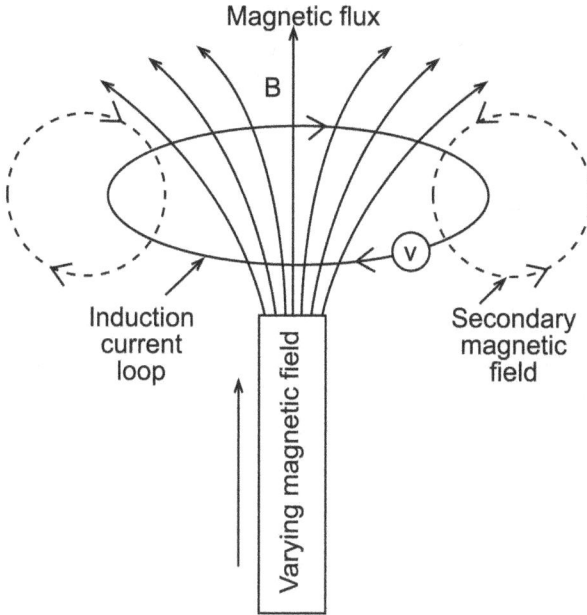

FIGURE 7.5 Faraday's law of electromagnetism; a moving magnet (B) produces an induction current (V).

magnetic field. The induced magnetic current in turn creates a secondary magnetic field around the loop wire, shown in the figure as two dotted circles. The interaction of magnetic and electrical fields during a time-varying sequence with an underground conducting body and subsequent distortion/anomaly in the field is the basis of the electromagnetic survey.

7.8 FIELD SURVEY METHOD

An electromagnetic subsurface survey is carried out either by placing the measuring equipment at the earth surface (land) or airborne.

The land equipment consists of two electrical circuits (primary and secondary) placed at the surface and a third is underground geological circuit. A primary current transmitter circuit (loop) consists of a battery, and another secondary electromagnetic wave receiver loop with a voltmeter. Both circuits are located at the earth's surface with appropriate distance apart. The size of the loops, the strength of the varying current and the spacing (spread) between the two loops are designed according to the object of the survey. Usually the loops are of a rectangular shape having variable dimensions depending upon the penetration needed.

A primary magnetic field (H_p) is generated by the current flowing in the primary transmitting wire loop (Maxwell's law). The generated primary electromagnetic wave spreads spherically all around. A major portion of the wave directly strikes the surface receiver loop. Some portion of the waves penetrates into the ground. As the radius of the spherical waves increases, the wave strength and amplitude decrease. The wave signals become weak, decay rapidly and ultimately vanish after propagating down to the limiting depth. The earth acts as a conducting medium for the electromagnetic wave from the transmitting loop to the receiver loop via subsurface conducting materials. The primary electromagnetic wave interacts with the subsurface conducting material of rock. An induction current is produced in the subsurface conducting rock body according to Faraday's law of electromagnetism. This induction current again becomes the source for generating a secondary electromagnetic wave (H_s). The secondary magnetic field waves are reflected and pass through the tilted angle secondary receiver loop placed at the surface. The receiver loop is an electrical circuit that picks up the maximum secondary magnetic field and converts it and records it as time-varying voltage signals.

The frequency and amplitude of the reflected secondary waves depend on the conductivity/resistivity contrast in the conducting rock body of anomalous rock. The anomaly in electromagnetic signals is attributed to the conducting body in the subsurface rock. The depth of electromagnetic wave penetration into the ground depends on the design of the loop, and the strength and frequency of the primary magnetic field. For the study of shallower depths, a small loop and spacing between the transmitter and receiver are needed. For the study of greater depths, larger loops are used, and the spacing between the transmitter and receiver loops is increased.

Airborne electromagnetic surveying (EM) is an obvious advantage over land surveying due to the time saved and the larger underground area covered, including surveying in harsh and difficult areas. Transmitter and receiver loops are mounted in a suitable place on the aircraft, or a transmitter loop is placed in the aircraft and another receiver loop is tied down to the tail of the aircraft. Another possibility is that the two loops are placed on separate aircrafts or a receiver loop is placed on the aircraft and the transmitter loop is placed on the earth's surface. A larger elevation of the aircraft covers a larger survey area, but weaker signals are received which affect the accuracy of measurements.

The EM survey data so obtained are interpreted to determine the geometry and depth of the underground conducting material.

7.9   INTERPRETATION OF EM DATA

For the determination of the depth of a subsurface targeted object, the following two procedures are used.

- Fix the positions of the transmitter and receiver loops at the surface and vary the frequency of the primary current.
- Fix the primary frequency and vary the spacing between the transmitter and receiver loops.

There is a limiting depth that can be achieved by propagating magnetic waves with a particular frequency. Beyond the limiting depth, the returning magnetic signal is not of sufficient strength to be detected at the surface. The depth (z) of an underground body is inversely proportional to the frequency of the primary electromagnetic field (f) and directly proportional to the resistivity (ρ) of the subsurface. A low-frequency wave (high wavelength) is likely to reach a greater depth at a constant-resistivity subsurface. But very low frequency signals are weak. The returning signals are difficult for the receiver to detect. A longer spacing between the transmitter and receiver facilitates deeper wave penetration. But the wave signals' strength decreases with increasing depth. A greater size of loops helps signals to reach a greater depth than small-sized loop. However, a greater sized loop generates weaker signals.

C. INDUCED POLARIZATION SURVEY

7.10   INTRODUCTION

Induced polarization (IP) is an electric phenomenon related to the accumulation and orientation of electric charge or ions on the surface of a conductor or semiconductor. Polarization in a conducting body is affected by applying an external electric field. A subsurface 'geo-electrolytic cell' provides an external field for the polarization of an object in underground conditions. Other terms used for IP are 'induced potential field' and 'interfacial polarization'. A polarized body acts like a capacitor or dielectric. A capacitor/dielectric temporarily stores charges in an electrical circuit, where the charge decays slowly after the withdrawal of applied current.

The induced polarization phenomenon is utilized for the geological study of subsurface materials. The method is based on the response of materials to the applied electrical field. IP surveying is similar to the method of electrical resistivity surveying. But the theory as well as the techniques differ considerably. Induced polarization depends on the subsurface electrochemical reactions and conductivity of the targeted material. The IP survey cannot be applied with confidence in oil/gas prospecting. Oil/gas response to electrical fields is poor, whereas conducting minerals are very sensitive to the electrical field, creating formidable anomalies, with respect to the bulk sedimentary rock.

7.11   ELECTROLYTIC CELL

An electrolytic cell consists of two parts, an external 'electrical circuit' and internal 'chemical circuit' (electrolyte and electrodes). An electrolyte is a conducting solution of soluble ionic minerals in water. The external electrical circuit consists of a battery and an electric wire connected to the two electrodes via a battery. The current flows in the external circuit via electric wire by the transfer of electrons from the positive electrode (anode) to the negative electrode (cathode). On the other hand, current flows in the chemical circuit of the electrolytic cell via electrolyte by the exchange (motion) of the cations (positive ions) and anions (negative ions) from the negative electrode (cathode) to positive electrode (anode).

7.11.1   Physical Electrolytic Cells

A typical electrolytic cell (electrolytic circuit) consists of two circuits, an external and an internal, and common two electrodes. In the external circuit the electrodes are connected to a battery and a voltmeter through a conducting wire. Externally current flows from anode to cathode solely by the movement of electrons.

In the internal chemical circuit the electrodes are immersed in an electrolyte solution contained in a cell. Under the influence of an external current, ionization (decomposition) of the mineral present in the electrolyte takes place into cations and anions. The electrochemical decomposition of ionic material creates its own current and electromotive force (EMF). An internal cell current flows from cathode to anode through the ion exchange process. Anions generated at the cathode, with an excess electron, travel to the anode. At the anode electrons are delivered to cause the current to flow externally. At the same time the anode generates cations that are transferred to the opposite direction through the electrolyte solution. Finally cations are deposited at the cathode. Ions are deposited at each electrode in equilibrium conditions. When such equilibrium is disturbed, this results in an uneven distribution of ions. A lesser or greater concentration of ions is found on one or the other electrode. Such a condition is known as 'over voltage'. Over voltage is the difference between the theoretical value (equilibrium) of the current and the excess current needed to drive the current. The electrode with excess ions is 'polarized' and the phenomenon is known as 'induced polarization'. Current at over voltage value is called the 'saturation current'. A polarized electrode acts like an electrical capacitor. A capacitor (condenser) temporarily stores electric charge in an electrical circuit. A polarized electrode continues to release storage energy (current) even after the current switch is turned off. The stored charge gradually decays over a period of time.

7.11.2   Geo-Electrolytic Cell

A geo-electrolytic cell simulates the conditions of the physical electrolytic cell described above. Conducting rock materials and electrolyte in the pore space of rock are necessary ingredients for forming a subsurface geo-electrolytic cell. The geo-electrolytic cell also consists of two parts, an external circuit at the surface and an internal underground circuit geo-electrolyte cell. The external circuit is similar to that in a physical electrolytic cell. Two electrodes (anode and cathode), connected by a battery, are fixed on the ground surface, keeping a suitable distance between them. This makes a current supply circuit into the underground. In between these, two other potential electrodes along with a voltmeter are fixed at the earth's surface, making a potential measuring circuit. In the external circuit, current flows by the transferring of electrons from anode to cathode.

An internal underground chemical circuit is set up by the geology of the area. The sedimentary rock may contain conducting ionic minerals and electrolyte (water) in pore space. A portion of the conducting minerals behaves as a cathode and another part as an anode within a sedimentary rock. Underground ionic current is set up due to electro geo-chemical reactions, and current flows from the cathode portion to the anode portion by ion exchange reactions in electrolyte solution. The cations move from the anode portion to the cathode portion and anions from the cathode portion to the anode portion, constituting an ionic current. Current flows by the exchange of ions in the pore space of the rock filled with electrolyte. The ionic current flows linearly in a homogenous uniform sedimentary rock in the presence of conducting materials. Like electrical energy, ionic energy (current) is used up to overcome the resistance offered by the rock and dissipates as heat. The ion exchange continues to occur till enough cations are accumulated at the cathode portion and the equilibrium is disturbed. Current flow is hindered and ultimately stopped. The interface is said to be polarized. Polarization is interpreted in terms of the geology of the conducting rock body.

All the components of a geo-electrolytic cell contribute to the polarization. Some are described here again.

7.11.2.1   Conducting Material and Applied Current

It is observed that when current flows in a non-conducting homogeneous sedimentary rock, the resulting subsurface voltage continues to increase linearly with applied current. If the current is switched off, the voltage drops immediately to zero. In the presence of semiconducting materials in sedimentary rock, the resulting voltage first increases rapidly compared to the homogeneous sedimentary rock. After further applied current, it decreases. If the voltage crosses over the homogenous sedimentary rock, then a steady state (over-voltage) is achieved. For conducting material, if the voltage increases rapidly and quickly, the polarized steady state (over-voltage) is achieved. Thus, it may be concluded that the rate of polarization of a material depends on its conducting property. A more conductive object is polarized more quickly than a semiconductor, and non-conducting material does not polarize at all.

7.11.2.2   Resistivity/Conductivity Contrast and Polarization Potential (EMF)

Polarization potential depends on the electromotive force (EMF) across the rock material. The potential or EMF is the force of the 'attracting ability' or the 'pushing ability' for an electron in any material, and thus is a measure of current flow. The EMF force depends on the rock resistivity/conductivity contrast. A conducting body and the surrounding rock media provide different resistivity/conductivity contrast. A large resistivity and conductivity difference between them favors early polarization of the materials. No resistivity/conductivity contrast exists in homogenous sedimentary rock, so there is no polarization in the homogeneous sedimentary rock.

7.11.2.3   Electrochemical Reactions

Ion exchange and electrochemical reactions affect the charge decay time. The electrochemical reactions are related to the origin and concentration of ions, the chemical composition of the rock and electrolyte. Therefore, polarization is also related to the exchange of ions, ionic current, ion diffusion, the nature of electrolyte, surrounding rock and deposition at the interface between the conducting materials.

7.11.2.4   Size, Shape and Distribution of Conducting Material

The size, shape, character and distribution of conducting material in the rock matrix affect the polarization. If the conducting material is a distinct and compact block, it is likely to be polarized before the well-spread and disseminated materials. The former requires less energizing current than the latter for polarization.

7.11.2.5 Dielectric Property of Material

The dielectric property is the most important factor in electrical measurements. The dielectric property is designated as the dielectric constant (k), and it is the prime factor for IP phenomena. The dielectric constant of a material is its ability to conduct an electric current as well as store the charge (current) and release it gradually. Therefore it is related to the polarizability of the body. It determines the material capacitance. The capacitance is the ability of the material to store a charge (current). Different matter has different dielectric constants according to its current conducting ability. The dielectric constant for a vacuum is 1; the dielectric constant for water is 80. The dielectric constant for wet rock is greater (50) than for dry rock (20).

Dielectric properties also play an important role in ground penetration radar surveying.

7.12 SURVEY METHOD AND RESULT

An AC or DC current is introduced in the earth through current electrodes for less than one minute. In a normal electric circuit, voltage grows and vanishes momentarily with switch on and off positions. In proper subsurface polarizing conditions, the voltage grows, attains a maximum value and then decays gradually, as indicated by the potential electrode circuit.

After turning the current on first, the voltage continues to grow linearly from the starting time (t_0) within the conducting homogenous sedimentary rock. Thereafter, a saturation state is attained, and no more current flows. The saturation voltage (V_{max}) signifies the polarized state of the underground conducting rock body. Initially the potential drops quickly after turning the current off. After a short time, the voltage decays slowly and ultimately attains zero value with time (t_d). The slow drop in voltage is due to the polarization state of the conducting body. The shape and area under the time/voltage curve are used to interpret the subsurface conducting rock body.

D. GROUND PENETRATION RADAR SURVEY

7.13 INTRODUCTION

Ground penetration radar (GPR) surveying is applied for many scientific, commercial and military purposes. The radar uses high-powered radio waves for the determination of the location, direction and velocity of aircraft, ships, spacecraft, road vehicles and missiles as well as for weather forecasting. The radio frequency range is also used for the broadcasting of radio and TV signals. It is successfully used for the geological surveying of the subsurface. Ground penetration radar (GPR) surveying closely resembles the seismological survey (Chapter 6). Seismological surveying uses the seismic wave (mechanical energy) whereas radar surveying employs electromagnetic radio frequency waves. They are invisible and travel with the speed of light in air (3×10^8 m/s).

GPR uses the radio wave portion of the electromagnetic radiation consisting of 10^2 to 10^8 Hz frequency and wavelength from 10^2 to 10^6 meters. The radio waves travel with a faster speed in matter than in air. The subsurface propagation of radar wave obeys the same law as seismic waves. The velocity of a radar wave (V) in the subsurface is controlled by the electrical properties of the rock materials, notably resistivity, conductivity and electrical permittivity. Electrical permittivity and resistivity are defined by the 'dielectric constant' of a semi-conductor (rock material). The dielectric constant is a measure of relative permittivity more than the resistivity. The dielectric constant of water is 80 and it is considered a highly electric conducting material, whereas rock materials have a dielectric constant of 4.0 to 50.0. The materials with a dielectric constant of less than 8 are considered poor conductors or good insulators. The presence of pore water and ionizable minerals in the rock makes the rock conductor. In homogenous subsurface rock the propagation of radar waves is linear, though of course diminishing in energy (amplitude) with the passage of time and depth penetrated.

At the interfaces of two layers or at the geological discontinuity where the incident radar wave strikes the boundary, reflection/refraction of the radar wave takes place. There is an electrical resistivity contrast at the geological discontinuity and at the two layers' interface. A contrast in dielectric constant at the interface results in the reflection of a portion of the radar wave.

7.14 SURVEY METHOD AND RESULT

Radar is a wireless electrical circuit. It consists of two antennas; one antenna is the radar wave transmitter (T) and the other is the wave receiver antenna (R). The radio waves are generated at the surface and are transmitted through the transmitter antenna. Some of the waves penetrate into the ground. The waves that are reflected back to the surface are detected by the receiver antenna and recorded. The receiver also records direct waves from the transmitter called 'surface waves'. The surface and reflected radar waves are recorded as amplitude versus time/distance traveled. The timing of the initial triggering (t_0) and the final receiving time (t_x) of the wave at the receiver antenna wave are recorded in milliseconds. The surface wave travels a very short distance, whereas the reflected pulse travels a two-way path, from the transmitter to the point of incidence at the subsurface interface and back to the receiver. Obviously, this travel time is greater than that of the direct surface wave.

A number of radar wave transmitting and receiving antennas are used to cover a wider subsurface area. The depth penetrated (z) by the GPR wave depends on the resistivity (ρ) of the underground rock and the frequency (f) of the radar wave. High penetration (z) is achieved by lower frequency and high rock resistivity.

A rock of greater electrical resistivity favors more propagation of radar waves into the ground, and also low-frequency radar waves achieve greater depth. A more conductive rock and high frequency are limited to a lower depth. Additionally depth determination is also related to the geometrical arrangement of the transmitter/receiver antennas at the surface. For shallower penetration determination, transmitters and receivers are placed on the surface along a line at a small distance. For covering a large survey area, the transmitters and receivers are moved apart on the survey line but the spread between them is kept constant. The radar waves reflected back from deeper rock (layer II) are weak, broad and lack sharpness and resolution compared to the signals from shallower rock (layer I).

E. SELF POTENTIAL SURVEY

7.15 INTRODUCTION

Self-potential is an example of passive electrical surveying. The survey method does not require an external current supply source (battery). The earth's natural potential (self-potential) and associated electrolytic current are generated in the subsurface by conducting materials. Electrochemical reactions at the interface of the conducting rock materials in the presence of electrolyte create potential difference (self-potential) and electrolytic current. The measurement of the natural potential is correlated with subsurface mineral materials.

7.16 ELECTROCHEMICAL CELL

The electrochemical cell is a device to generate electrical current solely by chemical reactions. An electrochemical cell is a chemical cell that does not require any external electric current (battery) to function, as needed in the case of 'physical electrolytic and geo-electrolytic cells' (discussed earlier).

7.16.1 PHYSICAL ELECTROCHEMICAL CELLS

The cell consists of an anode zinc electrode (oxidation), in a porous pot dipped into zinc sulfate solution, and a cathode copper electrode (reduction), copper dipped into a copper sulfate solution. Both electrodes are immersed in solutions of their salt. Thenode supplies the electrons to the outer circuit and the cathode accepts electrons from the circuits. Thus the flow of current in the external circuit takes place by the transfer of electrons from anode to cathode. The internal ionic current flows from the cathode to the anode by the exchange of anions (SO_4^{-2}) and cations (Cu^{+2}).

7.16.2 GEO-ELECTROCHEMICAL CELLS

Necessary ionizable minerals in presence of electrolyte are needed for electrochemical reactions. The occurrence of underground chemical reactions is the basis for geo-electrochemical cells. The underground self-potential and current generated by subsurface chemical reactions set up the underground geo-electrochemical cell. The subsurface rock containing a conducting body (ionizable mineral) and electrolyte in the pores may be divided into an upper anode-oxidation portion and a lower cathode-reduction portion. The current flows from the upper oxidation portion, due to the release of electrons, to the lower reduction compartment of the conducting body, at the same time generating cations. The lower cathode-reduction portion accepts elections and generates anions. Ionic current is generated through the exchange of anions and cations. A subsurface geo-electrical circuit is set up. Thus a self-potential is developed at both ends of the subsurface conducting rock body. The potential creates an anomaly, monitored at the surface against the back ground regional field. The anomalous potential is measured by the two non-polarized electrodes (not connected to a battery) and voltmeter set up at the surface. The shape of the curve obtained by plotting the anomalous potential (V) versus surface electrode distance (d) is related to the geology of the underground conducting body. By varying the distance between electrodes and placing them in longitudinal and transverse positions alternatively, a larger subsurface area is covered by the survey.

7.17   SURVEY METHOD AND RESULT

The survey method by this technique is a simple electrical measurement. Two electrodes are introduced into the ground with an appropriate spread (30–50 meters) in the line of survey. The electrodes are non-polarizable. The non-polarizable electrodes are pure metallic bars immersed into a solution of their salts, for example, a copper metal bar in $CuSO_4$ solution and zinc metal in $ZnSO_4$ solution. Both bars and solutions are contained in a porous pot. The porous pot allows the slow leakage of the solution into the ground. The potential measurement circuit is completed by connecting the two electrodes through a voltmeter at the surface.

The potential anomaly is created by the subsurface conducting body. The interpretation of a potential anomaly is similar to magnetic anomaly interpretation. With this potential anomaly along with the shape, height, area and half band-width of the anomaly peak, the size, shape, depth and quantity of the subsurface anomaly rock (conducting body) are estimated. The survey is limited to conducting bodies at shallow depths. Other natural and artificial electrical sources present in the area interfere and create noise signal.

F.   NATURAL TELLURIC FIELD

7.18   INTRODUCTION

Natural telluric (earth) current and telluric magnetism are natural fields of the earth. Both are present in and around the earth. These are low-frequency and low-energy waves. The telluric wave energy is utilized for the study of underground mineral and conducting bodies. Three kinds of natural telluric fields are identified:

- Telluric electric current field (frequency range from 10^{-1} to 10 Hz)
- Telluric magnetic field (frequency range from 10^{-1} to 10 Hz)
- Natural audio frequency magnetic field (AFMAG) (frequency range from 1 to 10^2 Hz)

The origin of the telluric current and magnetic field is the ionosphere. The ionosphere is the upper portion of the earth's atmosphere (90–500 km above sea level) containing charged (ionized) particles. The charged particles generate a magnetic field. The magnetic field starting from the ionosphere and reaching earth is known as the telluric magnetic field. The radiation from the sun and the earth's rotation leads to variation in the telluric magnetic field. The variation of earth's magnetic field also introduces telluric current in the earth. The variation of the telluric magnetic field induces an AC current of low frequency in the earth.

7.19 NATURAL TELLURIC CURRENT SURVEY

The telluric current penetrates from shallower to deep into the earth for several kilometers. The subsurface flowing telluric current is measured by a potential difference (V) recording circuit containing a voltmeter and two electrodes fixed at the earth's surface at suitable distance. The electrodes are non-polarizable metal bars, not connected to any battery. The electrodes are made from cadmium and zinc metal bars immersed in cadmium chloride and zinc sulfate solutions.

The survey is based on the electrical properties of conductivity and resistivity contrast of the subsurface. A uniform sedimentary rock exhibits a very low conductivity and uniform resistivity. The corresponding potential difference at the surface is uniform and constant, and no distortion (anomaly) is observed in potential. Whenever a conductivity/resistivity contrast exists in sedimentary rock there will be anomaly in potential difference. Uniform sedimentary rock containing a defined structure such as an anticline or salt dome exhibits electrical contrast. The resistivity contrast at the interface between the sedimentary rock and the structures deflects the underground telluric current. Therefore distortion (anomaly) in potential is observed at the surface. The observed potential anomaly is correlated with the geology of the anomalous structures of the underground rock. The presence of mineral ore in the sedimentary rock generates a formidable potential anomaly. Likewise the presence of an oil/gas reservoir results in anomalous potential.

7.20 NATURAL TELLURIC MAGNETIC SURVEY

Magnetic telluric surveying is based on the combination of the telluric current and telluric magnetic fields. It is similar to the electromagnetism method.

The natural telluric magnetic survey is difficult as both magnetic and electric fields are measured.

The measurements of telluric magnetism consist of two sets; one is the recording of the telluric current and the other is telluric magnetism. Both are measured simultaneously at the same spot.

- The underground magnetic field is measured by its inductive effect (current) on the electric loop placed at the surface or by use of a fluxgate magnetometer.
- The subsurface telluric current is measured at the surface, in the same way as stated above for the measurement of potential difference. The recording electric circuit consists of non-polarizable, porous electrodes (cadmium and zinc) immersed in their own solutions and a voltmeter.

Underground conductivity/resistivity contrast distorts the prevailing telluric current (potential difference) and magnetic field. These two anomalies are correlated with the anomalous conducting rocks. The electric field (E) is expressed as milli-volts and the magnetic field strength (B) expressed in nT. The depth (z) to which the magnetic telluric field penetrates in the rock is inversely proportional to frequency (f) and directly to resistivity (ρ_a) of the rock. Depth penetration increases with decreasing frequency of the telluric magnetic field (B). For greater penetration, low frequency is required.

7.21 NATURAL AUDIO FREQUENCY MAGNETIC FIELD (AFMAG)

The natural audio frequency magnetic field (AFMAG) is a natural alternating magnetic field (telluric electromagnetism). The frequency range ($1–10^2$) of this AFMAG electromagnetism is much less than the active artificially produced electromagnetism frequency range ($10–10^3$) used in survey. Therefore higher penetration depth is expected with the use of AFMAG.

The AFMAG is generated mainly from lightening, thunder storms, tornadoes and hurricanes in the upper or near surface (local) atmosphere. The audio electromagnetic generated waves spread spherically from the source point, between the upper atmosphere and below into the earth's surface. Some portion of the natural telluric magnetism penetrates into the earth's subsurface. The measurement technique for the audio natural electromagnetic field differs and is more difficult than the active electromagnetism survey. Two tilted signal receiver coils are used instead of one, to accommodate the fluctuating direction and intensity of the field. However there is no need for a primary transmitter loop for the generation of artificial electromagnetism, as here waves are produced by the natural processes of thunder storms, lightening, etc. Any moving magnetic field is associated with electrical as well as magnetic fields. A magnetic field has both horizontal and vertical components. When a horizontal magnetic field component passes through a subsurface conducting material, the axis of the magnetic field is deflected and an anomaly occurs. Measurement of the deflection of the axis is related to the geology of the conducting material. A major factor that affects the depth penetrated is the choice of frequency and electrical properties of the rock medium (apparent resistivity).

The following events are responsible for the generation of noise signals in AFMAG survey and need to be eliminated from genuine AFMAG signals:

- The interaction of charged particles (radiation) from the sun with the ionosphere and the earth's main magnetic field.
- The interaction of materials falling through the ionosphere.
- Manual activities involving electricity, atomic nuclear reactions, the aviation industry and other such installations.
- Noise signals generated by the measuring instrument.

7.22 CONCLUSION

The geo-electrical survey is the study of the interaction between electric current/magnetic field and the subsurface rock materials. The electric current and magnetic field may be manually applied from an external surface source (active) or may be inherent in the rock (passive). Four active survey methods are electrical resistivity, electromagnetism, induced polarization and ground penetration radar (GPR). Four passive survey methods are self-potential, natural telluric electrical current, natural telluric magnetic field and audio frequency magnetic field (AFMAG). Different rocks behave differently with respect to resistivity/conductivity contrast. Generally sedimentary rock is an insulator. In the presence of ferromagnetic elements and pore water with salts (electrolyte), the rock conducts electricity. Oil/gas is a bad conductor of electricity. In a resistivity survey current (AC or DC) is introduced in the ground through two current electrodes; the resistance offered by the rock is measured by the two potential sensing electrodes placed at the surface. When there is a geological discontinuity there is a resistivity contrast. Rock geology is inferred from the knowledge of apparent resistivity. A flowing current produces a magnetic field around itself and a moving magnet produces electric current, which forms the basis of active electromagnetic (EM) surveying. Electromagnetic waves are generated and introduced in the rock by a surface primary transmitter wire loop. The wave propagates, interacts with underground conducting rock material and generates secondary electromagnetism which is reflected back to the surface. The returning EM waves are recorded by a secondary tilted receiver wire loop, and the results are correlated with the features of the conducting subsurface rock. Induced polarization is the accumulation of charge (electrons or ions) around a conducting rock which acts as a polarized electrode in a subsurface geo-electrolytic cell. The geo-electrolytic cell is set up by introducing current by an external electric circuit into the ground. The current flows through external circuits and in solids by electrons transfer, whereas the subsurface conduction of current takes place in electrolyte by ion (cations and anions) exchange. The accumulated charge in the conducting rock continues and decays slowly even when the current is switched off. The accumulated charge decays with time and is measured; this is attributed to the rock geology. The subsurface propagation of ground penetration radar (GPA) waves obeys the law of seismic wave propagation. The velocity of GPA waves is controlled by the electrical properties of the underground rock. The GPA waves are reflected back to the surface from the geological discontinuity. The reflected wave is reduced in amplitude, frequency and strength and related to the interface geological discontinuity. A subsurface geo-electrochemical cell generates self-potential and electrolytic current by electrochemical reactions. The measurement of self-potential at the surface by the two non-polarized porous electrodes is correlated with the geology of the self-potential-generating rock. Telluric electrical and magnetic fields are generated in the ionosphere and penetrate several kilometers deep into the earth . The variation in the intensity and direction of waves is accompanied by variation (anomaly) of potential during underground propagation that is related to the geology of the path traveled path. Natural audio frequency magnetic& electric fields is generated in thunder storms and lightening. Both fields penetrates into the earth. When the AFMAG wave passes through a conducting rock, an anomaly in current (potential) occur and the axis of the magnetic field is deflected (anomaly) according to the geology of the rock.

8 Petroleum Geochemical Survey

8.1 INTRODUCTION

Geochemical survey is the study of the chemical nature of the earth. The earth as a whole may be regarded as composed of chemicals. The state and stability of a chemical depend on the surrounding environment. The study of earth in terms of chemicals is geochemistry, and the methodology used is known as the geochemical technique. If such technique is utilized for petroleum prospecting, the subject is called 'petroleum geochemistry' or 'petroleum geochemical survey'.

All geological materials are formed as result of chemical transformation over geological time. Petroleum is an organic substance; its subsurface formation equally owes to both organic and inorganic matter. Original organic matter is stored and processed in a subsurface factory (rock) consisting of inorganic materials. The essential requirements of a chemical conversion are a well-designed reactor, proper feedstock, temperature, pressure, catalyst, solvent and timing. For subsurface petroleum conversion all these factors are provided by the geological environment. Petroleum geochemistry is defined by many scientists in their own ways. Two standards definition are cited below:

> Petroleum geochemistry is a branch of earth science (geology) that uses the tools of chemistry principles for understanding and explaining the upstream geological process that leads to the deposition of inorganic and organic sediments, consolidation, maturation, generation and production of underground petroleum (oil and gas).
>
> Petroleum geochemistry is the branch of geochemistry, which deals with the application of chemical principles in the study of origin, maturation, generations, migration, accumulation and alteration of underground petroleum.

Based on the two definitions and explanations of geochemistry and geology, a diagram (Figure 8.1) is constructed to illustrate and compare the chemical process with the geological process leading to the generation of petroleum and dead carbon residue. The study of macroscopic structure and processes, both in the subsurface or on the surface of the earth, is a subject of earth science (geology). The geochemistry examines the microscopic details of the earth. It begins with atomic/molecule level, then proceeds further, to see the formation of elements, compounds, chemicals, minerals, rock and their interaction with environments. The earth or more precisely rock is composed of several chemicals (minerals). The state/stability of chemicals are defined by the environmental conditions. As the condition of the earth changes, the equilibrium conditions are disturbed. The chemicals try to adjust themselves in the new conditions through chemical

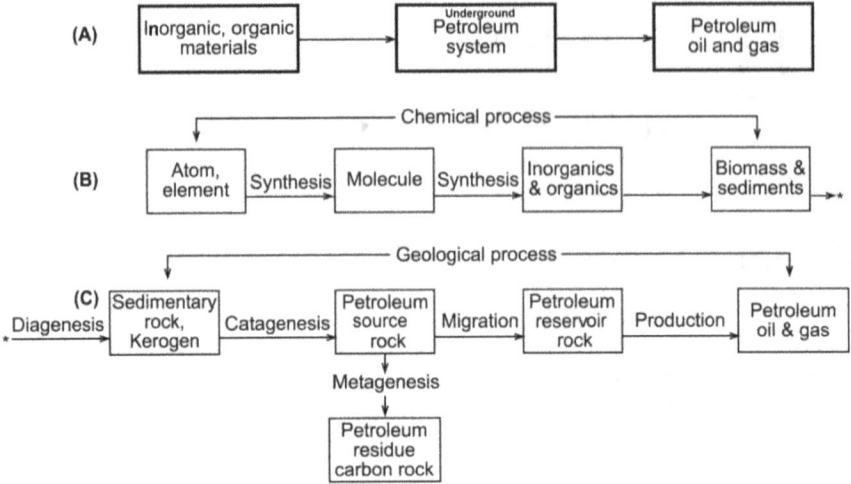

FIGURE 8.1 Path way from atom/molecule to biomass to petroleum oil/gas.

transformation. New systems are established with a change of environment. The whole rocky system is not in a static equilibrium, it is in a dynamic state. The geochemist focuses his attention on the extent of transformation of chemicals, from one form to another, under different environmental conditions. He also studies when and how the changes have occurred. Geochemistry deals with qualitative and quantitative interpretation and the total transformation mechanism of the geological events. Geochemistry includes the following subjects:

- Interactions of associated constituents (chemicals) within a rock system.
- Interaction of the earth's crust/rock with the hydrosphere (water).
- Interaction of the earth's crust rock with the atmosphere (air).
- Interaction of the earth's crust/rock with the biosphere (flora and fauna, plants and animals).
- Interpretation of geochemical data for petroleum prospecting.

Geochemistry supports other methods of prospecting to minimize the risk involved in the exploration, drilling and production of petroleum. Geochemistry is joined by other branches of sciences, particularly physics which has considerably helped in exploration, and production activities. Likewise biology has greatly contributed in understanding of the origin and transformation of organic matter into petroleum. The interaction of biology and geology has given rise to another branch of geology known as bio-geology. The combination of biology and geology has led to the emergence and development of subjects like paleontology and palynology. Bio fossils are used to understand the subsurface processes and events. Geology is the study of the earth's solid material. Some space is given here for the explanation of the term 'solid'.

8.2   STATE OF MATTER

Geology is concerned with solid matter and its interaction with different environments. It is appropriate to know and examine the solid at the microscopic level. Matter exists in three states, gas, liquid and solid. The 'state of matter' is defined only by stating the prevailing conditions of temperature and pressure. Matter changes from solid to liquid and gases and vice versa.

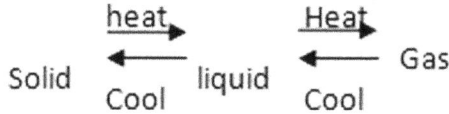

$$\text{Solid} \underset{\text{Cool}}{\overset{\text{heat}}{\rightleftharpoons}} \text{liquid} \underset{\text{Cool}}{\overset{\text{Heat}}{\rightleftharpoons}} \text{Gas}$$

For example, water below 0°C is solid (ice) and above 0°C it is liquid. Water above 100°C is gas (steam). Each form of water, that is solid, liquid and gas, possesses different characteristics and order. The three forms of matter merge into each other at critical temperature and pressure and become indistinguishable, one phase having critical volume.

8.2.1   ORDER-DISORDER

One of the major aspects associated with matter is the 'ordered' or 'disordered' state of the system. Order prevails in solids but not in liquids and gases. The cohesive forces in solids are strong enough to hold the solid particles in fixed locations. The solid has a fixed volume, definite shape and rigidity at given environmental conditions and is said to be in an 'order state'. The atoms, ions and molecules of a solid do not move; they are held together by chemical or physical bonds. The cohesive attractive forces among the liquid and gas molecules are small and negligible. The liquid and gas molecules are free to move and take the shape of the storage container. In gases and liquids, atoms, ions and molecules move randomly, rotate, vibrate and lack order. Gas and liquid are said to be in a 'disordered state'.

Complete order is only achievable at absolute 0 K (–273°C). At absolute zero the atoms, ions and molecules of a material are all at rest and in fixed positions. At any higher temperature, some kind of kinetic energy is imparted to the particles of the matter. Therefore, above absolute zero (0 K) temperature some disorder occurs in every solid. With an increase of temperature and kinetic energy, the movement of the particles increases, first within the solid structure. With a further rise in temperature, the solid particles lose their cohesive forces and turn into the more mobile liquid state. With a further rise in temperature, liquids become gases, losing total cohesive forces among the molecules.

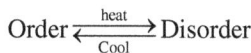

$$\text{Order} \underset{\text{Cool}}{\overset{\text{heat}}{\rightleftharpoons}} \text{Disorder}$$

Conversely the fusion (combination) of atoms, ions and molecules takes place with a decrease of temperature (cooling). This is the principle of solid crystallization from a solution or molten state.

The solids exists in two forms, namely crystalline solids and amorphous solids.

8.2.2 CRYSTALLINE SOLIDS

Crystallized substances are said to be true solids. In the crystalline solid, the particles (atoms, ions or molecules) are highly arranged in geometrical patterns, giving the crystal a specific shape, smooth surface and defined interfacial angle. A crystalline solid is a repeated three-dimensional network of atoms, ions or molecules; together they may be termed as 'particles'. The overall arrangement of particles is simply called 'crystalline form'. The following terms are used in crystallography to define a crystal.

8.2.2.1 Unit Cell

The individual repeating member of the crystal lattice network is the 'unit cell'. The unit cell is the smallest aggregate of particles repeating itself in the crystalline structure and is of a particular geometrical shape for example 'cubic unit cell'. The particles are placed on the corners (simple) or interior (body-centered) or on the sides (face-centered) of the cubic unit cells. The overall shape of the crystalline solid is the cumulative shapes of the unit cell.

8.2.2.2 Crystal Lattice

A crystal lattice is a grid or framework consisting of unit cells. For example 12 'unit cells' form a crystal lattice. The crystal lattice is a symmetrical three-dimensional arrangement of particles (atoms, ions, molecules) of the crystalline solid. The crystalline solid is highly symmetrical about a point or a plane or an axis of the structure.

8.2.2.3 Geometrical Shape

All the crystal solids are put into eight categories according to the position of particles and geometry of the unit cell: cubic, trigonal, tetragonal, orthorhombic, rhombohedral, hexagonal, monoclinic and triclinic unit cell structures.

8.2.3 AMORPHOUS SOLID

Atom, ion and molecule particles are randomly arranged in an amorphous solid. The particles lack symmetry and order. Glass, wood and plastics are amorphous solids. The particle disorder resembles the disorder of liquid and gas particles. Glass is actually not a solid. It is a super cooled liquid. A very old window glass can be observed to be thicker at the bottom than at the top, because of extremely slow downward flow of glass particles. A small quantity of glass material is found in association with igneous rock minerals.

8.3 TYPE OF BONDING & CRYSTALLINE FORM

A bond is a pair (two) of electrons joining two atoms. Different types of bonding give different classes of crystal. The following are well-known bonds and crystalline forms.

8.3.1 IONIC BOND/IONIC CRYSTAL

The ionic crystal is composed of positive and negative ions. The electro-static attractive forces between two oppositely charged ions (cation and anion) create a strong ionic bond. Sodium chloride is an example of an ionic crystal. The Na^+ and Cl^- ions surround each other in their fixed crystal lattice site (corner). The ionic crystals are hard and rigid solids with high melting points.

8.3.2   Van der Waals Bond/Molecular Crystal

Van der Waals is a physical bonding between the particles of the crystals. The particles are neutral molecules, not ions or atoms. It is also called a molecular bond. Molecular bonds are weak, and these crystals are of low melting point and unstable under stress and dynamic conditions. Van der Waals bonding is not found in earth minerals.

8.3.3   Covalent Bond/Atomic Crystal

Atoms occupy the corner of the crystal lattice. The atoms are bonded to each other by covalent bonds. A covalent bond between two atoms is formed by the atoms each sharing one electron with the other. A network of covalent bonds produces a macro-molecule crystal. Covalent bonds are strong. The crystals are hard, rigid and of high melting points. Examples of network covalent bonds are diamond and silica.

8.3.4   Metallic Bond/Metallic Crystal

The atoms in a metallic crystal are held together by a special type of bonding. A metallic bond is an electro-static force between the two atoms. The electro-static force is generated by a positively charged metal atom and a delocalized (detached) electron. A positively charged atom of the metal occupies the corner of the lattice. The metallic crystals are closely packed spherical atoms of metals. They are hard and high-melting-point solids. They are hexagonal closely packed or cubic closely packed and body-centered structures.

The different bonds described above impart and create different characteristics in the crystals. More than one type of bond may occur in a crystal. If a weaker bond, for example Van der Waals bonding, occurs in combination with more a stable ionic or covalent bond, the crystal is governed by the properties of the stronger and more stable bond in terms of high melting point and rigidity, etc.

8.4   ISOMORPHISM & POLYMORPHISM OF SOLIDS

The isomorphism and polymorphism phenomena are closely related to mineral crystals and solid state.

8.4.1   Isomorphism (Analogous)

Isomorphism (same form) is the similarity in crystalline structure between two or more minerals of different chemical composition. Minerals showing isomorphism contain different elements, but their structural formulae are similar. For example, the crystal structures of sodium nitrate and calcium sulfate as well as potassium hydrogen phosphate (KH_2PO_4) and ammonium hydrogen phosphate ($NH_4H_2PO_4$) are similar and are said to be isomorphous. Crystals of these substances are almost identical. Isomorphism is exhibited by those minerals whose anions and cations have the same size or nearly the same size and same number. Chemical nature is not important for isomorphism. The size of atoms or ions and the number of the ions are important. The size of an element (cation or anion) is expressed as the atomic radius. The atomic radius is the distance from the center of the nucleus to the boundary of the surrounding electrons.

8.4.2 POLYMORPHISM (ALLOTROPY)

Polymorphism (many forms) indicates the ability of a mineral to exist in more than one form or crystalline structure. For example calcium carbonate crystallizes in orthorhombic and trigonal structural forms. Each structural form of a mineral is known as a polymorph. When polymorphism occurs in an element, it is known as allotropy. 'Allotropy' is the existence of an element in more than one structural form, for example sulfur and carbon. Allotropic forms of sulfur are crystalline sulfur (α and β forms) and plastic sulfur. Carbon exists in different crystalline forms as diamond, graphite and charcoal. Polymorphic forms of silicon dioxide (SiO_2) crystals are quartz and tridymite.

Each crystalline structure of a mineral exhibiting polymorphism has distinct physical properties and crystal structure. Polymorphic substances are chemically the same (chemical composition and properties) but the physical properties differ. It is logical to conclude that the differences in physical properties arise from the different crystalline structure of the same atoms, ions or molecules. A mineral or any solid changes from one polymorph (structure) form to another under certain conditions of temperature, pressure and environment. The occurrence of different polymorphs in different places indicates the different conditions of the rock under which the crystal (polymorph) has formed.

8.4.2.1 Enantiotropy

Enantiotropy is defined as 'The relation of two different polymorphism forms that have a definite transition temperature and can change reversibly into each other'. A mineral can change from one polymorphic form to another at a certain temperature, provided the two forms have the same vapor pressure.

8.4.2.2 Monotropy

Monotropy defines the existence of polymorphs/allotropes of a substance or element in two forms; one form is stable and the other is metastable (theoretically unstable). The metastable form can change to a stable form at a suitable temperature. The stable form is not reversible to the metastable form. The vapor pressures of the stable and metastable forms are different. It is reported that diamond is the metastable form and graphite is the stable lattice of carbon. However the rate of change of diamond to graphite is very slow and unnoticeable.

8.5   MAGMA, CRYSTALLIZATION OF MINERALS

The major component of the magma in the earth's mantle is silicate minerals; it may also contain other minerals as minor impurities. Silicon dioxide is the basic monomeric unit of the silicate molecule along with other metallic oxides. On the basis of chemical composition three types of magma are identified. The first type, mafic magma, contains 45–50% of SiO_2. The intermediate type magma contains 55 to 65% SiO_2. The SiO_2 content of the third type, felsic magma, is 65–75%. The SiO_2 content of magma is 59.14 wt% on an average basis, followed by 15.34% Al_2O_3. Other significant oxides are calcium oxide (6.88%), iron oxides (5.08%) and magnesium oxide (3.49%) and several more.

'Crystallization' is the physical process of the separation and purification of dissolved solids from solution. The dissolved solids are separated, from a given solution, during the crystallization process according to their melting points in almost pure crystalline form. Fractional crystallization is the separation or solidification of minerals in solution or molten state according to the order of their melting points. From hot magma, as it cools, the first to crystallize is the highest-melting-point mineral, followed by progressively lower-melting-point minerals, along with the drop in temperature.

Crystallization is a natural as well as an artificial process. The process involves the slow cooling of a hot solution. The natural crystalline minerals are formed from the cooling of hot molten magma. It is a temperature-, pressure- and time-dependent process. The basic law governing crystallization is that the lower temperature and higher pressure favor solidification and crystallization of the dissolved solid minerals in solution. Conversely a higher temperature and lower pressure facilitate the dissolution of solids into hot solution and molten magma. Precisely the crystallization from hot magma is subjected to subsurface environmental conditions of temperature and pressure. Crystallization is change of state from liquid to solid state under definite environmental conditions. The molten magma solidifies on cooling.

As the magma cools, the dissolved minerals separate out as crystal solids with a definite geometrical shape. The formation of perfect crystal during cooling is impaired by many factors existing in magma, for example impurities, agitation, mobility and rapid change of temperature. The incorporation of the minor impurities in the bulk crystal turns the well-defined crystal into a complex mixture of components. The rate of magma cooling is not uniform, and how and when crystallization takes place cannot be stated clearly. The solid crystal and liquid of the mineral can coexist in the molten state. The major factors that control the crystal formation in magma are the variable temperature (300–1200°C), mobility of magma and presence of impurities.

8.6   SURVEY OF IMPORTANT MINERALS

Most of the minerals are crystalline, lifeless solid inorganic matter, with some exceptions. Their number is enormous in the nature, more than 3000, but only a few are important. Although rock is a complex heterogonous formation of various minerals, a mineral itself has an order, geometrical structure and homogeneous and chemical formulae. A mineral is composed of two chemical entities. A negative ion (anion) is an 'acidic free radical' (non-metallic) and a positive ion (cation) is a 'basic free radical' (metallic). The combination of the two ions produces a stable neutral mineral. On the basis of anion content, the minerals are divided into eight groups. The following discussion and list of the minerals along with their basic chemical units is given in Table 8.1.

8.6.1 Silicate Minerals $(SiO_4)^{x-}$

About 92% of the crust contains silicate minerals and the remaining 8% of the minerals of the crust are non-silicate inorganic minerals. Silicates are the most important and abundant minerals. They are complex compounds mainly consisting of silicon (Si) and oxygen (O) atoms together with other metallic elements. The silicate minerals undergo polymorphism or polymerization (repeated units of $(SiO_4)^{4-}$ to form macromolecules). Silicate minerals form rings, sheets or frame networks. Chemically they are represented by two different formulae, namely a 'combination formula' or a 'collective formula'. For example, the anorthite mineral is represented by:

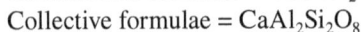

Combination formulae = $CaO \cdot Al_2O_3 \cdot 2SiO_2$
Collective formulae = $CaAl_2Si_2O_8$

The negative ion of the silicate is formed by the atoms of silicon and oxygen. The silicon atom occupies the space between four oxygen atoms. The positions of the four oxygen atoms are at the corners of a tetrahedron. The four negatively charged oxygen atoms of the tetrahedron (SiO_4) are bonded in different manners to different positively charged cation (mostly Ca, Al and Si) atoms to give various kinds of neutral silicates as given below.

8.6.1.1 Isolated Tetrahedral/Ortho Silicates $(SiO_4)^{4-}$

The isolated tetrahedral silicates contains discrete tetrahedral anions $(SiO_4)^{-4}$. Oxygen anions are linked to cations (metal) by bonding between them. Four monovalent or two di-valent metal cations react with tetrahedral anions $(SiO_4)^{-4}$ to form a neutral silicate molecule. Examples of these groups are sodium silicate Na_4SiO_4, barium silicate Ba_2SiO_4, beryllium silicate (Be_2SiO_4) and magnesium silicate (Mg_2SiO_4). The silicon to oxygen ratio is 1:4.

8.6.1.2 Double Tetrahedral/Pyro Silicates $(Si_2O_7)^{6-}$

The anionic ion $(Si_2O_7)^{6-}$ is formed by joining two tetrahedral units by one oxygen atom. An anionic ion $(Si_2O_7)^{6-}$ requires six mono-valent metal cations (M) to make a neutral silicate $(M_6Si_2O_7)$. The silicon/oxygen ratio is 1:3.5. An example is the hemimorphite silicate mineral, $Zn_4(OH)_2 Si_2O_7 \cdot H_2O$. It contains a hydroxyl group (OH) along with hydrated water.

TABLE 8.1
Common Natural Minerals

	Mineral Type	Mineral Base	Chemical Unit	Mineral Name
(a)	**Silicates**	Silica, alumina	SiO_2, Al_2O_3	Kaolite
	"	Silica, alumina	"	Shale
	"	Silica, alumina		
	"	Clay		
	"	Silica, alumina, potash	SiO_2, Al_2O_3, K_2O	Potash, feldspar
	"	Silica, alumina	"	Potash, mica
	"	Silica, alumina, soda	SiO_2, Al_2O_2, Na_2O	Sodalite
	"	Silica, alumina, calcium oxide	SiO_2, Al_2O_3, CaO	Anorthite
	"	Silica, alumina, iron oxide	SiO_2, Al_2O_3, e_2O_3	Hematite
(b)	**Oxides**	Silicon dioxide (silica)	SiO_2	Sand
	"	Silicon dioxide (silica)	"	Quartz
	"	Silicon dioxide (silica)	"	Sandstone
	"	Silicon dioxide (silica)	"	Flint
	"	Silica, marine lime	"	Kieselguhr
	"	Manganese dioxide	MnO_2	Pyrolusite
	"	Titanium dioxide	TiO_2	Rutite
	"	Tin dioxide	SnO_2	Cassiterite
	"	Cuprous oxide	Cu_2O	Cuprite
	"	Potassium oxide	K_2O	Potash
	"	Hydrogen oxide	H_2O	Water
	"	Magnesium oxide	MgO	Periclase
	"	Zinc oxide	ZnO	Zincite
	"	Manganese oxide	MnO	Psilomelane
	"	Lead oxide	PbO	Litharge
	"	Iron oxide	Fe_2O_3	Hematite
	"	Aluminum oxide	Al_2O_3	Corundum
	Complex oxides	Magnesium alumna	$Mg\,Al_2O_4$	Spinet
	"	Ferrochromium oxide	$FeCr_2O_4$	Chromite
	"	Beryllium alumina	$BeAl_2O_4$	Chrysoberyl
	"	Zinc alumina	$ZnAl_2O_4$	Gahnite
	"	Iron oxide	Fe_3O_4	Magnetite
	Hydroxide	Magnesium hydroxide	$Mg(OH)_2$	Brucite
	"	Aluminum hydroxide	$Al(OH)_3$	Diaspore
	"	Manganese hydroxide	MnO(OH)	Maganite
	"	Iron hydroxide	FeO(OH)	Geothite
(c)	**Carbonate**	Calcium carbonate	$CaCO_3$	Calcite
	"	Iron carbonate	$FeCO_2$	Siderite
	"	Strontium carbonate	$SrCO_3$	Strontianite
	"	Magnesium carbonate	$MgCO_3$	Magnesite

(Continued)

TABLE 8.1 (CONTINUED)
Common Natural Minerals

	Complex carbonate	Calcium magnesium carbonate	$CaMg(CO_3)_2$	Dolomite
	"	Calcium iron carbonate	$CaFe(CO_3)_2$	Ankerite
	"	Calcium zinc carbonate	$CaZn(CO_3)_2$	Minrecordite
	"	Barium calcium carbonate	$BaCaMg(CO_3)_2$	Brytocite
	Hydroxyl carbonate	Cupric hydroxyl carbonate	$Cu_3(OH)_2(CO_3)_2$	Azurite
	"	Cupreous hydroxyl carbonate	$Cu_2(OH)_2(CO_3)$	Malachite
	"	Lead hydroxyl carbonate	$Pb_3(OH)_2(CO_3)_2$	Hydrocerussite
	"	Zinc hydroxyl carbonate	$Zn_5(OH)_6(CO_3)_2$	Hydrozincite
	Hydrated carbonate	Hydrated calcium carbonate	$CaCO_36H_2O$	Ikaite
	"	Hydrated magnesium carbonate	$MgCO_35H_2O$	Lonsfordite
	"	Hydrated magnesium hydroxyl carbonate	$Mg_5(CO_2)_3(OH)_24H_2O$	Hydromagnesite
	"	Hydrated calcium carbonate	$CaCO_3H_2O$	Hydrocalcite
	"	Hydrated sodium carbonate	$Na_2CO_310H_2O$	Natron
(d)	**Sulfate (anhydrous)**	Calcium sulfate	$CaSO_4$	Anhydrite
	"	Barium sulfate	$BaSO_4$	Barite
	"	Strontium sulfate	$SrSO_4$	Celestite
	"	Lead sulfate	$PbSO_4$	Anglesite
	Hydrated sulfate	Hydrated calcium sulfate	$CaSO_42H_2O$	Gypsum
	"	Hydrated magnesium sulfate	$MgSO_4H_2O$	Kieserite
	"	Hydrated magnesium sulfate	$MgSO_45H_2O$	Starkeyite
	"	Hydrated magnesium sulfate	$MgSO_47H_2O$	Epsomite
	Sulfate hydroxide	Hydroxyl copper sulfate	$Cu_3(SO_4)(OH)_4$	Antlerite
	"	Hydroxy potash alumino sulfate	$KAl_3(SO_4)_2(OH)_6$	Alunite
	"	Hydroxy copper sulfate	$Cu_4(SO_4)(OH)_6$	Brochantite
	"	Hydroxy potash ferric sulfate	$KFe_3(SO_4)(OH)_6$	Jarosite
(e)	**Sulfide**	Iron sulfide	FeS	Pyrite
	"	Cupreous sulfide	Cu_2S	Chalcocite
	"	Cupric sulfide	CuS	Covellite
	"	Lead sulfide	PbS	Galina
	"	Silver sulfide	Ag_2S	Argentite
	Complex sulfide	Cupric iron sulfide	Cu_5FeS_2	Bornite

(Continued)

TABLE 8.1 (CONTINUED)
Common Natural Minerals

	"	Cuprous iron sulfide	$CuFeS_2$	Chalcopyrite
	"	Copper cobalt sulfide	$CuCO_2S_4$	Carrollite
	"	Argento arsenic sulfide	Ag_2AsS_3	Propucite
(f)	**Phosphate**	Ammonium magnesium phosphate	$(NH_4)MgPO_4\,6H_2O$	Strurite
	"	Lithium ferro phosphate	$LiFePO_4$	Triphylite
	"	Hydrated zinc phosphate	$Zn(PO_4)_2 4H_2O$	Phosphophyllite
	"	Hydroxyl calcium phosphate	$Ca_5(PO_4)_3(OH)$	Apatite
	"	Calcium fluoro phosphate	$Ca(PO_4)_3F$	Fluorapatite
	"	Calcium chloro phosphate	$Ca(PO_4)_3Cl$	Chlorapatite
	"	Calcium fluoro phosphate	$Ca_3(PO_4)_2CaFe_2$	Phosphate rock
	"	Cerium phosphate	$Ce(PO_4)$	Monazite
(g)	**Halite**	Sodium chloride	$NaCl$	Halite
	"	Potassium chloride	KCl	Sylvite
	"	Silver chloride	$AgCl$	Chlorargyrite
	"	Calcium fluoride	$CaF2$	Fluorite
	"	Silver bromide	$AgBr$	Bromargyrite
	Complex halite	Calcium hydroxy chloride	$Ca_2Cl(OH)_3$	Atacamite
	"	Sodium alumino fluoride	Na_3AlF_6	Cryolite
	"	Ammonium silicon fluoride	$(NH_4)_2SiF_6$	Cryptohalite
	Hydrated halite	Hydrated magnesium chloride	$MgCl_2 6H_2O$	Bischofite
	"	Hydrated potassium magnesium	$KMgCl_3 6H_2O$	Carnallite
(h)	**Evaporite (marine)**	Calcium sulfate	$CaSO_4$	Anhydrite
	"	Hydrated calcium sulfide	$CaSO_4 2H_2O$	Gypsum
	"	Potassium calcium magnesium sulfate	$KCaMg(SO_4)_6H_2O$	Polyhalite
	"	Hydrated magnesium sulfate	$MgSO_4 H_2O$	Kieserite
	"	Calcium magnesium carbonate	$CaMg(CO_3)_2$	Dolomite
	"	Calcium carbonate	$CaCO_3$	Calcite
	"	Magnesium carbonate	$MgCO_3$	Magnesite
	"	Sodium chloride	$NaCl$	Halite
	"	Potassium chloride	KCl	Sylvite
	"	Potassium magnesium chloride	$KMgCl_3 6H_2O$	Carnalite
	"	Potassium magnesium sulfate chloride	$KMg(SO_4)Cl3\,H_2O$	Kainite

(*Continued*)

TABLE 8.1 (CONTINUED)
Common Natural Minerals

Evaporite (fresh water)	Hydrated magnesium sulfate	$MgSO_4 7H_2O$	Epson salt
"	Sodium borate	$Na_2B_4O_7 10H_2O$	Borax
"	Sodium bicarbonate carbonate	$NaHCO_3 NaCO_3 2H_2O$	Trona
Evaporite (biological)	Calcium carbonate	$CaCO_3$	Limestone

8.6.1.3 Cyclo-/Ring Silicates $(Si_3O_9)^{6-}$

The cyclo-silicates are obtained when the two oxygen atoms of each SiO_4 anion unit share two corners of a tetrahedron, resulting in the formation of a closed ring anionic ion $(Si_3O_9)^{6-}$. The silicon/oxygen atomic ratio 1:3. An example is benitoite silicate, $BaTi (Si_3O_9)$.

8.6.1.4 Single-Chain Silicates $(SiO_3)^{2n-}$

In a single-chain silicate, two oxygen atoms are shared by each tetrahedron, resulting in the formation of a simple chain with anionic formula as $(SiO_3)^{2n-}$. Pyroxenes silicates are an example.

8.6.1.5 Double-Chain Silicates $(Si_4O_{11})^{2n-}$

Double-chain silicates having anion formula $(Si_4O_{11})^{2n-}$ are produced in which two simple chains are linked together by a shared oxygen. The silicon/oxygen atomic ratio is 1:2.8. An example is the amphibole mineral, $Ca_2Mg_5Si_4O_{11}(OH)_2$.

8.6.1.6 Sheet Silicates $(Si_2O_5)^{2n-}$

In the sheet structure, silicates are formed by the sharing of three oxygen atoms by each tetrahedron. The sheet is bounded by cations. An example of a sheet structure silicate is talc mineral $[(Mg_3(OH)_2(Si_4O_{10})]$.

8.6.1.7 Three-Dimensional Network Silicates $(SiO_2)_n$

The three-dimensional network silicates do not contain any metallic cations; they only have silicon and oxygen atoms. The group may be classified as silicate as well as oxide minerals. Three-dimensional frameworks of silicates are produced when all the four oxygen atoms of each tetrahedron are shared. The silicon/oxygen ratio is 1:2. Silica (SiO_2) is an example of a three-dimensional network silicate. With different structural forms, silica produces minerals like quartz and tridymite.

8.6.2 Oxide Minerals $(O)^{2-}$

An oxide mineral consists of a closely packed structure of oxygen atoms in which positively charged cations are located in the interstices (narrow space). Silicon dioxide (SiO_2) is the most important and abundant oxide existing in nature. The purest form of silicon dioxide is quartz. The next most important oxide is aluminum oxide (Al_2O_3), known as alumina. Both silicon and aluminum oxides are found in earth in different polymorphic forms. Sand is the crushed form of quartz. Sandstone consists of sand particles bonded together. Kieselguhr rock is made of silicon dioxide which is the remains of siliceous marine organisms. The bond between silicon and oxygen is covalent. Silicon atoms are bonded tetrahedrally to four oxygen atoms. Silicon dioxide is very hard, having a high melting point. It is the hardest substance having a crystalline structure used as a semiconductor. Aluminum oxides occur naturally in several crystalline polymorphic forms. The precious gemstones ruby and sapphire are polymorphic forms of alumina. Hematite and magnetite are well-known iron oxide minerals. Magnesium oxide (magnesia) is a white hygroscopic mineral. Pyrolusite` and ramsdellite are polymorphic minerals of manganese dioxide. In addition to these, several hundred natural oxides of various types are known. Oxides are formed either by covalent linkage or by ionic bond between oxygen free radicals $(O)^{-2}$ and one or more metallic free radicals (cations). The carborundum mineral is silicon carbide (SiC) and occurs in nature as traces.

8.6.3 Carbonate Minerals $(CO_3)^{2-}$

Numerous carbonate mineral deposits are known to exist at the earth's crust, surface and shallow depth. These are comparatively softer substances that react instantaneously with acid, producing carbon dioxide gas. Important minerals are magnesite ($MgCO_3$), calcite ($CaCO_3$) and dolomite ($CaCO_3 \cdot MgCO_3$). Carbonate minerals have chemical, biological and geological origins.

8.6.4 Sulfate Minerals $(SO_4)^{2-}$

Sulfate ions $(SO_4)^{2-}$ are represented by a structure in which four oxygen atoms occupy four corners of a tetrahedron with one sulfur atom at the center. This is similar to silicate ion $(SiO_4)^{4-}$. But the sulfate minerals do not undergo polymorphism or polymerization (repeated units of $(SO_4)^{-2}$ to form macromolecules, as silicate minerals do. The sulfate ion remains as a single unit. It is unable to form rings, sheets or frame networks as SiO_4^{4-} ions do. The sulfate mineral is formed by any one of the following phenomena.

- Oxidation of sulfide (S^{2-}) minerals and subsequent deposition as sulfate, during underground geo-thermodynamic circulation of formation water containing sulfide ions. Oxidation of minerals containing sulfide radicals, for example iron sulfide, to sulfate minerals in subsurface conditions.
- Water-soluble alkali or alkaline earth sulfates crystallize upon complete vaporization of sea water, forming solid sulfate deposits. The sulfate mineral is formed as evaporite deposits as hydrothermal veins in rock. A vein is a distinct sheet of crystallized mineral within rock. A vein is formed when a dissolved mineral is precipitated.

- With limited supply of water, a saturated solution of sulfate is formed, which is known as bittern or brine solution. Subsurface flowing water with soluble sulfate reacts with calcium ions in mud, clay and limestone to form metallic sulfate minerals such as alabaster, plaster of paris and gypsum ($CaSO_4 2H_2O$) minerals.
- 'Anhydrite mineral' is a sulfate mineral chemically known as anhydrous calcium sulfate ($CaSO_4$). The anhydride mineral deposit occurs in sedimentary strata from which sea water has been evaporated.

8.6.5   SULFIDE MINERALS $(S)^{N-}$

Sulfide mineral consists of sulfide radical $(S)^{n-}$ along with a metallic cation radical $(M)^{m+}$ to make a neutral stable mineral. The metal ion $(M)^{m+}$ is usually derived from a transition metal. Transition metals have variable valences and are more reactive. The chemical formula of the mineral is M_nS_m where M is a metal ion; m and n integers denote the variable valency of the sulfide and metallic ions. Sulfides of iron (pyrite), lead (galena), mercury (cinnabar), molybdenum (molybdenite) and copper (chalcopyrite) are a few examples of useful sulfide minerals. Sulfide mineral containing nickel, cobalt, zinc and silver are commercial minerals and are classified as ores. Ores are used to extract their respective metals. The sulfide minerals are mostly of igneous origin. They are black and exist in different crystalline structures. They exhibit high density and can be hard or soft materials.

8.6.6   PHOSPHATE MINERALS $(PO_4)^{3-}$

Phosphate mineral is a salt of phosphoric acid $\{H_3(PO_4)\}$. Phosphate ion $(PO_4)^{3-}$ is an isolated tetrahedral coordinated radical. The radical combines with h equivalent positive cations, so that a neutral and stable phosphate mineral is produced. Variable physical properties are witnessed in the phosphate minerals. The mineral is vitreous, dull, with moderate density and average hardness. The phosphorus element in the ion is replaceable by arsenic, vanadium and antimony atoms to produce arsenate $(AsO_4)^{3-}$, vanadate $(VO_4)^{3-}$ and antimonite $(SbO_4)^{3-}$ minerals. About 200 phosphate minerals are known, often associated with other minerals; halite (Cl, Br and F)$^{-1}$ and hydroxide (OH)$^-$ minerals. Some examples of phosphate minerals are as follows:

- Amblygonite (lithium, sodium, aluminum fluoride, hydroxide, phosphate).
- Anapaite (hydrated iron, calcium phosphate).
- Apatite (calcium chloride, fluoride, hydroxide, phosphate). Chloro-apatite (calcium chloro-phosphate) occurs in soil. It is a source of phosphorus food, consumed by plants. The fluoro-apatite (calcium fluoro-phosphate) mineral is found in animal bone. So the mineral is called 'biological apatite'.

8.6.7   HALITE MINERALS $(F, CL, BR, I)^-$

Halogen-containing minerals are known as halite. Halogens are a group of five chemically similar elements, recorded in the seventh group of the periodic table.

The periodic table is the arrangement of elements according to their atomic structure and characteristics. Naturally occurring minerals containing fluorine (F^-), chlorine (Cl^-), bromine (Br^-) and iodine (I^-) negative ions are termed as 'halite salts'. The fifth element of the halogen group is astatine, which is not found in any minerals. About 100 halite minerals have been identified, but only a few of them are significant.

The most common halite is known as common salt ($NaCl$). It is found as a solid in the subsurface and as dissolved material in water (lakes and oceans). A subsurface deposit of the salt is exploited by drilling and water injection to make an underground brine solution. Later the brine solution is pumped out and stored for crystallization. An important aspect of subsurface halite deposits is that the geological forces push them upward to form an arched structure known as a 'salt dome'. Salt domes are significant both for mining and underground petroleum systems. A major portion of reservoir formation water contains halite minerals. The composition of reservoir water depends on reservoir environment.

Lake/sea water is subjected to various geographical and weather conditions. Loss of water due to evaporation during hot weather and also a limited supply of fresh water during dry seasons disturb the solution equilibrium. Therefore dissolved salt begins to precipitate out as solid residue.

Fluorite (calcium fluoride) is a commercial mineral. Since it has commercial value it is also called 'ore'. The ore is also known as fluorspar. It mostly occurs in igneous granite rock. The ore is used to make ornaments, ceramic, glass and optical lenses. Calcium fluoride plays a vital role in the chemical industry for the production of hydrofluoric acid (HF), which is used to make many chemical products.

8.6.8 Evaporite Minerals

Evaporite rock does not contain silicon and aluminum minerals. Evaporite rock includes all the minerals discussed above, namely carbonate, sulfate, halite and anhydrite minerals. They are all solid materials. The term 'evaporite' is used for those minerals that have been separated either by precipitating out or evaporation from their solution in aqueous media. The aqueous media can be marine or land lake water. Evaporite mineral may be defined as an 'originally water-soluble mineral which is obtained as a result of precipitation, crystallization and evaporation from aqueous solution'.

The factors affecting precipitation and evaporation are temperature, pressure, humidity, weather, environment and water alteration. A limited water supply, shallow basin and lake, high temperature, arid conditions and quiet environment favor rapid water evaporation and crystallization of the dissolved solid. The separated solid minerals settle down in the bottom forming sedimentary rock. There are 70–80 evaporite minerals that have been identified, with some of economic importance, e.g. gypsum, common salt, potash salt from marine water and gypsum salt and borax from land lakes. The evaporite minerals can be divided according to their origin into the following types:

- Evaporite from marine water.
- Evaporite from terrestrial (land) lake water.

- Evaporite from biological processes. Limestone may be of biological origin in addition to inorganic origin. Limestone is thought to be formed from the remains of shelled sea fauna settled down at the bottom. Over geological time the deposits turn into limestone.

Evaporite salts are of different solubility behavior in water media, which affects precipitation and separation. More solubility leads to less precipitation, and less solubility favors quick separation. On this basis, among the evaporites, the first to separate is carbonate because it is the least soluble. The last to separate are magnesium and potassium chloride because they are most soluble in water. Carbonate may start separation as quickly as at 10% evaporation, and chlorides may not separate even at 90% water evaporation. Complete evaporation leads to the formation of evaporated solid sedimentary rock, which continues to change and alter due to external factors, wind, water and underground stresses. Anhydrite and gypsum can take or lose water of hydration. Subsurface halite minerals can flow and appear at the top of sedimentary rock. The moving halite is known as diapirs. Anhydrite, gypsum and halite form cap rock. Halite deposits can even come out at the surface and form commercial deposits for salt mining.

8.7   GEO-PHYSICOCHEMICAL FACTORS INFLUENCING MINERALIZATION

Mineralization is broad term; it is used in geology, metallurgy and soil science. The mineralization process is the deposition or formation or introduction of a particular mineral or ore or nutrient or fossilization of organic matter into sedimentary rock or soil. It should not be confused with sedimentation. Sedimentation is a complex geological process involving the weathering, erosion, deposition, transportation and settling of mineral. The sedimentation process begins with tiny mineral particles and leads to the formation of a solid formidable framework of sedimentary rock. Sedimentation is slow process; it can only be expressed in geological time scale. Mineralization is related to the following points:

- The formation of minerals and ore in soil and subsurface conditions.
- The subsequent deposition of minerals and ores.
- The process of the fossilization of organic matter in the inorganic matrix.
- The conversion and oxidation of complex organic matter to simple inorganic compounds and soil micronutrients.

The relevant geo-physiochemical conditions for the formation of a mineral and subsequent deposition are provided by the lithosphere, hydrosphere, atmosphere and biosphere. Here mineralization is taken up by discussing the role of geo-physicochemical factors, namely ionic potential, hydrogen ion concentration (pH), redox potential and the colloidal system in the process.

8.7.1   Ionic Potential

Ionic potential is one of the main factors that provides the necessary conditions for the chemical conversion of one mineral to another in aqueous media. The ionic potential of an element is defined as the ion charge (z) divided by ion size (r) in terms of its radius that is (z/r). The ratio (z/r) is a measure of charge density at the surface of the ion. The denser the charge, the stronger the bond formed by the ion with an oppositely charged element. The ionic potential (z/r) or ionic charge density gives the ion the strength that will determine how it will be electro-statically attracted to an opposite charge or how it will be electro-statically repelled by a like charge. Accordingly all the elements are divided into three groups, low, intermediate and high ionic potential. Ionic potentials along with their groups are tabulated in Table 8.2.

TABLE 8.2
Ionic Potential of Elements

Group	Element	Ionic Potential
Low ionic potential group 1	Cs^+	0.60
	Rb^+	0.68
	K^+	0.75
	Na^{2+}	1.0
	Li^+	1.5
	Ba^{2+}	1.5
	Sr^{2+}	1.8
	Cu^{2+}	2.0
Intermediate ionic potential group 2	Mn^{2+}	2.5
	La^{2+}	2.6
	Fe^{2+}	2.7
	Co^{2+}	2.8
	Mg^{2+}	3.0
	Y^{2+}	3.3
	Lu^{2+}	3.5
	Sc^{2+}	3.7
	Th^{4+}	3.9
	Ce^{4+}	4.3
	Fe^{3+}	4.7
	Zr^{4+}	5.1
	Bc^{2+}	5.7
	Al^{3+}	5.9
	Ti^{4+}	5.9
	Mn^{4+}	6.7
	Nb^{5+}	7.5
High potential group 3	Si^{4+}	9.5
	Mo^{6+}	9.7
	B^{2+}	13
	P^{5+}	14
	S^{6+}	20
	C^{4+}	25
	N^{6+}	38

Modified: Brain H. Mason, *Principles of Geochemistry*, John Wiley and Sons, New York, 1966. (Table 6.2, page 162).

Ionic potential determines the stability of the mineral or the formation of a new mineral. Great variation in ionic potential is observed from 0.6 $(Cs)^+$ to 38 $(N)^{6+}$. On the basis of ionic potential, elements are divided into three groups:

- Group 1. Low ionic potential and low z/r ratio. Forms weak cation–anion bonds. Thus less stable salts and minerals are formed. Low-ionic-potential ions produce water-soluble neutral minerals when reacted with anions. They remain soluble even at low temperatures. For example, sodium, potassium and calcium minerals are readily soluble and remain in solution during the sedimentation and weathering process and hardly take part in deposition and sedimentation.
- Group 2. Intermediate ionic potential and z/r ratio. Strong cation and anion bonds are formed; they produce stable solid minerals which are precipitated by hydrolysis in aqueous media. For example, iron, aluminum and cobalt are separated as solid hydroxide in aqueous media. Cations like (Al^{+3}) have sufficient charge density to make a strong bond with anions like (O^{-2}) or $(OH)^-$ resulting in the formation of insoluble minerals Al_2O_3 or $Al(OH)_3$.
- Group 3. Very high ionic potential (z/r) ratio. Examples are cations of non-metallic elements such as sulfur, boron, nitrogen and phosphorus. Cations of high ionic potential make complex soluble compounds by reacting with corresponding anions; on decomposition they form insoluble minerals.

8.7.2 HYDROGEN BONDING IN WATER

Water plays an important role in the mineralization process. The properties and reactivity of a mineral are affected by aqueous medium. In addition to ionic bonding, the reactivity is dependent on the hydrogen bonding of water molecules. Water behaves like a dipole, having two poles $(H^+ \ \& \ OH^-)$; this creates hydrogen bonding. The molecules of water are strongly bonded by cohesive forces due to dipoles (hydrogen bonding). The strong cohesive forces among water molecules make it behave differently compared to the compounds of comparable structure like hydrogen sulfide, phosphorus oxide and ammonia.

The strong cohesive forces in water are also reflected by its high surface tension (72.75 dyne/ cm at 20°C) compared to the lower surface tension (26.77 dyne/cm) for carbon tetrachloride and benzene (28.5 dyne/cm) at 20°C. The dissolution of many minerals in water is due to hydrogen bonding because the cohesive force of attraction among a mineral's atoms is greatly reduced in water medium.

Water's homologous (similar structure) substances, for example hydrogen sulfide and ammonia, are gases. Water boils at 100°C and freezes at 0°C. The high boiling point of water is due to hydrogen bonding. Water has a high dielectric constant (80) and high surface tension (72.75 dyne/cm). It has a high heat of vaporization of 9730 cal/mole. The unique properties of water are attributed to the strong hydrogen bonding in water molecules. The hydrogen of one molecule is bonded to the oxygen of another water molecule.

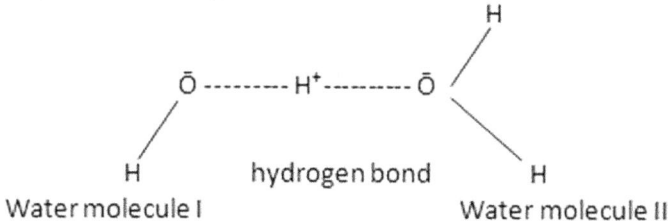

Water molecule I   hydrogen bond   Water molecule II

The bondage (linkage) of water molecules through hydrogen bonding creates unique water properties. Water molecules are linked together through hydrogen bonding; similar to monomeric molecules combined together in macromolecules. Other homologues of water do not form hydrogen bonding; molecules exist as discrete individual units. The hydrogen bonding in water gives extra cohesive forces among the molecules of water. Water dissolves much ionic substance because of hydrogen bonding and high dielectric strength. Hydrogen bonding in water is a weak physical interaction; under some environmental stress, the hydrogen bonding is broken and a new equilibrium is established. Water undergoes hydrogen bonding with other molecules containing charged functional groups, creating new types of compounds.

8.7.3   Hydrogen Ion Concentration in Water

Water is ionized into cations [H^+] and anions [OH^-]. The concentrations of both the cations and anions are equal, which is 10^{-7} mole/liter at 20°C in pure water. Thus water is neutral. Practically, hydrogen ion concentration [H]$^+$ is expressed as logarithmic scale of [H]$^+$ ions as follows:

$$pH = \log_{10} [H]^+ = \log_{10} (10^{-7})$$
$$pH = 7$$

The pH value of neutral water is 7. Actually water is either acidic or basic depending on the pH value or hydrogen concentration expressed as mole per liter. The pH value is the inverse of the hydrogen ion concentration.

pH = 1 → {H^+] = 10^{-1} (mole/l) → strongly acidic
↓
pH = 7 → {H^+] = 10^{-7} (mole/l) → neutral
↓
pH = 14 → {H^+] = 10^{-14} (mole/l) → strongly alkaline

Aqueous media having pH values of 1–6 are acidic and 8–14 are alkaline. Every chemical reaction taking place in an aqueous medium is dependent on the pH value of water. A specific pH value favors a particular chemical reaction; that chemical reaction is not possible at another pH value. For example, the precipitation of iron as $Fe(OH)_3$ from its solution is only possible above pH 8.5 (alkaline medium). Hydroxide is not precipitated below this pH value. The solubility of iron hydroxide in aqueous medium increases rapidly with the decrease of pH value (more acidic). The solubility of iron hydroxide in acidic media is about 10^5 times greater than at pH 8.5 (alkaline media). River and stream water have higher concentrations of dissolved iron, because rivers are acidic. The acidic character of rivers is due to the dissolved carbon dioxide and oxides of sulfur and nitrogen. As soon as the acidic water of a river reacts with the alkaline ocean water, iron begins to precipitate out as solid hydroxide. Therefore the iron content of ocean water is much less than that of river water. Clay mineral is a combination of Al_2O_3 and SiO_2; their respective proportions depend on the pH value. In acidic media (pH less than 4) the mineral will be rich in silica and between 5 and 9 pH value the clay will be rich in alumina.

8.7.4 OXIDATION–REDUCTION REACTION (REDOX POTENTIAL)

Oxidation–reduction reactions (redox potential) bring about certain chemical transformations. The redox potential helps in the formation of minerals along with influencing the mineral solubility, insolubility, separation and precipitation, under a given set of conditions prevailing in an aqueous environment. An atom, or ion or element or molecule of a mineral has a natural tendency (stored energy) to undergo oxidation–reduction reactions in aquatic conditions. Suppose a bar of zinc metal is dipped into water, containing copper sulfate solution. There will be a mutual exchange of electrons between the zinc metal and copper sulfate solution. The zinc bar will dissolve in solution, releasing electrons:

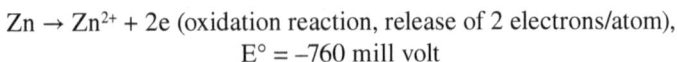

$$Zn \rightarrow Zn^{2+} + 2e \text{ (oxidation reaction, release of 2 electrons/atom)},$$
$$E° = -760 \text{ mill volt}$$

Copper ions in solution will capture the electrons:

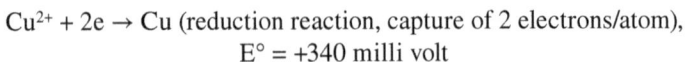

$$Cu^{2+} + 2e \rightarrow Cu \text{ (reduction reaction, capture of 2 electrons/atom)},$$
$$E° = +340 \text{ milli volt}$$

The net reaction is:

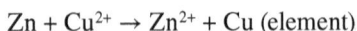

$$Zn + Cu^{2+} \rightarrow Zn^{2+} + Cu \text{ (element)}$$

The net result of the combined oxidation–reduction reaction is the liberation of Cu from the copper sulfate solution. Oxidation is the loss of an electron or increasing the oxidation state (number) of the Zn, from 0 to 2+. The oxidation number of a neutral element is zero. A reduction reaction is the gain of an electron or decreasing the oxidation state, for copper from 2^+ to 0. A chemical that has gained an electron is called an oxidizing agent, while the atom or molecule that loses an electron is known as the reducing agent. The tendency to lose or gain electrons is called the 'oxidation–reduction reaction' or 'oxidation-reduction potential' or 'redox potential'. The redox potential of any reaction is a comparative term. It has been standardized with respect to the reference hydrogen molecules in water, under standard conditions of temperature (25°C), partial pressure of hydrogen (one atmosphere) and one molarity (1 M) hydrogen ion concentration in aqueous media. Under the standard conditions the following oxidation–reduction reaction (redox potential) takes place:

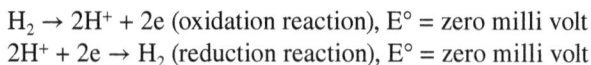

$H_2 \rightarrow 2H^+ + 2e$ (oxidation reaction), $E° =$ zero milli volt
$2H^+ + 2e \rightarrow H_2$ (reduction reaction), $E° =$ zero milli volt

Hydrogen molecule (gas) loses electrons (oxidation) and the hydrogen ions in water accept electrons (reduction). The redox potential observed under the specified conditions has been arbitrarily fixed as zero (0). The unit of potential is milli-volt (mV), and under standard conditions potential is denoted by the symbol $E°$.

The theoretical value corresponds to the discharge/liberation/deposition/evolution of ions and electrons. However, additional potential or 'over potential' is needed for the discharge of electrons and ions. The redox reaction depends on the nature of the chemical species but also on the characteristics of the aqueous medium. An aqueous mineral solution having high redox potential has a tendency to oxidize a lower redox potential mineral. It means the higher the material's positive potential, the greater the affinity it shows for electrons, and it acquires a higher capacity to oxidize by capturing electrons from another chemical. Likewise a lower redox potential species has a lower affinity for electrons. It has a higher tendency to release (donate) electrons and is a reducing agent.

Oxidation-reduction potential indicates the concentration of electrons. On the other hand, the pH indicates the concentration of hydrogen ions in water. The redox potential is related to the hydrogen ion concentration. A decrease of 1 pH brings about a 58-mV increase in redox potential. An increase of pH to the alkaline range (high pH) decreases potential. The activity of electrons, the standard redox potential ($E°$), is therefore related to the activity of the hydrogen ion concentration or the pH value. The stability of the minerals is related to the hydrogen ion concentration and redox potential (pH–$E°$) relationship. The relative values of pH and $E°$ determine whether a mineral will be soluble in a given aqueous condition. Both pH and $E°$ values facilitate the ionization of atoms and molecules. Ionic forms are reactive species that determine the extent of redox reaction and conversion to new species.

Once again consider the oxidizing and reducing strength of copper and zinc elements and their ions in the light of redox potential. The ionic forms Cu^{++} and Zn^{++} can act as oxidizing agents (accept electrons). The cation Cu^{++} is a more oxidizing agent than the cation Z^{++}. The $E°$ value for Cu^{++} (+340 $E°$ milli-volt) is much higher than for Zn^{++} (–760 $E°$ milli-volt). Zn element is the more reducing agent (electron donor). Similarly any chemical will oxidize other compounds in a downward trend of decreasing redox potential. The standard potential of Fe^{++} ions in an acidic solution (low pH) is +770 $E°$ milli-volts. But when the pH is raised to the alkaline range (high pH), the potential drops to –560 $E°$ milli-volts. In the alkaline medium the Fe^{++} is oxidized to Fe^{+++} to form $Fe(OH)_3$, which is precipitated out as solid mineral.

8.7.5 COLLOIDAL SOLUTION AND SEDIMENTATION

A colloidal solution or colloidal dispersion is the intermediate stage between a true solution and a suspension of a substance or mineral in water. The subject has already been discussed with reference to oil/gas generation in Chapter 2. The behavior and role of colloidal minerals in aqueous solution are taken in connection with the mineralization process. Colloidal particles of the mineral in aqueous medium usually carry a charge. A colloidal solution of minerals is unstable. It is stable under given conditions. The dispersed colloidal particles may dissociate to smaller particles to form a true solution or further coalesce/aggregate to form a bigger suspended particle big enough to settle down as insoluble solid matter. The stability of a colloid depends on a number of factors, for example, the charge on the colloid particle, the presence of electrolyte and fluctuation in physico-chemical conditions in the medium. Some colloids are specifically important in the sedimentation process. They are retained for a longer period of time and are transported longer distances to be discharged later from the colloidal solution.

The role of colloidal particles in the mineralization process is elaborated by citing the example of silica salute. At low concentrations, silica forms a true solution. In true solution silica behaves as an ionic compound. It is precipitated according to the ionic potential of the aquatic medium. With an increase of concentration followed by the aggregation of smaller silica particles into bigger particles, a colloidal system of silica is formed. Apart from the concentration effect, another factor that helps to aggregate the molecules is the adsorbent capacity of silica. Silica adsorbs other chemical species in water through physical and chemical bonding. Physical bonds or the Van der Waals bonds are weak bonds. The chemical bonds are strong, formed by the transfer or sharing of electrons, creating a valence bond or ionic bonds of the absorbate (chemical species) and adsorbent (silica). Thus many minerals are removed through adsorbing by silica from the solution and re-precipitated at other location. Silica is not the only absorbent mineral; other chemical species show adsorbent tendencies. Temperature and concentration both affect the adsorbent capacity. Low temperatures and high concentrations of the absorbent facilitate quick and more adsorption; high temperatures and lower concentrations of the absorbent favor desorption.

8.8 INTERACTION OF THE EARTH'S CRUST WITH SURFACE CONDITIONS

The earth's crust is like a thick wrapping paper on the surface of the earth. The crust is defined as the distinct outermost solid lithosphere of the earth's globe. The earth's crust is important to mankind. The crust is directly linked to the atmosphere, hydrosphere and to all living species (biosphere) of the earth. The lithosphere consists of sedimentary, metamorphic and igneous rocks. Sedimentary rock was formed about 600 million years ago. Before that igneous rock covered the earth's surface and was exposed to the hydrosphere (water) and atmosphere (air). Igneous rock was formed under conditions of high temperature

(700–1200°C) and pressure (4000–6000 psi). The constituents of the igneous rock were formed under severe subsurface conditions of temperature and pressure. The constituents cannot remain stable and in their original form at the relatively much lower temperature and pressure conditions of the surface of the earth. At the earth's surface in the presence of ample water and oxygen, the mineral constituents of igneous rock undergo alteration. The altered minerals further undergo transformation by reacting with themselves and with water/oxygen to generate new chemical species. The new chemicals are stable under the prevailing conditions and are subjected to change with alteration in environmental conditions. The newly formed chemicals initiate the sedimentation process and are the precursor of sedimentary rock at the surface of the earth. The earth's crust surface and environmental conditions consist of the following parts:

- Lithosphere (rock)
- Hydrosphere (water)
- Atmosphere (air)
- Biosphere (life)

The formation and alteration of the lithosphere (rock) have already been described in previous chapters.

8.9 HYDROSPHERE

Water originated on earth due to the condensation of water vapors from the early atmosphere. Water also formed from underground volcanic activities. All the water present on the earth as oceans, seas, snow and ice, fresh water (lake, river and spring), atmosphere water vapors, clouds and underground is known as the hydrosphere. Water is freely and slowly exchanged among these reservoirs, going from one source to another, maintaining their individual characteristics in each phase. The ocean is the major reservoir of water.

Water Reservoir	Volume %
Ocean and sea	98.30
Snow and ice	1.66
Lake, river, spring	0.035
Water vapor	0.0001
Underground water	--
Total	99.9951

In addition to the major sources, 'ground water' is also important. It originates from any of the mentioned sources but undergoes drastic variation in properties. Water is necessary for the survival of all life. All geological, biological, physical, chemical, environmental, natural and artificial activities are greatly influenced by water.

8.9.1 OCEAN WATER

Ocean water covers about 70.8% of the earth's crust. The water layers from the sea bottom up to the water surface, on the basis of thickness, can be divided into three sections:

Layers	Thickness (m)	Volume %
Upper layer	0 (surface)–500	2.0%
Middle layer	500–2000	18.0%
Deep layer, trench	2000–10,000 (sea floor)	80.0%

Ocean bottom sections, on the basis of topography, continental margin and the deep sea plane, have been discussed in Chapter 4.

8.9.1.1 Ocean Temperature

The average temperature of the ocean water surface is around 25°C. The temperature is maintained by incoming solar radiation. With depth, the water temperature decreases rapidly, because sun radiation cannot penetrate. The temperature near the ocean bottom is as low as 0 to –5°C. The temperature and density of the ocean water in different regions of the earth do not vary substantially, whereas the temperature and density difference in the terrestrial regions is quite significant.

8.9.1.2 Density and Circulation of Ocean Water

The density of ocean water is reported between 1.025 to 1.028 g/cc. The density depends on the temperature of water. Variation of density due to water circulation is also observed. Usually cold water is circulated from the geographical poles bottom to the equator, and warmer water from the equator surface to the cold geographical poles. This forms a water cycle in the ocean.

8.9.1.3 Dissolved Gases in the Ocean

The ocean water surface is in contact with atmospheric gases. The exchange of gases between oceanic gases and atmospheric gases takes place. The concentration of dissolved gases in the ocean varies with environmental conditions. Nitrogen (8.4–14.5 ml/l at 20°C) dissolves in water, but it does not play any role in chemical reactions. The content of dissolved helium and argon inert gases is 0.0002 ml/l. Dissolved oxygen (0.0–9.0 ml/l) is needed by organisms and is also used in the decomposition of organic matter. Carbon dioxide (34.0–56.0 ml/l) dissolves in four forms, CO_2, CO_3^{2-}, HCO_3^- and carbonic acid, and it has different functions. It is needed for biological activities and sedimentation. Hydrogen sulfide (0.0–22.0 ml/l) is present as ions in ocean water and as free gas in stagnant shallow water.

8.9.1.4 Chemical (Ionic Form) Composition of Ocean Water

Ocean water contains more dissolved salts (ionic form) compared to river water as evident from the fact that total dissolved solids in sea water are about 3.5% , whereas the dissolved solids in river water is much lower, compared to sea water, from 0.001 to 0.5.0 % depending on the geographical location. The major dissolved salts in sea water are of sodium (30.5%), potassium (2.0%), magnesium (3.5%) and calcium (1.2%) elements; in addition to these, there are more than 100 dissolved salts found in the ocean in trace quantities. Ocean water is termed 'brackish or salty' water. In spite of water exposure and interaction with the earth's atmosphere, crust and different water sources, the chemical composition of ocean water is almost constant. The composition does change appreciably; it has the capability to maintain the balance through its own processes of transferring ionic constituents among different sources.

8.9.1.5   The Total Salinity

The term 'water salinity' is used to express dissolved solid material in water. Water salinity is defined as the 'total weight in grams of solid material dissolved in one kilogram ($^o/_{oo}$) of ocean water or grams of solid dissolved in 100 grams water (%)'. The composition of dissolved solids is expressed in terms of percentage g/100 ml (%,) or part per thousand weights g/kg) ($^o/_{oo}$) or mg/L (parts per million, ppm). Normally the total solids are expressed as equivalent oxides of all the bicarbonate, carbonate and sulfate materials and as equivalent chlorides of all the fluoride and bromide materials.

The total salinity of sea water is almost constant at 35 ($^o/_{oo}$), with minor variations from 34.5 to 35.5 ($^o/_{oo}$). The variation is due to the weather conditions prevailing in rainy regions and hot high evaporation rate zones. The origin of salinity, that is the dissolved solids in the oceans, is from two sources.

- Weathering of rock at the surface of the earth
- Weathering bottom floor of the ocean

The weathered materials at the earth's surface are carried into the ocean by flowing water (rivers, streams, rain). The igneous and magma activities in the adjacent mantle in contact with ocean water have their share of total dissolved solids in the ocean. The additions of solids in the ocean have continued over geological time. One can expect that the ocean would have become a mass storage of such material. But the fact is otherwise. A balance in solids is maintained in ocean water by nature. The total dissolved solid materials remain at 35.0 ($^o/_{oo}$) in the ocean.

8.9.1.6   Balance of Solid Material in Ocean Water

The total materials that are added continuously are being withdrawn or removed from the ocean simultaneously, through physico-chemical and biological processes comprising the following phenomena:

- Adsorption/desorption
- Precipitation
- Biological processes

Certain substances, whether liquid or gaseous or solid, when they come in contact with other solids, are adsorbed on the surface of the solid. The adsorption may be either a physical attraction or chemisorption by chemical bond. The metal ions in ocean water are removed by adsorption on the surface of the sediments. The adsorption capacity of fine-grained rock is more than that of course-grained rock. The deeper rocks have fine texture, so they adsorb more trace metals than the shallow course-grained rock. Simultaneously, desorption of certain elements from sediment also occurs. An exchange of metals ions takes place between sediment and ocean water. The exchange depends on the size and ionic charge of the metal ions. Heavy metal ions, for example barium and calcium, are removed from the ocean water by exchange with lighter ones, sodium and potassium, on the sediment surface.

Numerous dissolved metal salts are precipitated from the ocean water as solid materials and are deposited in the bottom as sediments. For example, calcium chloride is precipitated as calcium carbonate, in the presence of carbonic acid. Iron, chromium and aluminum are precipitated as insoluble hydroxides. In general, heavy metal ions are more easily precipitated as insoluble materials than the lighter metal salts. Manganese and iron transition metals are removed as insoluble complex hydrated ferromanganese compounds and settle down at the bottom.

The most significant processes for maintaining the total salt content of sea water are the biological processes. Calcium ions are removed and metabolized as solid calcium carbonate (calcite). Calcium carbonate forms the protective shell of the organism. After the death of the shelled organism, the calcite settles down to form coral reef. Elements constituting micronutrients for living organisms are easily removed as food. Iron and copper compounds are removed and consumed by marine organisms. Similarly other soluble food materials containing phosphate and nitrate ions are consumed by organisms. At the death of the organism, some of these elements become part of the sediments and some are transformed to water as soluble compounds.

8.9.2   ICE AND SNOW

Ice and snow are the second biggest source of water after the oceans/sea. Ice is frozen water and snow is the frozen atmospheric water vapor formed by the crystallization process in cold conditions (reduced temperature) of atmosphere. Snow and ice can be a single crystal or an agglomerate of crystals. The density of water and ice exhibits special features and plays an important role in the ecosystem. A solid form of any substance is heavier than its liquid form, but the solid form of water (ice) is lighter than liquid water. The density of ice (solid) is lower than the density of liquid water below 4°C. The density of water is at a maximum (1.0 g/cc) at 4°C and 760 mm (atmospheric pressure). The density varies according to environmental conditions. The density of ice is always below 1.0 g/cc at any other temperature above or below 4°C. The abnormal behavior of ice and water serves useful purpose. Ice being lighter than water below 4°C, floats at the surface of the water. Below the floating ice, there exists warmer water that did not solidify. The conditions provide a warmer flowing water environment, necessary for the flora and fauna to survive in the ocean. There is a marked difference in the density of ice from different sources. Glacier ice has lowest density 0.84–0.85 g/cc among all types of different ice from different sources. The density of snowflakes is very low, 0.004–0.30 g/cc, almost comparable to the density of gases. It is due to the presence of trapped air in the snow. Snow and ice originate from three sources. The first is frozen water droplets in the atmosphere as snowflakes and hail drops. The second source is the frozen and frosted water in rivers, lakes, seas and glaciers. Ice may also be found as frozen water in the interstitial of soil sediments, beneath the earth's surface.

8.9.3 LAKE, RIVER, SPRING-FRESH WATER

Life survives on 'fresh water'. Other terms used for fresh water are 'sweet water' and 'terrestrial water'. Fresh water is of low salinity compared to brackish water. Fresh water contains micro quantities of nutrient elements needed for life. Fresh water constitutes very little of the total water content of the earth. The survival of mankind depends on fresh water. Fresh water sources, especially rivers, play an important role in sedimentary rock and petroleum basin formation as well as in the enrichment of the earth's soil for agricultural purposes. The flowing water, rivers and streams, brings a lot of weathered, eroded and decomposed land materials and offloads the materials in low-lying land (basin) and oceans. The materials are not only of large quantity but also of diversified nature and variety. Wide variations in the concentration of ions between rivers and ocean water are observed. The concentration of Na^+ and Cl^- ions is much lower in river water than in the ocean. The kind of dissolved salts in river water shows from which region the river has originated and flowed. The river water contains trace amounts of sparingly soluble minerals of iron, aluminum and silicon which owe their origin to weathered sedimentary and igneous rocks. Sea water is devoid of these elements. The ionic composition indicates that more soluble salts are found in ocean water. Sparingly soluble salts are abundant in river water. The calcium content of river water is very high (12.5%) compared to sea water (1.2%).

8.9.4 WATER VAPOR

The presence of water vapors in the atmosphere is due to the evaporation of water. Evaporation occurs from different water sources, namely oceans, rivers and lakes, etc. Water vapor is also introduced in the atmosphere by the perspiration of animals and transpiration of plants. The rate of evaporation of water depends on many factors such as temperature, relative humidity, surface area and the concentration of dissolved solids. Higher temperatures and less humidity accelerate evaporation. The rate of evaporation from sea water is about 2--3% less than the fresh water rate, because of higher salinity in the ocean. The amount of evaporated water from the ocean is far greater than the small quantity of evaporated fresh water. The contribution from ocean water to the atmospheric water vapor is around 85% of the total, whereas fresh water sources provide 15% of the vapor in the atmospheric pool.

8.9.5   UNDERGROUND WATER

Underground water is found below the earth's surface and is present in the pore void space of soil and rock. The water is called pore or interstitial water. The quality and quantity of underground water depend on the geographical location and rock structure. Underground water quality is related to the total dissolved solids. As depth increases, the quantity of dissolved salts also increases. For example, salt content of 187.0 g/l at a ground depth of 1500 meters increases to 230.0 g/l at a depth of 2000 meters. The water is characterized on the basis of total dissolved solids, fresh water (<1.0 g/l), brackish water (1–10 g/l), saline water (10–100 g/l) and brine water (>100 g/l). An aquifer is an underground permeable rock formation having a sizable amount of fresh water.

8.10   EARTH'S ATMOSPHERE

Atmosphere is a gas mixture surrounding the earth. Earth's atmosphere is a mixture of gases along with liquid and solid material as aerosols that are retained by earth's gravity and surround the globe. Aerosols are fine droplets of liquid and fine dust particles. The atmosphere plays an important role in sustaining life of organism. Some of the components of the atmosphere are exchanged with the ocean and earth. The atmosphere controls the geological processes of the erosion, weathering and abrasion of rocks. Solar energy first passes through the atmosphere before reaching earth. The atmosphere acts as a filter and prevents the harmful solar radiation reaching earth and allows the useful radiation in different parts of the earth for the survival of organisms.

8.10.1   INITIAL STAGE OF THE ATMOSPHERE

The atmosphere appeared along with the creation of the earth about 4.7 billion years ago by one of the following cosmic processes:

- Direct condensation of cosmic incandescent gases and aggregation of cosmic dust particles (cosmic dust cloud).
- Separation from solar nebula. The sun was also formed earlier by cosmic dust particles (cosmic dust cloud).

Both the earth and the atmosphere were created from the 'cosmic dust cloud'. The cosmic dust cloud was created from the 'radiation dust' consisting of particles, atoms, elements, molecules and ions. The cosmic dust cloud was mainly composed of hydrogen, methane, ammonia, water vapors and some noble gases; together they formed the early atmosphere and earth. The earth and atmosphere were indistinguishable. During the aggregation of the particles of the cosmic dust cloud, some of the particles were squeezed and condensed into solids to form earth. Un-aggregated gases formed the atmosphere. The atmosphere was devoid of oxygen but rich in hydrogen, presenting a reducing environment. Non-reactive and light gases, noble gases, diffused out from the atmosphere and were lost in cosmos. Reactive and heavier gases remained with the solid mass of the earth that later on continued to be released from the earth, taking part in the development of the atmosphere.

8.10.2 MIDDLE STAGE OF THE ATMOSPHERE

The middle stage lasted between 4.7 and 2.0 billion years ago. Some oxygen was produced in this stage, but the atmosphere remained predominantly reducing. During this period, the earth's structure was fully developed as the distinct crust, mantle and core parts of the earth. Volcano activities were developed that were associated with the expulsion of dissolved gases from the mass of the earth. That was a real step for the separate formation of the atmosphere. Volcanic gases like hydrogen, nitrogen, hydrogen sulfide, carbon monoxide and some water vapors moved toward the surface of the earth and the atmosphere. The later period of the middle stage saw the formation and release of oxidized gases such as carbon dioxide, sulfur dioxide and abundant water vapors. Still the atmosphere remained reducing and lacked free oxygen.

8.10.3 PRESENT STAGE OF THE ATMOSPHERE

The present atmosphere began to appear from 2.0 billion years ago. In this stage, there was a gradual increase of free oxygen in the atmosphere. The atmosphere turned from a reducing to an oxidizing environment. Two processes were responsible for the mass production of oxygen. The first process was the photo chemical decomposition of water vapor in the atmosphere, and the other process was the appearance of photosynthesis. The water molecules were decomposed into oxygen and hydrogen by the ultraviolet radiation of the solar energy. The mass production of oxygen occurred with the advent of the photosynthesis process about 2.0 billion years ago. In the presence of solar energy and green pigment of plant leaves, water and carbon dioxide reacted to produce carbohydrates and oxygen.

8.11 CONSTITUENTS OF THE ATMOSPHERE

Three types of constituents are present in the earth's atmosphere, which are gases, water vapors and dust particles.

8.11.1 GASES

The main natural components of the atmosphere are nitrogen gas (78.1 vol %)and oxygen gas (21.0 vol %). The remaining about 1.0% of the atmosphere is composed of numerous (more than 30) trace chemicals and gases. Other significant components of the atmosphere are argon (0.9 %), carbon dioxide (0.03 %), neon (0.002 %), helium (0.0005 %), methane (0.00015 %) and krypton (0.0001%). Oxides of nitrogen, hydrogen, ozone and xenon gases are found in trace quantities less than 0.00005 %. Some of these constituents are variable. They change according to the climatic conditions. Helium is one of the lightest and natural gases. Being the lightest, the gas escapes from the atmosphere to the upper cosmos and is added to atmosphere continuously by natural processes. Helium gas is coproduced with natural gas in certain regions of the earth. Helium gas is produced by the radioactivity of uranium and thorium isotopes. Argon is produced by the radioactive isotope of potassium. The water vapors are an important constituent. Considerable variation occurs in the water vapor content (0.0–5.0 %) of the atmosphere. It depends on the location and climatic conditions. Sodium chloride in vapor form is found near the ocean and decreases on land away from ocean. Bromine and iodine originate from ocean water and contaminate the atmosphere in minor amounts.

Additionally human commercial and industrial activities are introducing gases which are adversely affecting the atmosphere's composition. Sulfur and nitrogen compounds are known as contaminants and pollutants. They are introduced by human activities in variable amounts. The concentration of sulfur dioxide is very low but the combustion of fossil fuels is responsible for the increase of sulfur dioxide in the atmosphere. The combustion of fossil fuels increases the concentrations of carbon dioxide, carbon monoxide, oxides of nitrogen, oxides of sulfur and unburned hydrocarbons in the atmosphere. Fluorine comes from mining and industrial sectors. Plants and animals are also responsible for the consumption and production of carbon dioxide.

8.11.2 ATMOSPHERIC WATER VAPOR

Nature maintains the gaseous composition of the atmosphere almost constant except the water vapors. The water vapors in the atmosphere vary from place to place and time to time. They vary between 0.0 and 5.0%. The presence of water in the atmosphere affects the climate. Mountain and desert areas of the earth are almost devoid of water vapor. Dry weather prevails there. In the dry and desert areas, a wide difference in temperature occurs between night and day time. The raining tropical areas are humid (more water vapor) and warm. The temperatures remain almost unaltered during day and night in humid areas, because water vapor absorbs enough solar energy during the day and continues to release heat during the night. With elevation, the concentration of water vapor decreases. At a certain height the water vapor is almost nil. As the atmospheric elevation increases, the temperature drops, so any water vapor present is condensed to liquid and falls down. Fog, mist and clouds are dispersed water vapors in the atmosphere.

8.11.3 DUST PARTICLES

Atmospheric dust is suspended minute particles of solid material in the air. The origins of dust are the erosion and weathering of rock, desert wind and also human industrial activities. The concentration of dust is at a maximum in the vicinity of its origin (commercial, industrial and mining activities). The dust particles reduce the hot solar radiation reaching the earth's surface. Dust also decreases the intensity of reflected radiation from the surface of the earth. The dust particles in air act as a nucleus for the condensation and agglomeration of water vapor to form water drops that fall down as local rain. Smoke is dispersed minute particles of carbon in air.

8.12   ATMOSPHERIC ZONE

The atmosphere consists of layers of different gases with different temperatures surrounding the earth. The total thickness of the atmospheric layer is not well defined. The upper layer is rarefied, thin and of low density compared to the lower layer. The lower layer adjacent to the earth's surface is compressed, thick and of high density. With an increase of height/elevation, the density, water vapor, temperature, pressure and constituents of the atmosphere decrease. There are no sharp boundaries between the layers; overlapping exists. The layers or zones may be constructed on the basis of the constituents, the decrease in pressure, density and water vapor and the fluctuation in temperature with height. The atmospheric layers are structurally divided into six zones, namely the troposphere, stratosphere, mesosphere, thermosphere/ionosphere, exosphere and solar wind zone. The salient features of the six atmospheric zones are summarized below.

8.12.1   Troposphere Zone

The troposphere is adjacent to the earth's surface. The thickness of the troposphere is variable. From equator it covers an elevation of about 16 km and decreases to a height of 8 km at the poles. Also the thickness changes with the season. It is less thick in winter and more in summer. The troposphere is the most turbulent zone of the atmosphere. It is the densest part of the atmosphere, and 80% of the total atmospheric mass is found in this layer. Most of the heavy constituents of the atmosphere are concentrated in this zone, for example dust particles, water vapors, carbon dioxide, oxides of nitrogen, argon, pollutants besides nitrogen and oxygen. These constituents absorb the heat and are responsible for the increase of temperature or global warming. The concentration of heavy constituent decreases with elevation, and correspondingly the temperature also decreases with altitude. The temperature range of the zone is (–40)–25°C. It is estimated that the temperature drops by 6.4°C for every 1 km in height. At a height of 16 kilometers the pressure is reduced to 0.0001 atm.

8.12.2   Stratosphere Zone

The stratosphere is the next highest zone of the atmosphere between 16 and 56 km. The transition between the troposphere and the stratosphere is known as the 'tropopause'. The stratosphere is a quieter layer of the atmosphere. The main constituents of the layer are nitrogen, oxygen and ozone gases. In the stratosphere, ozone plays an important role. The ozone absorbs all the harmful ultraviolet radiation from the sun and prevents the rays from striking the earth. Therefore, due to the absorption of ultraviolet rays, the temperature of the zone rises from –40 to –25°C. The water vapor content and density are almost negligible. The pressure is reduced to less than 0.00001 atm in the zone.

8.12.3 MESOSPHERE ZONE

The next highest zone of the atmosphere is the mesosphere. It lies between 57 and 90 km in altitude and the temperature ranges from –25 to –80°C. The boundary between the stratosphere and the mesosphere is called the 'strato-pause'. In this zone minor concentrations of oxygen, nitrogen and ionized oxygen and oxides of nitrogen are present. Since the constituents of the zone do not absorb solar radiation, the temperature of the zone decreases with elevation from –25 to –80°C. The pressure continues to decrease with elevation.

8.12.4 THERMOSPHERE/IONOSPHERE ZONE

The thermosphere starts from 91 km and extends in an upward direction up to 500 km. The transition between the mesosphere and the thermosphere is known as the 'meso-pause'. Extremely low concentrations of oxygen, nitrogen and oxides of nitro-gen exist in molecular and ionic forms. The ions are capable of absorbing some portion of the solar radiation; therefore, the temperature within the zone increases with altitude from –80 to –40°C. Pressure is reduced to nil in this zone. The International space station orbits in this layer between 350 and 500 km. Ionosphere

8.12.5 EXOSPHERE ZONE

The exosphere is the outer layer of the atmosphere above 500 km. This layer contains extremely low concentrations of hydrogen and helium. The zone is extremely rarefied; almost vacuum. In the lower part of the layer, oxygen and nitrogen in low concentrations are present. Gas constituents are far apart from each other. No cohesive force exists between them. The gases freely escape from this zone into the upper solar wind zone beyond 10,000 km. Most of the satellites orbiting earth are located in the exosphere. The zone is devoid of any pressure but the temperature rises from –40 to 20°C. Sun is very bright in the zone. This zone is considered as outer most layer of the atmosphere and atmosphere ends here.

8.12.6 SOLAR WIND ZONE

The solar wind zone is the beginning of the vast cosmos space beyond 10,000 km. Its border with the lower exosphere is ill defined. The transition exists between the upper part of the exosphere at about 10,000 km and the solar wind. The exosphere merges into the solar wind zone and becomes indistinguishable. The solar wind zone consists of charged particles emitted from the surface of the sun.

8.13 ENERGY BALANCE OF THE EARTH

The earth receives tremendous amounts of solar energy in the form of electromagnetic radiation. The earth must use or reject the energy in order to maintain energy equilibrium and sustainability. The heat input by the solar radiation must equal the heat output by the earth. The atmosphere plays an important role in maintaining the heat balance of the earth. Part of the incoming energy is absorbed and the remainder is reflected back into space. Thus, the earth maintains a heat balance and maintains the optimum climate conditions necessary for life on the earth. The energy balance is supported by variation in the temperature as well energy distribution. Great fluctuations are observed in temperature in various zones of the atmosphere. The energy distribution occurs as follows:

- About 50 % of the total solar energy is reflected back by the atmosphere into space before reaching the earth.
- 15 % of the energy is reflected back by the earth's surface.
- About 28 % of the radiation is utilized for the heating and evaporation of the earth's surface water.
- About 5 % of the radiation is absorbed by the land surface.
- About 2 % the radiation is utilized by the plants and animals of the earth.

The above heat balance has been compiled on an average basis considering the earth as whole. Different parts of the earth receive and reflect different amounts of energy. This heat imbalance on the earth also serves a useful purpose. Warmer areas near the equator and colder regions in the north and south poles trigger atmospheric (air) and oceanic water currents that keep the climate suitable for living.

The sun radiates energy like a black body at very high temperatures (6000 K). A black body completely absorbs all light and emits thermal radiation. A major portion of the sun's radiation energy consists of visible electromagnetic radiation (about 45% of 700–250 nm wavelength), and infrared radiation (about 47% of 700–2000 nm wavelength). Energy transfer from one place to another occurs by conduction and convection currents. A convection current is created by the circulation of hot and cold gases and water. The transfer of heat by conduction takes place by atoms, molecules and ions of the materials.

8.14   BIOSPHERE/LIVING ORGANISMS

All matter on the earth can be divided into two systems; one is the 'organic biotic living system (biosphere)' and other is the 'inorganic abiotic non-living system (atmosphere, hydrosphere and lithosphere)'. Living systems are those which are produced, metabolized, respire, respond to stimuli, grow, move, decay and die. Non-living bodies are lifeless. They cannot breathe, reproduce, grow or respond to stimuli. The biosphere is the region of the earth where life in any form may exist. The earth's regional factors provide the necessary elements that support the production and sustenance of life. The biosphere is the precursor of petroleum.

8.14.1   Habitat for Living Organisms

Living organisms include all forms of life, plants, animals, microorganisms, insects, spores, flora, fauna and bacteria of all kinds. The living habitat is the natural home for all plants and animals on earth. The biosphere includes the living habitat of all organisms. The lithosphere, hydrosphere and atmosphere from about 8.0 km below sea level to 6.0 km meters above sea level are the habitat of all kinds of organisms. Bacteria can survive the harsh conditions of oil-water brine in underground sedimentary rock, whereas the polar bear lives in cold conditions below –50°C. The zone of life is more scattered and varied in nature compared to the zone of the hydrosphere, atmosphere and lithosphere. A life region exists where some kind of life-supporting factors are available. Life (flora and fauna) on the earth started much later, as the necessary conditions for life were created a long time after the major geological events. Life-supporting factors are as follows:

- Supply of oxygen
- Availability of water
- Suitable climate conditions (temperature, pressure and humidity)
- Supply of energy source (food)
- Supply of food nutrients
- A suitable habitat in the hydrosphere, atmosphere and lithosphere

When the above-mentioned life-supporting factors were available, early life (primordial bacteria) emerged around 3.5 billion years ago. Earth's oldest life was the prokaryote (incomplete life) organism, namely cyanobacteria; afterward life began to develop and establish itself on the earth.

8.14.2 Comparison of the Biotic and Abiotic Spheres

The biotic-sphere or biosphere is the living habitat of all kinds of living organisms (life) in the abiotic (lifeless) hydrosphere, atmosphere and lithosphere. Apparently there is a tremendous difference between biotic and abiotic matter. The weight and volume of biotic matter are insignificant compared to the volume and weight of the hydrosphere, atmosphere and lithosphere. Although there is a much smaller amount of biotic life, the effect on the earth is enormous. Biotic interactions with the abiotic sphere bring about large numbers of transformations and alterations. Biotic life needs the abiotic environment for survival and sustainability. Biotic life cannot exist without the support of the abiotic earth.

The biotic environment is subjected to addition, alteration, variation, transformation and extinction within short periods of time compared to the abiotic environment. The lifecycle of a living organism is much shorter, almost nothing, compared to the abiotic sphere in geological time. It may be noted that the geological process involving inorganic abiotic activity takes a long geological time, whereas living organism (biotic) activities are visible within short periods of time. For example the formation of carbonate rock and siliceous deposits by secreting and dead organisms occurred over geological time.

The abiotic environment is relatively static compared to buoyant and dynamic biotic life. Even within biotic species great difference occurs. The productivity, survival and forms of life are different from one biotic species to another and from one place to another. Again this diversity in biotic species is more visible on land than in aquatic environments. The types and forms of living organism species are manifold on land compared to the hydrosphere and atmosphere. Life on land is restricted to a selected surface area. There are vast uninhabited areas of dry desert, high mountain regions and ice-cold polar regions on land. There seems to be uniformity in species and productivity in hydrosphere aquatic life compared to land life. Most of the living organisms in the hydrosphere are concentrated within the zone of underground up to the depth of 200 meters, where solar radiation can penetrate and the photosynthesis process is possible. Below that depth, solar rays cannot penetrate, so the photosynthesis process is not possible. Life below 200 meters deep is rare; only anaerobic bacteria survive. Aquatic living organisms far exceed land organisms in number. The interaction of biotic life with other living species along with the interaction with abiotic matter is known as the 'ecosystem'. Ecology is the study of the whole biosphere.

8.14.3   Bio-Geo-Chemical Cycle

Bio-geo-chemical cycle system study that how the biological, chemical and geological process works so that the necessary ingredients of life are assured and sustained. The study basically includes the supply and demands of fundamental elements of life, which are, hydrogen, carbon, oxygen, nitrogen, sulfur, phosphorus and other elements. The bio-geo-chemical cycle in ecosystem terminology is part of the 'ecological cycle'.

The bio-geo-chemical cycle describes how the basic materials required for life in the biosphere are regenerated and recycled after their use so that there is a continuous supply and life is maintained on the earth. The life-supporting ingredients are water, energy elements, nutrients, micronutrients and the abiotic environment. The ingredients are connected to each other through the 'bio-geo-chemical' cycle. The habitat of living organisms supports bio-geo-chemical functions. The habitat is a bio-geo-chemical pool. A bio-geo-chemical pool is a place where the necessary components are easily available for life. That place may be the ocean, atmosphere, subsurface or the earth's crust.

The bio-geo-chemical pool is of two types. One is an active pool and the other is a storage pool. For example, the photosynthesis process, where carbon dioxide, water, sunlight and plant life are freely available, is an active pool. The fixation of carbon (carbohydrate formation) and circulation of carbon dioxide are continuous phenomena. An example of a 'storage pool' is the formation of coral reef by carbonate-emitting organisms over geological time. Carbon in carbonate rock is an example of a storage pool. The carbon is released from the carbonate rock as carbon dioxide gas, which takes a long geological time. Fossil fuels are another example of a carbon storage pool. The weathering and erosion of rock are an example of a storage pool. From this storage, necessary micronutrients are incorporated into the atmosphere, soil and ocean for living organisms. In turn the remains of dead organisms recycle these microelements to the sediment, soil and rock. The exchange of materials from an active pool is rapid and at a high rate. The rate of material transport from a storage pool is a slow and usually occurs through geological time.

Gases play a vital role in the biogeochemical cycle for providing necessary ingredients for life. The bio-geo-chemical cycle process of carbon, oxygen, nitrogen and sulfur elements takes place in a gaseous state. Even the water cycle also occurs through water vapor (gas) in the atmosphere. All these gaseous elements belong to an active pool. The cycle of solid bio-geo-chemical components mostly belongs to storage pools, for example carbon, iron, phosphorus and calcium.

8.14.3.1 The Water Cycle

Water is the most important of all the material necessary for life. The ocean is the bulk source of water; from here it goes to the atmosphere as vapor. From the atmosphere the water falls to the earth's surface as rain. It may be noted, although land area is much smaller than the aquatic surface, the land receives a greater amount of rain water than the ocean's surface. Rain water is drained back to the original source (ocean) through rivers, streams and by other channels. The water cycle is not only important for maintaining water balance in all the sources of water, but also plays an important role in the ecosystem and geological formation of the earth's crust as follows:

- The water cycle maintains a balance in nutrients in the soil and ocean for the consumption of living organisms. Water acts as a storage pool for many dissolved, suspended materials for soil and rock formation and food nutrients for living organisms.
- Water in the atmosphere as vapor, oceans, rivers, streams and lakes is open, mobile and accessible. The water from these sources is known as an 'active pool'. All these sources constitute about 97% of the total water present. The remaining 3% constitutes the storage pool, and is stored in underground rock reservoirs and as solid glaciers and ice caps at the north and south poles. Underground water, ice caps and submerged glaciers have limited accessibility and interchangeability; such water is termed as a 'storage pool' or inactive pool'.
- Water helps to eliminate or reduce the material that may be harmful to the living organisms on land. The water provides necessary conditions for necessary physico-chemical-biological reactions and the metabolism process in the ocean.
- Plant transpiration releases about 90% of their water into the atmosphere.

8.14.3.2 Nitrogen Cycle

Living organisms need nitrogen in substantial amounts to synthesize amino acids, proteins and other life compounds. More than 79% nitrogen is available in the atmosphere. The abundant atmospheric nitrogen cannot be utilized by organisms for their metabolism processes. Nitrogen is a non-reactive inert element. Therefore, nitrogen has to be 'fixed' into food or chemicals in such a way that it can be used by organisms or as fertilizer or as soil nutrients. Some bacteria living in plants can transform atmospheric nitrogen directly into ammonia. The ammonia is then transferred to amino acids. Proteins are made from amino acids. Protein is a constituent of all living organisms. Another type of bacteria convert the ammonium ion (NH_4^+) into nitrite ions (NO_2)$^-$ and nitrate ions (NO_3)$^-$. Nitrate mineral in the soil becomes a nutrient for plant life. Nitrogen in the atmosphere is converted to oxides of nitrogen by lightning and solar radiation in the atmosphere. The nitrogen oxides are brought to the earth's surface by rain drops, where they are converted to soluble nitrate minerals. Artificial fixation of nitrogen is carried out by interacting atmospheric nitrogen and hydrogen gas in the ratio of 1:3, to produce ammonia (NH_3). Ammonia is used to manufacture urea and other nitrogen fertilizers needed in agriculture.

8.14.3.3   Oxygen Cycle

Oxygen is life. It is essential for all living organisms. The oxygen 'consumption cycle' consists of respiration by animals, combustion of fossil fuels and decomposition of organic matter. The 'supply cycle' of oxygen is the photosynthesis process which consumes carbon dioxide and releases oxygen. Photochemical decomposition of water vapor in atmosphere produces oxygen. All these processes maintain consumption/generation sustainability of oxygen on earth for all living organisms. The 'storage pools' of oxygen are oxides of minerals. But on weathering or decomposition the minerals do not form oxygen directly. Instead carbon dioxide and water are formed. Carbon dioxide is used in the photosynthesis process to produce oxygen and energy material carbohydrate.

8.14.3.4   Phosphorus Cycle

Phosphorus is an essential constituent of animal life. Animal teeth and bones as well as some lipids contain phosphorus. Additionally, phosphorus is a component of nucleic acid and the cell membrane. Phosphorus is incorporated as phosphate in sediments and sedimentary rock by the remains of decomposed dead animals. Rock transfers the phosphorus to the soil to act as a nutrient for the plant. The phosphorus is cycled from plant to animal, from dead animal to rock and soil and again from soil to plants to animals. The same phosphorus cycle occurs in both land and aquatic environments.

8.14.3.5   Carbon Cycle

Carbon is abundant both in organic and inorganic forms. Storage pools of carbon are carbonate mineral and fossil fuel. Active pools are atmosphere (carbon dioxide) and sea water (dissolved carbon dioxide). Carbon recycling has already been described in detail in Chapter 2.

8.15   BASIC STRUCTURE OF LIVING ORGANISMS

The building blocks or the small units of all animals and plants are enormous numbers of living cells. A living cell simulates the conditions of a living species. A cell has a definite structure, composition, activity, respires, performs a particular function, moves independently and duplicates/replicates responses to stimuli, decays and ultimately dies.

8.15.1 Evolution of Living Cells

The evolution of living organisms or living cells succeeded the chemical evolution. The chemical evolution succeeded the matter (lifeless) evolution in earth's history. The chemical era provided the necessary ingredients to begin life on earth. From inorganic chemicals, first chemical polymers then biological polymer macromolecules were produced. The biological polymer further evolved to build the living cell, the basic structural unit of life. This stage is known as 'biological evolution'.

8.15.1.1 Bio-Chemical Species

Ammonia, phosphoric acid, carbon dioxide and water are the seed chemicals for the production of chemical polymers that later turn into biological polymers, proteins and nucleic acid. Both chemical and biological evolution had occurred, first in aquatic media, mostly in the ocean. At that time, biopolymers along with inorganic chemicals and polymers were dispersed randomly in the ocean. The further transformation of polymers by biochemical processes resulted in highly ordered biological polymers which were in a dynamic equilibrium state in the prevailing environment. The biopolymer is considered as the initial stage of the living cell.

8.15.1.2 Gelatinous Semi-Fluid Drop

The original biopolymer species, protein and nucleic acid, were soluble in water. With further polymerization and increase of molecular weight the solubility decreased. The macromolecules formed a colloidal solution. The colloidal solution paved the way for cell aggregate formation. Protein and nucleic acid polymers are polar (dipole) in nature, similar to the dipole water molecule. Therefore, strong interactions between water and polymer macromolecules occurred. The agglomeration of large numbers of colloidal macromolecules resulted in the formation of gelatinous semi-fluid in the ocean water. The semi-jelly fluid was distinguishable in water with a sharp boundary, floating on the water as droplets. The droplets fused together to give a bigger gel-like drop. The drop represents the clustering of polymers of different types of macromolecules.

8.15.1.3 Living Cell Evolution

The cluster of polymers continues to interact with the inorganic and organic constituents of ocean water. There was free exchange of chemicals between the cluster polymer and the surrounding water. Therefore, a dynamic equilibrium was established, that manifested a living character. This was initial stage of life. When the dynamic equilibrium condition was disturbed, the living cell of macromolecule died away. In the meantime, the emergence of enzyme proteins (catalytic activity) accelerated the exchange of chemicals between cellular clusters of polymer drops (gelatinous semi-fluid) and water. That was really a biological process instead of a chemical reaction. Now the cell cluster of polymer proteins is able to synthesize its own enzyme with the available material present in the environment. This led to the continual generation of unit cells, the building block of life. Any cell that could not produce its own constituent and enzyme died out. Protein facilitates the creation of polynucleotide biopolymers in the presence of catalytic enzymes in the unit cell. Now the cells are fully developed. The cells are capable of multiplying themselves and grow. The old cells die. Therefore, a sustainability is maintained of living organisms. Protein biopolymers are associated with the metabolism and growth of the cell. The polynucleotide polymers are responsible for the hereditary and reproduction system in the cell.

8.15.2   Living Cell Structure

Biological polymers themselves are lifeless materials in discrete and separated form, but they do grow or decay. After the incorporation of biopolymers in the cell they assumed the role of living organism. In the cell the biopolymers are arranged in an orderly manner with fixed geometry. In the cell they can interact between themselves and exchange material with the surroundings. They are full of life, grow and die.

Two types of living cells are identified; one is the prokaryote and the other is the eukaryote. Prokaryotic cells are a primitive form of life and are simpler and smaller than eukaryotic cells. Prokaryotic cells are formed in lower forms of life that is, bacteria, etc. Eukaryotic cells are associated with high forms of animal and plant lives. Eukaryote cells are bigger and more complex cells than the prokaryote cell. All unit cells are independent bodies with life and include necessary organs to perform different functions as described below.

8.15.2.1   Membrane & Cell Wall

The outermost part of a living cell is its membrane. A cell membrane or plasma membrane or biological membrane acts as a selective permeable membrane. It controls the material that goes in or out of the cell. The membrane helps to establish potential energy inside the cell. The membrane consists of two layers made of proteins and phospholipids. The membrane of a plant cell has an additional outer layer called the wall membrane, or cell wall. The cell wall is made of lipo-polysaccharide polymers.

8.15.2.2   Cytoplasm & Cell Life

Most of the inside space of the cell is occupied by a gelatinous semi-fluid substance known as 'cytoplasm'. The cytoplasm responds and reacts to external stimulation. This is a characteristic of life. The cytoplasm is placed adjacent to the inner surface of the membrane. The cytoplasm contains micro-tubes or micro-filament known as a cytoskeleton. The cytoskeleton provides the structural framework of the cell and maintains the shape and resists deformation in the structure.

The cytoplasm is a colorless, semi-permeable, gelatinous substance. The cytoplasm contains colloidal fluid containing granular particles known as cytosol and also consists of water, proteins, electrolytes and cell nutrients. The granular part of the cytoplasm is known as organelles. Many types of organelles are identified. Each type performs different functions. Organelles may be treated as organs of the cell that perform different functions. Organelles are similar to animal organs like the heart, kidney, liver and lungs that perform different functions. Organelles perform many functions. One type of organelle synthesizes certain compounds that on decomposition provide necessary energy in the cell. Unwanted substances introduced into the cell are decomposed and destroyed by different types of organelles.

8.15.2.3 Nucleus of the Cell

The nucleus of the cell is situated inside the cytoplasm. One of the most important constituents of the nucleus is nucleic acids. It is the most complex part of the cell and is composed of substances that control the hereditary and genetic characteristics. The nucleus is again surrounded by a special membrane made of phosphide polymeric compounds. The phosphide membrane has small openings that act as inlet and outlet valves. Both inlet and outlet valves are called 'annuli'. The annuli allow required and needed nutrients to go in and out of the nucleus. With elapse of time, every cell undergoes subdivision, multiplication and extinction. The fully grown mother cell gives rise (division) to a daughter cell. The cell grows through a biological process called 'metabolism'. The metabolism process is of two kinds. The metabolism process that breaks down large molecules to provide energy for the cell is known as the 'catabolism process'. Second, the 'anabolism process' uses energy to create complex molecules to perform other biological functions. In the process of cell division and mother/daughter events, the important aspect is the equal distribution of a nucleus substance known as the 'chromosomes'. The chromosomes control the hereditary characteristics. Chromosomes are a complex substance of deoxyribonucleic acid (DNA) and proteins (histamine). The complex substance constitutes a material known as 'genes'

Sequences of mother/daughter division, growth, reproduction, decay and extinction are natural phenomena for the alteration in the living cell. Environmental, physical, chemical and biological conditions affect the living cell and bring about changes. Certain physical and chemical agents (mutagen) are capable of changing the genetic substance (DNA). The change or alteration in genetic message is known as a 'mutagenesis process'. The mutagenesis process alters or modifies the DNA genetic message brought forward by genes. Genetic study is an engineering science and technology subject.

8.16 ELEMENTAL COMPOSITION OF LIVING ORGANISMS

There are enormous numbers of different kinds of organisms living in different geographical conditions on the earth. It is expected that their composition varies greatly from species to species. However on an elemental basis all organisms contain the same types of elements in varied composition. The elemental composition of a living organism consists of carbon, hydrogen, oxygen, nitrogen and phosphorus along with minor quantities of other non-metallic and metallic elements. The elements play different roles in organisms for metabolism processes. The elements can be classified into three groups as follows:

- Energy elements group. Hydrogen, carbon, oxygen, nitrogen and phosphorus.
- Macronutrient elements group. Calcium magnesium, potassium, sodium and sulfur.

- Micronutrient elements group. Iron, copper, manganese, aluminum, zinc, cobalt, silicon, selenium, chlorine and iodine.

First, the energy elements are the main elements of every organism. The second group of elements are known as macronutrients and are prominent in the animal kingdom. The third group of elements necessary for all life is known as micronutrients. Carbon is mainly present as carbohydrate, nitrogen as amino acid and proteins. Elements like phosphorus, calcium and silicon form the bony skeletal structure of organisms. Sodium, potassium and chlorine are electrolytes and play their role in osmotic pressure in living cells. Magnesium is found in enzymes and leaves (chlorophyll). Sulfur is found in some amino acids. Micronutrients are required in very small quantities by the organism, but play useful roles in metabolism. Nearly all living organisms contain some metal in one form or the other. The metals are concentrated in the residue of the remains of the dead organism.

Elements are not present as such in a pure state. Different chemical bonding of these elements gives rise to different types of chemical compounds in living organisms. Different compounds perform specific functions in the body. The percentage composition of elements and chemical compounds differ considerably from one type of living organism to another.

8.16.1 Distribution of Elements

The distribution of elements in individual chemical compounds (carbohydrates, proteins, lipids and cellulose) of the organism was already given in Chapter 2, Table 2.1. Living organisms, both animals and plants, are mainly composed of these four groups of compounds besides water, bony skeleton and minor components. All the elements in organisms are distributed in these four types of compounds. The elemental composition of petroleum was also included in the table for comparison with other organic matter. The elemental composition of lipids is closely related to petroleum; other components of organic matter deviate considerably. Among the elements, oxygen and carbon dominate in organisms and so in each type of organic compound present in the organism. Oxygen is an integral part of water as well as of all other macromolecules found in organisms. The element carbon is attached to all the chemical compounds in different structural forms. Hydrogen is another element common to all organic compounds of organisms. Sulfur, nitrogen and metallic elements, though present in only small quantities in proteins and lipids, play important roles in organisms. Phosphorus and calcium are integral part of bony animals. In addition to all this, a large number of trace elements are present in organisms and perform vital functions.

8.17 CHEMICAL COMPONENTS OF LIVING ORGANISMS

To quantify all the components (organic compounds) found in living organisms is difficult; the average % compositions of the main types of chemical compounds (components) in animals and plants follow different patterns. The compound composition differs considerably among big and micro species of both animals and plants. Water is the principal component of all living organisms. In marine organisms, water may reach up to 99% of the total weight of the species. Water is vital for life. Its peculiar properties help to perform various functions in the organism. Water can undergo hydrogen bonding among water molecules as well as with other chemical constituents having a charged functional group such as in proteins and nucleic acid. Hydrogen bonding plays a critical role in the stability of these macromolecules. Beside water, all living organisms contain organic and organo-metallic compounds in widely varying proportions. The organic compounds are dominant in all living organisms but in certain bony animals the organic compounds may be as low as 30% of the total weight. The remaining (70%), composed of water and bone and teeth, all are inorganic compounds. The chemical compound composition of animal organisms is more complex than in the plant kingdom. The major components of animal life are proteins (55–70%), lipids (10–18%) and carbohydrates (15–30%); nucleic acid (2–3%) is present in minor quantities. Plants are mainly composed of carbohydrates (40–70%), lipids (5–20%), cellulose (50–70%), lignin (2–30%) and proteins (2–30%). Higher plants contain major portions of cellulose, lignin and tannin (2–3%) compounds. The percentage composition of the compounds is quoted on a dry weight basis. The range of water and mineral content in animals is 50–70% and in plants is 10–90%. The variation of the percentage of chemical compounds depends not only on each species but also on their living habitat.

8.17.1 PROTEINS

Protein is a vital constituent of all living organisms. Protein is also called polypeptide, since the chain is linked by peptide bonds.

8.17.1.1 Characteristics of Proteins

Proteins are soft and flexible substances. Proteins exhibit variable characteristics, depending upon the molecular weight and structure of the individual protein. The molecular weight of protein varies in a wide range from 5000 to several million. Proteins are the major source of nitrogen in organisms. One of the important biological functions of the living organisms is that they response to stimuli; proteins respond to shock. Protein helps the living organism to exhibit certain types of movement.

Proteins are of two types; one is fibrous and the other is globular. Fibrous proteins support and connect the tissues, muscles, skin and hair of the body. Fibrous proteins are comparatively stable and non-reactive. Globular proteins maintain and regulate the biological functions in living organisms. Another class of proteins contains a non-protein group as well in their molecular structure. The non-protein groups can be either carbohydrate (glucose) or lipid or nucleic acid or a metal ion. Accordingly proteins are named as carbohydrate-protein, lipo-protein, nucleo-protein and metallic-protein. Non-protein groups impart additional biological functions in the protein. Some proteins also act as catalysts. Such a protein is known as a catalytic enzyme. It catalyzes certain biological metabolism processes. Enzymes are more active and flexible than proteins.

8.17.1.2 Protein from Amino Acid

Protein is a biopolymer and is synthesized both naturally and artificially. Amino acid is the raw material for making proteins. Amino acid was initially synthesized during the 'chemical evolution' period of the matter era. It was formed by the interactions of primitive hydrogen, methane, ammonia and water molecules in the presence of atmospheric electric discharge. Amino acids are compounds having a carboxyl group (–COOH), amine group (–NH$_2$) and inert alkylene group (–CH$_2$–).

Proteins/polypeptides are formed by the polymerization and condensation of more than one amino acid molecule, through the peptide linkage (–CO–NH–) with the elimination of water. The polymerization and condensation are chemical conversions that take place in anhydrous medium (non-aqueous). Two amino acids form a di-peptide protein. Three amino acids form a tri-peptide. Several amino acid molecules produce polypeptide/protein. The mass natural conversion of amino acid to biopolymer proteins occurred in the ocean in the presence of a reaction promoter through the biological evolution process. Hence the terms 'bio-amino acid' and 'biopolymer' are used. In the ocean, the condensation and polymerization of amino acids took place in the presence of some reaction promoter such as hydrogen cyanide. Hydrogen cyanide absorbed the eliminated water molecules from the condensation products, and thus helped the reaction to proceed to completion with the formation of polypeptides. It should be mentioned that, on hydrolysis, proteins are converted back to amino acids in the presence of water and under suitable temperature conditions.

8.17.2 Carbohydrates

Carbohydrates are also known as polysaccharides or simply saccharides. The meaning of saccharide is 'sweet'. Carbohydrates are a series of compounds containing carbon, hydrogen and oxygen atoms and hydroxyl (–OH), aldehyde (–CHO) and ketone (=CO) functional groups. On the basis of hydroxyl functional groups, the carbohydrates are defined as polyhydroxy compounds having aldehyde and ketone groups. Carbohydrate is an essential part of all living organism. It provides energy and fiber. Carbohydrates are polymeric compounds and on hydrolysis yield simple monomeric carbohydrates. Fructose (honey sugar) and glucose (grape sugar) are simple monomeric carbohydrates. They do not yield simpler carbohydrate molecules on hydrolysis. Fructose and glucose are the simplest molecules, so they are termed as monosaccharide and monomers. A disaccharide, for example cane sugar, on hydrolysis yields two mono-saccharides, glucose and fructose. A tri-saccharide carbohydrate, on hydrolysis, produces three simple carbohydrate monomers. A carbohydrate that yields a number of simple carbohydrates (monomer) is known as a polysaccharide. Polysaccharides do not have a sweet taste. The general chemical formula of carbohydrates is denoted by $(C_6H_{10}O_5)_n$, where 'n' is an integer. If 'n' is greater than 10, two important polysaccharides, namely starch and cellulose, are produced.

Starch and cellulose are complex carbohydrates/polysaccharides. Starch is the main source of carbohydrate in food. It is mostly found in seeds, cereals and roots of plants. Starch is a colorless amorphous powder. The molecular weight of polysaccharide ranges from 20,000 to 150,000. A polysaccharide macromolecule contains between 300 and 2500 glucose monomer units. Cellulose is the essential part of the cell wall of trees and land plants. Higher plants synthesize large amounts of cellulose. The marine fauna algae and seaweed do not contain cellulose. Wood is 50% cellulose. Cotton is 100% cellulose.

8.17.3 Lipid Compounds/Oil, Fat and Wax

Oil, fat and wax together are called lipid compounds. They are comparatively inert and stable. Lipids are the constituents of original organic matter that appear in petroleum oil without much alteration. The lipids play an important role in geochemical survey and are termed as 'geo-chemicals'. Geo-chemicals are used to trace the history of organic matter in subsurface conditions over a geological time period.

Lipids represent large groups of compounds with widely different chemical compositions and structures. Broadly they are defined as the organic matter in organisms not soluble in water but soluble in organic non-polar solvents such as ether, carbon tetrachloride and chloroform. Excess food taken by organisms, especially carbohydrates, is converted into lipid and fat. Lipid becomes the store of chemical energy and carbon in the body of animals. Lipid and fat form a protective layer on the vital organs of the body. Thus the organs are protected from shock and temperature variation. Roots, seeds and fruits are the large storage of lipids in plants. Lipids are divided into three main classes:

- Simple lipid, oil, fat and wax
- Derived lipid, terpenes, steroids and fat-soluble vitamins
- Compound lipid, namely phospholipids, sphingolipids, glycolipids and porphyrins

8.17.3.1 Simple Lipid/Oil, Fat and Wax

Simple lipid derived from land/marine vegetation/plant sources contains about 80% paraffin hydrocarbon containing oxygen atoms. Oil, fat and wax produce fatty acid and alcohol on hydrolysis. Oil and fat are naturally occurring mixtures of triesters of long chain carboxylic acid with glycerol (glycerin). Hence oil and fat are termed triglycerides. Glycerol is a polyhydric alcohol containing three hydroxyl (OH) groups. If all three hydroxyl groups are esterified with the same kind of acid, the resulting triester is known as 'simple glyceride'. When the three hydroxyl groups are esterified by different carboxylic acids, the formed triglyceride is known as 'mixed triglyceride'. Carboxylic acids are long chain hydrocarbons (paraffins or olefins) containing carboxylic acid group (–COOH). Carboxylic acid may be saturated or unsaturated organic molecules, depending on the source of lipids. Long chain saturated hydrocarbons are solid wax, for example paraffin $C_{21}H_{44}$ is solid wax (mp 41°C). The unsaturated olefin hydrocarbons irrespective of chain length are always liquid or gas. Similarly carboxyl acid containing unsaturated hydrocarbon is always liquid or gas irrespective of chain length. For example, $C_{21}H_{42}$ is liquid (mp 3°C). Marine organisms contain unsaturated hydrocarbons (liquid), whereas land animals contain exclusively saturated (fat) hydrocarbons. Land plants contain both saturated (wax) and unsaturated hydrocarbons.

Marine organisms need liquid lipids which are lighter, for food storage, insulation and buoyancy in water. Land plants and animals needs solid wax and fat (heavy lipids), for protection from external adverse environmental conditions.

Plant organisms synthesize an odd number of carbon atoms in the hydrocarbon chain, generally from C_{25} to C_{37}. The most dominant carbon atoms are C_{27}, C_{29} and C_{31}. The marine plants synthesize smaller odd chain length hydrocarbons from C_{15} to C_{21} compared to the land plants. They contain high percentages of unsaturated hydrocarbons from one to six olefin bonds. The most dominant are C_{19} and C_{21} carbon atoms. Waxes are solid and characterized by their melting point that ranges from 35 to 110°C. Waxes are lubricating materials; they produce a sliding surface. They are like oil and fat, and do not dissolve in water, but are soluble in organic solvents. Waxes are synthesized and maintained mostly by land plants and animals. Waxes are chemically a mixture of long chain fatty acids, free fatty acid and long chain monohydric alcohols. Waxes contain generally even number carbon atoms in the range from C_{16} to C_{38}.

8.17.3.2   Derived Lipid, Terpene, Steroid and Oil-Soluble Vitamins

An example of derived lipids is terpenes. They are oily substances and are essential parts of plants. Therefore they are called 'essential oils'. Terpene oils have definite color, odor and particular flavor. The terpenes are a group of substances and together they are called 'terpenoid'. The general chemical formula for terpene is represented by isoprene unit $(C_5H_8)_n$. It is poly-olefinic hydrocarbon or poly-isoprene. It may be an oxygen derivative in the form of alcohol (–OH), aldehyde (–CHO) and ketone (=CO). The simplest terpene is the myrcene containing two isoprene units $(C_5H_8)_2$. A single unit (C_5H_8) does not exist. Unsaturated olefin hydrocarbons are very reactive. They undergo polymerization and cyclization reactions to produce a variety of compounds. Isoprenes are unsaturated hydrocarbons and reactive. They form long chain or cyclic compounds known as 'isoprenoids'. Isoprenoids may be simple molecules or very complex compounds. Poly-isoprene (natural rubber) is obtained from the oily substance latex. The latex is formed in the bark of many trees of tropical regions. Crude latex contains oil, fat, carbohydrate, proteins and natural rubber.

Steroids are fatty substances and include cholesterol, sex hormones and bile salts. A steroid contains five isoprenoid molecular units and is tetra-cyclic terpene.

Oil-soluble vitamins, A, D, E and K, are isoprenoid derivatives and considered as oil and fat substances. All of them are produced by selected animals and vegetables. Vitamin A is a group of unsaturated organic compounds, namely retinal, retinol, retinoid, retinoic acid and beta carotene. Vitamins D, E and K are calciferol, tocopherol and phylloquinone respectively.

8.17.3.3   Compound Lipids

Compound lipids contain more than one functional group. A few are mentioned below:

- **Phospholipid** is a phosphorus-containing compound lipid. Important group members are lecithin and cephalin, consisting of di-glyceride phosphate and a simple molecule such as ethanol amine and choline. It mostly occurs in plant and animal membranes.

- **Sphingolipids** are more complex lipids. They consist of a combination of carbon 18 unsaturated long chain aliphatic amino alcohol (sphingosine) and fatty acid.
- **Glycolipids** are a large group of sphingolipids substance containing one or more molecules of sugar (glucose or galactose). Both sphingolipids and glycolipids are widely distributed in plants and animals.
- **Porphyrin group.** The porphyrin group represents a functional group of hetero-multi-cyclic compounds containing four substituted pyrrole subunits with many functional groups. The four pyrrole subunits are interconnected by methine (=CH–) linkages at their α-carbon position. The two central nitrogen atoms of the pyrrole are coordinated to a metal, constituting a metallic porphyrin group. The type of substitution of pyrrole rings determines the kind of porphyrin compounds. The porphyrin functional group does not exist itself. The porphyrin group forms many compounds. The most well-known porphyrin compounds are chlorophyll (plant green pigment) and heme (hemoglobin red pigment).
- **Chlorophyll porphyrin compound.** Chlorophyll is a complex lipid hydrocarbon containing a metallic-porphyrin group. They play an important role in geochemical study. They are constituents of the green pigment of plant leaves. The chlorophyll helps the photosynthesis process. They are groups of complex compounds containing several functionalities in one molecule. Corner carbon atoms of the pyrrole sub-unit of the porphyrin basic unit are substituted by several functional groups such as oxide (=O), hydroxyl (–OH), methyl (–CH_3), ethyl (–C_2H_5), methine (= CH-), oxymethyl (–OCH_3), carboxyl (–COOH) and by isoprenoid unit (–C_5H_8). The two central nitrogen atoms of the pyrrole are coordinated to a metal, most probably magnesium. Other metals such as vanadium or nickel are also possible.
- **Heme porphyrin compound.** Heme is another porphyrin compound. The central basic unit of heme is similar to porphyrin, containing four pyrrole rings connected by several functional groups in the same way as in the case of chlorophyll. There is a difference in porphyrin substitutions. The corner carbon atoms of the four pyrrole sub-unit rings are substituted differently in heme than in chlorophyll. The substituted isoprenoid unit is of comparatively smaller molecular weight than in the chlorophyll. The two nitrogen atoms of the inner pyrroles are coordinated with a central iron atom in porphyrin called heme. The heme is the red pigment of the hemoglobin.

8.17.4 LIGNIN

Lignin is an important chemical component of wood. It is not found as geochemical in petroleum crude oil but a common component of coal. Lignin is the most resistant compound of any component of plants. The lignin residue preferentially forms peat and coal during the peatification and coalification processes. Lignin acts as a binder of cellulose fiber in plants.

Lignin is a polymeric compound consisting of monomers linked together to form macromolecules of molecular weight from 3000 to 10,000. Lignin is not a uniform homogeneous substance and not of well-defined chemical composition. The monomer of lignin polymer is not simple, contrary to the monomers of other polymeric compounds which are normally simple molecules. The monomeric unit or the building block of lignin contains one substituted aromatic ring along with one heterocyclic ring containing oxygen atoms. On decomposition lignin forms a number of phenol compounds, namely phenol, o-cresol and p-hydroxy-anisol.

8.17.5 Tannin Substance

Tannin substances are amorphous powder. They are found in the bark of the plant and protect the plant. The chemical composition of tannin varies widely. The building blocks (monomer) of tannin substances are gallic acid, ellagic acid and flavone. All these compounds contain substituted aromatic and heterocyclic rings. Tannin substances are capable of converting the gelatin of hides and skins into insoluble material. The conversion is known as the 'tanning process'. The tanning process converts skin into brown-colored leather.

8.17.6 Nucleic Acid

Nucleic acids are complex molecules present in the nucleus of living cells, with a wide range of molecular weight from thousands to millions. The nucleic acids are repeating nucleotide monomer compounds. A nucleotide is not a simple monomer as other common monomers. It is a complex monomer with an appreciable molecular weight. A nucleotide monomer is formed from three basic chemicals:

- Purine or pyrimidine base (heterocyclic compounds)
- Carbohydrate (sugar)
- Phosphate ester

First, the above three chemicals are used to synthesize nucleotide monomer. Nucleic acid is formed by the polymerization of nucleotides. Nucleic acids are of two types. One is deoxyribonucleic acid (DNA). The DNA carries genetic information. The daughter cell receives the same DNA structural information from her mother cell. The other type of nucleic acid is called ribonucleic acid (RNA). The RNA is involved in the biosynthesis of proteins in the body and induces the DNA information in the specific proteins. Though the quantity of nucleic acids is much lower, they play an important biological role in the living organism. However their contribution to the formation and preservation of the total organic matter is negligible.

8.18 GEOCHEMICAL SURVEY EXAMPLES

Petroleum is a mixture of organic and organo-metallic compounds and is associated with inorganic sediment and water. Oil/gas is analyzed through many classical and modern techniques. For example, organic components of oil/gas are analyzed by a combined technique of pyrolysis, gas chromatography and mass spectroscopy. Metallic components are analyzed by atomic absorption/emission spectroscopy and inductively coupled plasma spectroscopy. Non-metallic elements carbon, hydrogen, sulfur, nitrogen and oxygen are analyzed by commercially available elemental analyzers. Isotopic concentration is measured by mass spectroscopy. Certain physical parameters and indexes are used to find the maturity of source rock.

8.18.1 GEOCHEMICAL PROSPECTING AND OIL/GAS SEEPAGE

Oil/gas seepage has already been discussed in Chapter 4. An example of direct geochemical application is the survey of the seepage of oil/gas at the earth's surface. The survey data are interpreted for oil/gas prospecting. The survey works well, if the underground petroleum reservoir rock is located at a shallow depth just beneath the seep spot at the surface. The survey cannot be applied for deeper deposits. The geochemical survey of seeped hydrocarbons does not lead to definite oil/gas findings. However the seeped oil/gas analytical results are used as an exploratory technique along with other geological survey and geophysical methods.

A seep is the leakage of oil/gas from underground deposits and their appearance at the earth's land surface and ocean floor. It is reported that the total quantity of seep under water is almost twice as much as occurs on the land (continents). Several seeps have been identified in the continental shelf and adjacent margin, originating from the sea bottom below or from nearby petroleum deposits on land. The seepage of underground oil/gas in the sea floor is similar to volcanic activity forming a cone-shaped mound of sediments. From the mouth of the mound, oil/gas along with sediment erupts. The sediment spreads around the mound to different extents, depending upon the pressure of the expelled fluid. The sediment settles down at the bottom of sea and oil/gas is separated. The separated oil/gas travels up through the water and appears at the water surface. Gas is released at the water surface as bubbles and oil as droplets. Both oil and gas are exposed to atmospheric conditions for evaporation and weathering. Oil from seep may contain crude oil with a complete range or a portion of crude oil depending upon the geological nature of the subsurface deposit and migration path way. Another form of gas leakage is bio-gases. The subsurface bio-chemical decomposition of organic matter produces bio-gases that appear at the surface. There is no oil associated with bio-gases. The quantity of seeped oil/gas is not enough. Therefore neither the seeped oil nor gas has been found to have any commercial application worth mentioning. Only some deposits of natural asphalt, formed as a result of seepage and subsequent evaporation and weathering, are commercially utilized on a limited scale on a regional basis.

8.18.1.1 Sampling of Leaked Oil/Gas

The survey method involves taking oil/gas samples from the leakage point at the surface and the underground area beneath the leakage point. The sampling of sediments and water is carried out at regular intervals of depth. The sampled oil, gas, water and sediment are analyzed to get the following geochemical information.

- The chemical composition, physical properties and quantity of leaked oil/ gas are determined.
- Adsorbed oil/gas in pores of the sediments and oil/gas dissolved in the water sample.
- Trace and radioactive elements in sediments and water.
- The presence of bacteria that survive on hydrocarbons and are responsible for the formation of carbonate rock.

8.18.1.2 Artificial Leakage

For the last 150 years the petroleum industry has developed quickly. In addition to the natural seepage there is a tremendous increase of oil and gas leaks from various man-made installations and machineries such as land and sea tankers, storage tanks, transportation, pipelines and any other equipment using petroleum. It is estimated that millions of tons of oil are being leaked by human activities. The seepage of oil/ gas from natural and artificial means is a case of concern from the environmental point of view.

8.18.2 Physical Alteration and Fractionation of Oil/Gas

The seeped oil/gas encounters different environments, both in the subsurface path way and the land or water surface. Thus the seeped oil/gas is subjected to natural physical and chemical alteration, first at the subsurface conditions, and later on under the atmospheric environment.

The alteration or degradation of oil/gas is a continuous process. An alteration of oil/gas in subsurface conditions is a slow process, whereas at the earth's surface exposed to atmospheric conditions the degradation of oil/gas is rapid. Seeped oil/gas, whether at the earth's surface or in ocean water, is first physically separated according to its volatility and solubility. The evaporation of more volatile components of the oil is a rapid process under atmospheric conditions. The oil gets rid of its volatile components in a very short time. Only thick, heavy, nonvolatile oil and tarry residue is left at the surface that further continues to degrade slowly or is washed away by rain water.

Hydrocarbon gases are sparingly (less) soluble in water. So the gases appear at the water surface without any change. However associated non-hydrocarbon gases, hydrogen sulfide and carbon dioxide are washed away as dissolved gases in water. Some of the oil components are comparatively more soluble in water than the others. Soluble oil components, mostly aromatic hydrocarbons and hydrocarbon derivatives containing sulfur, nitrogen and oxygen atoms, form solutions with water and are retained in the water column. Lighter weight compounds are more soluble than the heavier. Un-dissolved oil containing saturated hydrocarbons appears at the water surface as droplets. The same solubility behavior applies to oil seep at the land surface. The aromatic hydrocarbons and sulfur, nitrogen and oxygen derivatives of hydrocarbons in land are preferentially washed away by running rain water toward low areas.

8.18.3 Chemical Alteration, Weathering of Oil

Chemical alteration, weathering, degradation and decomposition of oil are used here in the same sense. The exposure of oil/gas to atmospheric oxygen accelerates the degradation process. After physical evaporation, the residual thick oil is subjected to chemical degradation under atmospheric conditions. The floating oil droplets on the water surface drift to the shore and coalesce to form thick oil. Under atmospheric conditions the weathering is very rapid; in the presence of oxygen and aerobic bacteria the oil is quickly converted to asphaltic material. With progressive weathering the oil becomes more insoluble, thicker and solid. The weathering process is slow in subsurface conditions. Oil takes geological time to become an asphaltic substance. The contributory factors to the thickening of oil are summed up as follows:

- Atmospheric oxidation of oil leading to the formation of asphaltic materials, through polymerization and condensation reactions.
- Microbial degradation and oxidation, with the evolution of gas and water and thickening of oil leading to the formation of asphaltic substance.

The end result of atmospheric oxidation and microbial degradation of the oil is the formation of black asphaltic and pyro-bitumen (insoluble bitumen/asphalt) substances. Asphalt is a black viscous liquid (semi-solid) soluble in carbon disulfide, whereas pyro-bitumen is an amorphous solid and is insoluble in carbon disulfide.

8.19 GEOCHEMICAL METHOD, EVALUATION OF SOURCE ROCK/KEROGEN

Geochemical study is a series of activities; area location for sampling points, sample collection, preserving the samples in suitable containers for onward transmission, performing the appropriate analytical method in the laboratory, recording of analytical data and processing and interpreting the collected data for oil/gas prospecting. Modern geochemistry is an interdisciplinary subject involving thorough knowledge of electronics, physics, chemistry, biology, mathematics and computers. Geochemical assessment has been greatly benefited by sophisticated computerized analytical and data processing machineries. Geochemical analytical methods give qualitative and quantitative information about the point of the rock from where the sample has been taken; sample from source rock, migration path way and from reservoir rock. The collected data are correlated to identify the source rock, estimate the type and quantity of organic matter, maturation index of organic matter, potential timing of petroleum migration from source rock, follow the migration path way and evaluate and characterize the petroleum fluids in reservoir rock. Geochemical methods also follow the path way of oil alteration, tertiary migration, leaks, seeps and the formation of new oil accumulation outside the source and reservoir rocks.

8.19.1 Sample, Container, Quality and Quantity

The need for proper sampling, sample preparation and sample preservation cannot be over-emphasized. Sampling from each selected point in the whole subsurface/surface petroleum system is to be carried out. Proper sample preparation is vital for geochemical analysis. It should be representative of the bulk material. A representative piece/portion of material is taken from the bulk amount/storage/rock to carry out geochemical analyses.

Plastic bags for solids, glass bottles for liquids and steel cylinders for gas sampling are employed. The sampling container is well cleaned to avoid any contamination. After obtaining the sample, the container is sealed. A sealed sample is free from adverse exposure to the environment. The volatiles are preserved. The sampled container should be stored in cold storage immediately. It is always hard to get a duplicate sample once it is consumed in the laboratory. Therefore it is necessary to obtain a sufficient quantity of the sample at the time of sampling, so that the sample remains in sizable amount after testing. There may be a requirement to retain the sample for a fixed period of time, say from six months to a year, for rechecking.

8.19.2 Kerogen Sample Preparation

Qualitative and quantitative assessment of kerogen at various stages of diagenesis, catagenesis and metagenesis is an important aspect in geochemical prospecting. Kerogen is the immediate precursor of petroleum. A kerogen sample is obtained from a portion of the core of the representative source rock. The kerogen organic matter is isolated from the inorganic rock matrix by selective extraction. The required quantity of the rock is crushed and pulverized. The pulverized rock is first decomposed by hydrochloric acid to remove carbonate minerals. The carbonate free sample is treated with hydrofluoric acid. The hydrofluoric acid dissolves almost all the minerals except organic matter (kerogen) and some resistant minerals. The extracted kerogen may be subjected to further purification by centrifuging and magnetic separation. Any bituminous substance (thick oil) associated with kerogen is removed by solvent extraction using carbon tetrachloride or chloroform. Insoluble kerogen is obtained as a solid black mass. The hydrocarbons extracted from the kerogen/rock sample are analyzed for their constituents. The properties of the extracts differ from one sample to another. The difference in properties is correlated with the stage of kerogen maturation.

After obtaining a suitable representative and properly prepared sample of kerogen, it is subjected to one or a combination of geochemical analytical techniques, to obtain useful data that may be related to the maturity. The maturity of kerogen and source rock is a measure of their state for the generation of oil and gas and an indication of potential hydrocarbons. First the purified kerogen mass is visually examined for color, texture and state, and then subjected to analysis by the chosen classical or modern instrumental method.

8.20   ANALYTICAL TECHNIQUES IN GEOCHEMISTRY

To follow the chemical transformations of original organic matter to kerogen from the initial digenesis stage through the catagenesis stage and finally to metamorphosed carbon in the metagenesis stage, is one of the main purposes of geochemical surveying. The organic matter/kerogen is subjected to various physico-chemical, biological and geological transformations. Initially the kerogen is an amorphous solid, reactive, unstable and immature due to the presence of active functional groups. The early diagenesis process is accompanied by the loss of hetero atoms (O, N and S) in the form of evolved gases, ammonia, carbon dioxide, hydrogen sulfide and water. The loss of hetero atom is followed by condensation and polymerization to a bigger molecule. During catagenesis, macromolecules formed are cracked (broken) to form the bituminous molecular substances. The residual kerogen becomes condensed and cyclized. Near the end of catagenesis or the beginning of metagenesis, the paraffin groups (methyl, ethyl, etc.) attached to the aromatic or naphthenic rings are cleaved, leading to evolution of mostly methane gas. The remaining residue becomes more condensed and aromatized. The kerogen is now an uncreative, stable and fully matured crystalline solid. This is the stage of maximum maturity and stability of kerogen, free from bitumen; it can only be altered under severe metamorphism.

The kerogen is evaluated by various chemical analytical methods, from classical wet chemistry to modern sophisticated techniques such as high-performance liquid chromatography, gas chromatography and mass spectroscopy, but only a few are of interest to geochemists. It is not required, from a petroleum prospecting point of view, to follow the detailed chemical changes occurring in kerogen during all the stages. The important aspect is to measure the thermal maturity of kerogen. Maturity is defined by the following parameters:

- The quantity and quality of oil/gas that has already been generated in the source rock.
- The quantity and quality of total oil/gas potential of source rock.
- Source rock deposition and accumulation environment.
- Origin of oil/gas.

To address the above points, some specific and simple geochemical analytical measurements are made. There are two different approaches as follows:

- One is a direct method. It involves the chemical treatment of the kerogen/source rock sample to arrive directly at the useful maturity information.
- The other is an indirect method. The method involves, first, measuring some key parameters of the kerogen/source rock sample and interpreting them to find the maturity of the kerogen sample.

8.21   DIRECT METHOD, PYROLYSIS OF KEROGEN

Pyrolysis is the direct method for determining the thermal maturity of kerogen/source rock. The extracted kerogen or direct source rock samples may be used in pyrolysis. The laboratory pyrolysis process conditions facilitate rapid, in a few minutes to a few hours, maturation of organic matter in kerogen and its conversion into oil/gas and solid un-reactive crystalline kerogen residue. The same stage of kerogen maturation in the subsurface geological conditions is achieved in millions of years (geological time). Pyrolysis is the thermal decomposition of kerogen/rock sample at elevated temperatures. It is carried out normally under an oxygen-free atmosphere or under a limited supply of air to avoid drastic oxidation, or it may be performed under complete oxidizing conditions to fully convert the total carbon of the sample into carbon dioxide and water. The pyrolysis is also known as cracking processes. The process is the breaking of organic macromolecules into smaller molecular fragments. The temperature of pyrolysis and corresponding volatiles give a complete picture of potential and already generated oil/gas in the kerogen. The pyrolysis temperature (T_{max}) of kerogen increases with maturation. However there is a vast difference in laboratory pyrolysis conditions and the subsurface geological environment. They may lead to different kinds of chemical transformation of kerogen. The bitumen (oil/gas) obtained from kerogen under subsurface geological conditions may not be the same as obtained by the pyrolysis method.

8.21.1 LABORATORY METHOD

A small quantity of pulverized kerogen or source rock sample is heated to an elevated temperature under a programmed rise in temperature and under inert atmosphere or in complete oxidizing conditions. The evolved volatiles (oil/gas) are assessed by the following methods:

- Modern instrumental analytical technique. Pyrolysis under inert atmosphere is carried out by passing a slow stream of helium or nitrogen over the sample. The volatiles from the pyrolyzer are directly fed to a gas chromatograph (GC) for the identification and quantitative estimation of the individual oil components. The evolved gases from the pyrolyzer furnace first pass through the capillary column where the vaporized components are separated according to their volatility (boiling point). The separated and eluted volatile components enter the flame ionization detector in sequence for the identification and estimation of individual components. The electrical signals from the detector are fed to a processor and chart recorder for tracing a graph as a series of peaks.
- Alternatively the pyrolyzer may be coupled with a gas chromatograph–mass spectrometer (GCMS) to give structural knowledge of the components in addition to identification and quantitative estimation. The GCMS technique separates the components in the capillary column as in the GC method. The separated and eluted volatile components enter the ionizer chamber (detector) of the mass spectrometer in sequence for the identification of the molecular structure of individual components.

8.21.2 PROCEDURE

A pulverized kerogen or rock sample is subjected to pyrolysis. The laboratory pyrolysis is carried out by taking 100–200 mg of pulverized kerogen/source rock sample in a crucible and placing it in a temperature-controlled heating furnace called a pyrolyzer furnace. A lower quantity of sample is required for kerogen, and a higher quantity is required for a source rock sample. Typically the temperature in the pyrolysis furnace is increased at the rate of 25°C/min up to 400°C in the first stage under inert conditions of helium or nitrogen and then in a second stage from 401 to 550°C also under inert prescribed conditions. In the third stage (optional) the temperature may be raised from 550 to 850°C, under complete oxidation conditions, to fully decompose the organic matter. The volatilized hydrocarbons are detected and measured by the coupled gas chromatograph or GCMS.

8.21.3 GAS CHROMATOGRAM/MASS SPECTRUM INTERPRETATION

The graph (gas chromatogram) is a plot of the detector response shown as amplitude or peak height versus increase in temperature. A mass spectrum is a plot of the detector response (amplitude, peak height) versus atomic mass number of individual chemical components of the volatile/cracked oil/gas. The area under the peak or the peak height is proportional to the concentration of the component. If the peak area is properly calibrated with the standard sample, the concentration of respective components in the sample is determined. The graphs obtained through GC or GCMS analyzers are used to compute all the volatiles and thermally cracked oil/gas that have evolved from the kerogen. The qualitative information of the evolved oil/gas components is obtained through respective temperatures and peak height/area as follows:

- Pyrolysis under an inert atmosphere from 200°C to 400°C. The peak/peaks between these temperatures represent any already generated free but trapped oil in the sampled kerogen. The trapped oil from the kerogen is removed by thermal distillation at a temperature between 200 and 400°C. Kerogen may or may not be associated with free oil; it depends on the kerogen maturity stage.
- Pyrolysis under inert atmosphere from 401°C to 550°C represents the hydrocarbons including dry gas that have been generated by the thermal pyrolysis of kerogen molecules. The pyrolysis (cracking) temperature (T_{max}), at which the maximum amount of oil/gas is obtained, increases with the maturity of the kerogen. With maturation the kerogen becomes a more condensed, cyclized and compact organic mass. It requires a higher temperature to volatilize and crack the organic macromolecules of the kerogen.
- The cracked oil represents the potential oil in the kerogen that will be produced in source rock, over geological time. Some carbon dioxide gas is also produced as a result of cracking.
- Pyrolysis under oxidizing conditions from 551 to 850°C. By 550°C, all the potential hydrocarbons are removed from the kerogen. Only highly

condensed aromatic ring compounds remain. Ring compounds are stable and nonreactive; do not yield any oil/gas on further cracking. Heating the sample above 550 up to 850°C simply releases two kinds of carbon dioxide: residual carbon dioxide from the decomposition of residual oil and the other due to the decomposition of carbonate mineral.

Pyrolysis and evolved volatiles give a complete picture of the thermal history and potential chemical transformation of kerogen. They tell clearly how much oil has been generated and how much is expected in the future from a given kerogen/source rock. The pyrolysis temperature may be correlated to the three geological processes as follows:

Pyrolysis temperature for diagenesis stage kerogen: 200 to 400°C.
Pyrolysis temperature for catagenesis stage kerogen: 401 to 500°C.
Pyrolysis temperature for metagenesis stage kerogen: 501 to 550°C.

The summary of expected quantitative yield of the volatiles up to the pyrolysis temperature of 500°C is as follows:

Type I kerogen. It produces about 80% paraffinic distillate oil. The oil is of high pour point, low sulfur, high wax and few ring compounds.
Type II kerogen. The volatile yield is 50–70% of mixed crude oil. The oil is naphthenic with high sulfur and low pour.
Type III kerogen. Generates mostly gases and little oil.

8.22   INDIRECT GEOCHEMICAL METHOD FOR KEROGEN EVALUATION

Indirect geochemical methods involve the identification of some key property of the kerogen that has undergone significant alteration under subsurface conditions in geological time. The significant alteration is related to kerogen maturity. Kerogen properties vary considerably from one source rock to another, and within the source rock properties may vary from point to point. The parameters of kerogen reported in the following discussion represent typical values. Some selected methods that establish the kerogen maturity are given below.

8.22.1   Elemental Composition

Elemental compositional analyses may be applied to all types of hydrocarbons (oil, bitumen and kerogen). The path way for the conversion of organic matter and kerogen to petroleum could be assessed initially by recording the changes in elemental composition.

8.22.1.1   Determination of Elemental Composition

The elemental composition of kerogen and oil is determined either by classical chemical methods or by modern elemental analyzers. The working principal of an elemental analyzer is to completely burn the crushed solid (kerogen) or oil sample by placing about one gram of the sample in a crucible placed in a furnace. The evolved carbon dioxide, nitrogen dioxide, sulfur dioxide and water vapor are directly related to the nitrogen, sulfur, hydrogen and total organic carbon contents (TOC) of the kerogen. The evolved gases and water vapor are directly fed to an absorption column of the chromatograph instrument, along with a slow stream of an inert helium gas. The gases are eluted from the column according to their retention times and adsorption-desorption capacity on the column material. The eluted gases pass through a flame ionization detector (FID) or thermal conductivity detector (TCD). The anomalous detector signals due to evolved gases are sent to the processor which directly gives the carbon, nitrogen, sulfur and hydrogen percentages in the kerogen/oil. However, sulfur is determined separately due to technical reasons. It is determined by a sulfur-determining instrument. The kerogen sulfur is burnt in a furnace and the evolved sulfur dioxide is fed to an infrared (IR) detector and the anomalous signal due to gas is sent to a data processor. In the classical method the sulfur is determined by burning the kerogen/oil sample in a steel bomb filled with oxygen in the presence of water and hydrogen peroxide. The evolved sulfur trioxide is absorbed in aqueous medium. Sulfur is precipitated as sulfate by adding a barium chloride solution. The barium sulfate is gravimetrically measured and converted to sulfur equivalent. It is difficult to measure oxygen directly. The oxygen is determined indirectly according to the following formula:

$$\%\text{Oxygen} = 100 - (\%C + \%H + \%N + \%\text{Ash})$$

Ash is the residue left after the complete burning of the kerogen sample in a furnace at 750°C. The ash consists of mostly metallic oxides. The ash represents the total metal content consisting of the inorganic metal drawn from sediment and organo-metallic compounds of the of kerogen. The individual metals can be analyzed from the ash by using atomic absorption or atomic emission spectrophotometers. Oil-soluble metals are present in trace quantities.

8.22.1.2 Determination of Elemental Composition of Gases

The elemental composition of gases is determined from the data obtained by gas chromatography (GC) and gas chromatography–mass spectrograph (GCMS). The elemental composition is computed from the knowledge of individual components of the gas sample. Standard gas samples with known composition are available for calibration and comparison with the sample components.

8.22.1.3 Comparison of Main Elements of Petroleum/ Organic Matter/Kerogen

Carbon (83–87%) and hydrogen (11–14%) are the main elements in petroleum. The major elements in organic matter are oxygen (60–65%), carbon (18–20%) and hydrogen (10–12%) excluding the bony skeleton of the organism. The most important single element of petroleum is carbon but the most important single element of organic matter is oxygen. Higher amounts of nitrogen (2–4%) are found in organic matter than in petroleum (<0.5%). The sulfur content of organic matter is 2–4% and in crude it varies from <0.5 to 6.0%. The higher amounts of sulfur in crude oil is due to the presence of sulfur mineral (iron pyrite) in sedimentary rock.

The elemental composition of three types of kerogen is given in Table 2.2. Kerogen has variable elemental composition: type I, carbon (67–70%), hydrogen (10–8%) and oxygen (0–5%), type II, carbon (75–85%), hydrogen (5–8%) and oxygen (0–2%) and type III, carbon (90–95%), hydrogen (2–4%) and oxygen (less than 0.5%). It is clear that as maturation proceeds, the carbon content of the kerogen increases and the concentrations of all other elements decrease.

A comparison between the elements of organic matter and petroleum gives useful information. The data suggest that major reactions in kerogen for the formation of petroleum are de-hydrogenation, de-oxygenation, de-nitrogenation and de-sulfurization. These are the initial reactions occurring in the diagenesis stage with the loss of hetero atoms (O, N, S) and the evolution of H_2O, CO_2, NH_3 and SO_2 . This is followed by cyclization, aromatization and condensation in the catagenesis/metagenesis stages and further loss of hydrogen and oxygen within the kerogen macromolecule. Cyclization and aromatization are in fact the carbonization of kerogen that is the increase of carbon content. Dehydrogenation of kerogen due to thermal cracking results in the corresponding increase of carbon and hydrogen atoms in petroleum during the 'oil window'. Thermal cracking further leads to the cyclization and aromatization of the kerogen organic mass. It is clear that the path way of transformation from organic matter to kerogen to oil/gas could be followed by observing the elemental composition of these substances.

8.22.2 Total Organic Carbon

The elemental compositions of carbon, oxygen and hydrogen are initially useful for determining the potential of kerogen for oil/gas generation. But the 'total organic carbon' content is especially used to determine kerogen potential for oil/gas generation. The carbon content is expressed in three ways:

- Total carbon content of sample
- Total inorganic carbon
- Total organic carbon

Total carbon content (TCC) is the sum of 'total inorganic carbon' and 'total organic carbon'. The 'total inorganic carbon' comes from inorganic carbonate and carbonic acid in the sample. The 'total organic carbon' (TOC) originates from the organic matter in the sample. The TOC is obtained by subtracting the inorganic carbon from the total carbon content of the sample. Total organic carbon is determined by several geochemical methods involving physical, chemical, spectroscopic and petrological techniques.

The capacity of source rock for the generation of oil/gas depends on the quantity and quality of the associated organic matter apart from geological factors. The quantity of organic matter and kerogen is best expressed as TOC, which gives a good indication of the potential for oil/gas generation. TOCs are of three types. Each type has its own significance in contributing to the quality of the kerogen:

- **Converted carbon.** Oil/gas that has been generated in source rock, but not expelled from the source rock.
- **Potential carbon.** Carbon that has potential for conversion to oil/gas in the kerogen. It depends on the chemical structure of the organic matter which is likely convertible to oil/gas.
- **Residual carbon.** Residual carbon is that organic carbon which is unfit for generating oil/gas. It is highly aromatized, cyclized, polymerized and condensed rings system. It is termed 'dead carbon'.

A minimum 3.0% of TOC (potential) in source rock is considered as a good indication for generating commercial quantities of oil/gas. The total potential organic carbon of the kerogen is expected to decrease because of conversion to oil/gas and expulsion from source rock with maturation during diagenesis, catagenesis and metagenesis, and correspondingly residual and inorganic carbon are expected to increase.

8.22.3 Atomic H/C Ratio and Atomic O/C Ratio

The path way of transformation of organic mass to bitumen/oil/gas could be followed by noting the decrease in the atomic ratios of the H/C and O/C of kerogen. At the expense of a decrease in the H/C ratio of kerogen, a corresponding increase in the H/C ratio of petroleum is observed. Cyclization, decarboxylation, de-nitrogenation and de-sulfurization during the diagenesis, catagenesis and metagenesis processes in organic matter/kerogen correspond to increases of hydrogen and carbon content in petroleum. Petroleum has higher hydrogen (11–14%) and carbon (83–87%) contents than any of its predecessor organic mass.

The H/C ratio of lower molecule weight hydrocarbons is higher than that of the larger molecular weight substance. Methane's (CH_4) atomic H/C ratio is 4.0 and this decreases with the rise in molecular weight. The atomic H/C ratios of ethane (C_2H_6), propane (C_3H_8) and hexane (C_6H_{14}) are 3.0, 2.7 and 2.3 respectively. Cyclic compounds have lower H/C ratios. The atomic H/C ratio of benzene (C_6H_6) is 1.0, whereas the H/C ratio of naphthalene ($C_{10}H_8$), two benzene rings fused together) is 0.8. The range of the H/C ratio of organic matter, the precursor of petroleum, is 1.2–1.6. The extent of change in this atomic ratio depends on the quality of the original organic matter at constant geological conditions. The quality and quantity of generated oil/gas depend on the quality and quantity of original organic matter in the sedimentary rock. Organic matter with a higher H/C ratio is likely to generate more oil/gas than organic matter with a lower H/C ratio. Organic matter first transformed to kerogen. On the basis of its H/C ratio the kerogen organic mass is divided into two types. One is saprogenic type I kerogen, having a high atomic H/C ratio (1.4–2.0), and the other is the humic kerogen type III organic substance having a lower H/C ratio (0.90–0.95). The saprogenic substance is derived from organic matter rich in oil, fat, wax and lipid content. Saprogenic organisms are spores, algae, flora and fauna preserved in aquatic and shale mud environments. Saprogenic substances generate oil of a paraffinic nature. Humic organic substances are generated by land plants. The humic substances produce coal or very thick black asphaltic oil, depending on geological conditions. It is possible that both saprogenic and humic matter is deposited in the same sedimentary rock and generates type II kerogen (H/C ratio 1.0–1.3). The generated oil will have the character of both the saprogenic and humic substances. The kerogen becomes progressively cyclic, devoid of paraffinic side chain, and accordingly the H/C ratio is decreased. The H/C ratio of the residual kerogen type IV is reduced to 0.3–0.5. Typical values of the O/C atomic ratio for type I, II and III kerogen are 0.15, 0.20 and 0.25, respectively, in the initial stage and reduce to a common value of 0.025 at the end of metagenesis. Type IV reworked kerogen has an O/C atomic ratio of more than 0.25 in the initial stage.

8.22.4 Van Krevelen Diagram

The van Krevelen diagram (VKD) predicts the petroleum-generating capacity of source rock organic matter and maturation of kerogen. The van Krevelen diagram elaborates the kerogen maturation. The maturation process is associated with changes in the chemical structure of the kerogen, the generation of hydrocarbons, oil, wet and dry gases and non-hydrocarbon gases. The van Krevelen diagram is based on elemental analysis and calculated H/C and O/C atomic ratios. The diagram is a plot of the H/C ratio versus the O/C ratio of kerogen I, II, III and IV. Type I is saprogenic, type II is mixed (saprogenic + humic), type III is humic and type IV is residual kerogen. The maturation path way of kerogen and the transformation to oil/gas/residual carbon during diagenesis, catagenesis and metagenesis can be followed with the help of the diagram (VKD). The characteristics of kerogen are related to the relative distribution of the main elements, hydrogen, oxygen and carbon, in the kerogen, during the maturation process. There is a rapid loss of hydrogen (H/C ratio with respect to carbon) for type I kerogen, followed by type II and III. The hydrogen content is slightly altered in type IV kerogen. Type IV kerogen has an already meager amount of hydrogen atoms (<2%). Also, there is a decrease in the O/C atomic ratio (deoxygenation), first slowly and then quickly, relative to the carbon, with the evolution of CO_2 and H_2O. The oil/gas produced from the kerogen has a higher hydrogen content. Organic matter, having a higher content of hydrogen (higher H/C atomic ratio) atoms, is likely to produce more oil/gas. The kerogen type I derived from organic matter containing oil, fat, wax, spores, algae, flora and fauna preserved in an aquatic environment has the highest H/C atomic ratio. Kerogen type I generates more oil/gas. Type IV kerogen is devoid of sufficient hydrogen and is of low H/C ratio; there is only minor evolution of gases. It is almost equal to residual carbon, having no capacity to generate oil and very little dry gas. In fact kerogen type IV is reworked from types II and III kerogen. Type II is also oil-prone kerogen whereas type III(humic) is gas-prone kerogen. Four distinct phases are witnessed during kerogen maturation resulting in the progressive generation of (1) water and carbon dioxide, (2) oil/gas, (3) wet gas and (4) dry gas. A summary of the change of H/C and O/C ratios from initial diagenesis to the end of metagenesis is as follows:

	Digenesis (H/C)	Metagenesis (H/C)	Catagenesis (H/C)
Type I	2.0–1.4	1.3–0.6	0.5–0.3
Type II	1.3–1.1	1.0–0.6	0.5–0.3
Type III	0.95–0.90	0.8–0.6	0.5–0.3
	(O/C)	**(O/C)**	**(O/C)**
Type I	0.15–0.10	0.09–0.06	0.05–0.025
Type II	0.20–0.15	0.14–0.06	0.05–0.025
Type III	0.30–0.25	0.24–0.06	0.05–0.025

Characteristics of all types of kerogen continue to change with maturation until a comparative stable dead carbon is produced of fixed H/C (0.3) and O/C (0.025) atomic ratios. It is apparent from the VKD that all the four types of kerogen try to attain stability under equilibrium geological conditions during metagenesis. The stability and equilibrium are indicated by the converging of all four kerogen characteristics to a point of the same H/C and O/C ratios. At the end of maturation and during metamorphism, the four types of kerogen merge into each other and become indistinguishable, representing an inert crystalline refractory solid mass of graphitic carbon.

A summary of the approximate quantities of the volatiles evolved up to pyrolysis temperature of 500°C is as follows:

- Type I kerogen produces about 80% of oil. The oil is of a high pour point, low sulfur, few ring compounds, waxy and paraffinic in nature.
- Type II kerogen. The distillate yield comprises 50–70% mixed crude oil with little gas. The crude oil is naphthenic with high sulfur and a low pour point.
- Type III kerogen generates mostly gases and little oil.

8.22.5 HYDROGEN INDEX (HI) AND OXYGEN INDEX (OI)

The potential oil-generating capacity of kerogen can be predicted with the help of two additional terms, namely 'hydrogen index (HI)' and 'oxygen index (OI)'. The hydrogen index is defined as the milligrams of petroleum hydrocarbon (HC) generated in one gram of TOC. Similarly OI is defined as the milligrams of CO_2 evolved in one gram of TOC. Quantitative estimation of hydrocarbons and CO_2 is carried out by a pyrolyzer coupled with gas chromatography. Both HI and OI are related to the H/C ratio and O/C ratio. A plot of HI versus OI is similar to a van Krevelen diagram (VKD). So the HI/OI plot is called a modified van Krevelen diagram (MVKD). The HI of kerogen progressively decreases from 900 mg HC/g TOC to 100 mg HC/g

TOC during maturation. Similarly OI progressively decreases from 200 mg CO_2/g TOC to 25 mg CO_2/g TOC during maturation. Noting the HI and OI at different intervals allows the path way for the transformation of kerogen to be followed. The range of hydrogen index could be related to the oil, wet gas, dry gas and residual carbon generating capacity of kerogen as elaborated as follows:

- Oil-prone kerogen has a HI above 800.
- Kerogen type IV has a hydrogen index less than 150.
- The oil-producing range of HI is 800–550, wet gases 550–350, dry gas 350–100 and dead carbon below 100.

The hydrogen index of kerogen is drastically reduced from the diagenesis, catagenesis and metagenesis stages. At the end of metagenesis, only very nominal residual hydrogen is found in kerogen.

8.22.6 VITRINITE REFLECTANCE

Vitrinite is maceral, one of the components of coal. Macerals are organic compounds in coal analogous to the minerals of rock. Vitrinite is a vitreous substance, shiny like glass, having a high calorific value associated with volatile matter. The vitrinite reflectance method is frequently applied for the maturity determination of coal and petroleum kerogen. It is an optical method. Originally the method was developed for coal evaluation and its rank (maturity) determination. Vitrinite compounds are also found in petroleum kerogen, so the method was extended for the determination of kerogen maturity. Vitrinite reflectance is defined as 'An average of a series of measurements of the percentage of incident white light reflected from the vitrinite macerals at a 500× magnification in the oil immersion'.

8.22.6.1 Coal/Kerogen Constituents

Coal consists of three types of distinct microscopic organic units called macerals along with inorganic materials. 'Maceral' particles are building blocks of coal similar to 'mineral' particles in rock. The difference in optical properties of the three maceral groups is utilized for their identification in kerogen and coal. The three organic maceral groups of coal which are also found in kerogen are as follows.

- **Liptinite macerals** are an amorphous component of coal/kerogen derived from waxy, resinous and algal organisms and represent types I and II kerogen. Their vitrinite reflection is the lowest among the three types of macerals.
- **Vitrinite macerals** of coal and kerogen originate from the deoxygenated macromolecular organic compounds derived from cellular woody plants. They are equivalent to type III kerogen.
- **Inertinite macerals** in the coal/kerogen are formed from highly degraded and reworked plant residue deposits. The macerals are found in type IV kerogen. The vitrinite reflectance of the group is the highest.

8.22.6.2 Technique for Vitrinite Reflectance Measurements

The kerogen sample is prepared for measurement by pulverizing the solid and compressing to form a pellet. The kerogen pellet sample is irradiated with incident light. Numbers of reflections are noted from different points of the kerogen pellet sample. The average vitrinite reflectance intensity is reported as $\%R_0$. The intensity of reflected light from the prepared surface of the kerogen sample increases with the increase of kerogen maturation. With maturity, the glossy, vitreous and crystalline character of the kerogen also increases.

Figure 8.2 is the graphical representation of reflection data by 16 rectangular bars each representing a reflection point on the sample and the corresponding intensity ($\%R_0$) of reflected light from each reflection point. The reflection is due to the individual vitrinite maceral present in the sample. Wide variation in the intensity (indicated by the height of the bars) of the reflected ray occurs within a sample from point to point. Obviously refection data with minimum deviation are desirable. Deviation in reflection data is attributed to light scattering from the inertinite/liptinite macerals of the sample. An average vitrinite reflectance dataset compiled from a number of reflections with minimum deviation is taken for computation and practical application. The average reflection data determine the quantity of vitrinite in the kerogen/coal sample and give satisfactory evidence for the determination of the maturity of kerogen. With the increase of maturation of kerogen/coal rank, the intensity of vitrinite reflectance increases.

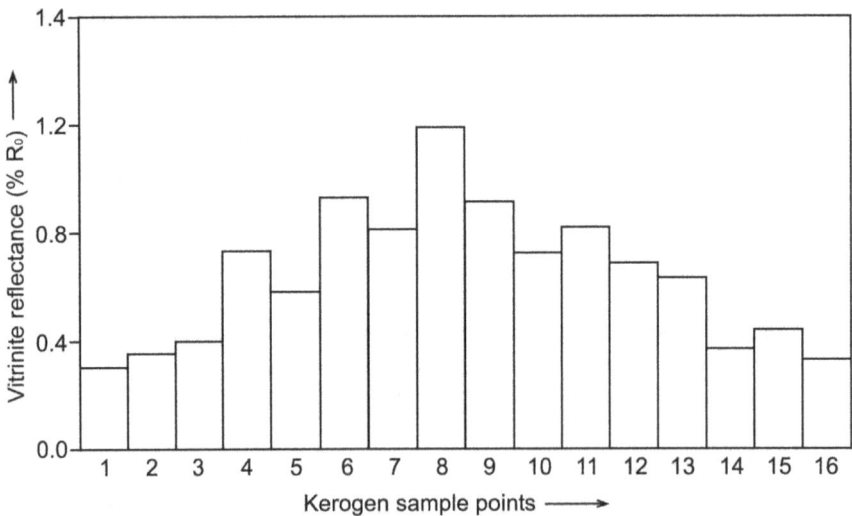

FIGURE 8.2 Graph showing vitrinite reflectance ($\%R_0$) as rectangular bars for 16 reflecting points on the kerogen sample. Source: Modified from John M. Hunt, *Petroleum Geochemistry and Geology*, W.H. Freeman and Company, New York, 1996 (Figure 10.25, page 369).

A summary of change of vitrinite reflection ($\%R_0$) during three geological stages is as follows:

	Diagenesis	Catagenesis	Metagenesis
Vitrinite reflection	0.2–0.6	0.8–2.0	3.0–5.0

8.22.7 THERMAL ALTERATION INDEX (TAI)

The thermal alteration index is the visual examination of color and correlation to determine the maturity of kerogen. The term 'thermal alteration index (TAI)' may imply that some kind of heat is applied to kerogen, but it is not so. Hardly any external thermal energy is applied to kerogen except the natural energy along with other geological factors. The TAI measurement is based on a color scale. The thermal alteration index (darkening of color) increases with geological time due to the increase of maturity of kerogen. The thermal alteration index method involves the visual examination of the kerogen sample and noting the color. During the early stage of diagenesis, liptinite macerals retain their original color of plants species. The colors of spores, pollen, cuticles, algae and amorphous organic matter are green and yellow. During the catagenesis stage, the color progressively turns from lighter to darker that is orange, brown and finally black. In the metagenesis stage, the colors of all the macerals turn to black irrespective of their origin and extent of maturation. At this stage it is extremely difficult to distinguish the kerogen types because all the three macerals attain similar characteristics. The macerals are all dark, and kerogen seems to have merged into one kind. This is the final stage of maturation.

The 'TAI color scale' has been given numbers arbitrarily from one to five, corresponding to colors from lighter to darker. A higher number denotes a higher maturity of kerogen. Darkening and the corresponding TAI of kerogen increase with depth and maturation. In addition to thermal maturity, the TAI gives some idea about the quality of organic matter of the kerogen. A summary of the change of the thermal alteration index during three geological stages is as follows:

	Diagenesis	Catagenesis	Metagenesis
TAI scale	1.0–2.5	2.6–3.8	4.0
TAI color	Green, yellow	Orange, brown	Black

8.22.8 FLUORESCENCE METHOD

Certain substances when struck by incident radiation of a shorter wavelength (visible or ultraviolet light) emits light of longer wavelength radiation with fluorescence. The fluorescence is the emission of light from a particular type of material because of incident rays. Fluorescence stops when the incident light is switched off. The emitted light is another kind of electromagnetic radiation having a higher wavelength than the incident light. Fluorescence is exhibited by material having an aromatic structure or chromophore groups with unsaturated chemical bonds. A chromophore is a chemical group responsible for the color of the organic substance. Organic compounds having saturated bonds (paraffin, naphthene) do not exhibit fluorescence.

Fluorescence is used to distinguish potential source rock from others. The kerogen sample is irradiated with visible blue light (470 nm) or ultraviolet radiation. The electrons in the unsaturated bond are excited and are raised to higher unstable energy levels. The unstable electron falls back to the ground state by emitting the excess energy in the form of fluorescence electromagnetic radiation. The wavelength of the emitted light is higher than the original incident visible to UV light. The intensity of emitted light and change in wavelength depend on the nature of organic compounds in the kerogen. An immature kerogen sample gives stronger fluorescence (more intensity in the emitted light) than a mature kerogen sample. Immature kerogen has more chromophore groups than the mature sample. With the increase of maturity there is a loss of chromophore groups in kerogen, and the emitted fluorescence intensity decreases accordingly. In the initial stage of kerogen (diagenesis) the fluorescent method works better than the vitrinite reflectance method. In this stage poor vitrinite reflectance signals are observed. This is because of low vitrinite content and the lack of crystalline character. The fluorescent method gives strong signals with high liptinite content. The color and wavelength (nm) of the emitted fluorescent light during three geological stages are as follows:

Diagenesis: blue and green. Wavelength 470 and 500 (nm).
Catagenesis: golden, yellow, dull yellow, orange and red. Wavelength 540, 600, 640, 660, 680 (nm).
Metagenesis: no fluorescence

8.22.9 BIOCHEMICAL MARKERS IN PETROLEUM

Biochemical markers are natural geo-chemicals found in source rock, soil, organic matter, kerogen, surface and subsurface petroleum oil/bitumen and coal. They are chemically stable, moderate to high molecular weight organic compounds with extremely low volatility. Inertness and non-volatility and compact structure help them to survive over geological time. Biomarkers are inert compounds compared to other organic compounds of petroleum. However they do undergo minor alteration within the subsurface conditions. At the same time biomarkers differ in stability within their member group. Some biomarkers are altered less and have survived for longer periods of geological time than others. Typically more stable biomarkers retain all or most of the original molecular structure of the specific living organism from where the biomarker originated. Markers are higher molecular weight members of paraffinic, aromatic, terpenes, steroids and porphyrin compounds. The distribution and identification of biomarker in solid source rock and in petroleum oil can be followed by geochemical analytical techniques and modern analytical instrumental analysis such as GCMS and HPLC. The biomarker information leads to the following geological information.

- Knowing the type and structure of biomarkers, the kerogen and oil are assessed.
- Knowing the total distribution of biomarkers in oil, the characteristics of the source rock could be worked out.
- Characteristics of biomarkers can help to predict oil-prone source rock or a gas-prone source rock.
- Knowledge of biomarkers is used to predict the depositional environment of kerogen source rock such as marine, lacustrine, fluvial, deltaic or terrestrial.
- Biomarkers help to determine the deposition environment in terms of lithology (shale or carbonate).
- Biomarkers help to determine the age of the source rock.
- Biomarkers give evidence for the thermal maturity of source rock.
- Biomarkers give information about the petroleum basin.
- Biomarkers are used for the identification of bio-geo-chemical processes in subsurface rock.
- The concentration of biomarkers decreases in kerogen with maturity and ultimately vanishes in the metagenesis stage on the formation of dead carbon and graphite.

Two examples of common biochemical markers found in petroleum are 'sterane' and 'hopane' hydrocarbons. Sterane is high molecular weight four cyclic compound derived from steroid or sterol. Hopane is also high molecular compound derived from triterpene.

The biomarkers are characterized on the basis of their origin, type, molecular size and molecular structural shape. This requires advanced knowledge of organic chemistry and highly sophisticated instrumentation.

8.22.10 Isotopes in Geochemical Survey

Atoms of an element with the same atomic number (numbers of electrons) but different atomic weights (neutron + proton) are isotopes.

Carbon forms two isotopes. Each atom has an atomic number 6 and atomic masses of 12 and 13 denoted as ^{12}C and ^{13}C respectively. The range of ^{13}C isotope in nature is much smaller compared to that of ^{12}C. It is between 1.02–1.13%. The carbon isotopic ratio ($^{13}C/^{12}C$) has been used to determine maturation, generation and alteration in oil/gas and the evolution of the source rock depositional environment, natural gas, oil, bitumen and kerogen. The carbon isotopic atomic ratio gives useful information about the gas origin. It can be decided whether a gas is biogenic or thermal or from the catagenesis or metagenesis stage. Thermal gas has a higher abundance of ^{13}C than biogenic.

8.23   OVERALL CORRELATION OF KEROGEN PARAMETERS

The correlation of kerogen characteristics with other relevant geo-physico-chemical parameters is given in Table 8.3. All the parameters which are used for the evaluation of kerogen are related to each other.

- Among all the parameters, only the pyrolysis method gives information directly about the quality, quantity and maturity of the kerogen in spite of some limitations. Pyrolysis is the direct and forced approach for kerogen maturation evaluation. The pyrolysis temperature (T_{max}) of kerogen increases with maturation from diagenesis and catagenesis to metagenesis. The temperature of pyrolysis and corresponding volatiles and cracked products give a complete picture of the already generated and potential oil/gas in the kerogen. The pyrolysis temperature is correlated with the three geological processes. It may be added that laboratory conditions of pyrolysis for the generation of bitumen rely mainly on artificial thermal input, devoid of the subsurface geological environment for the maturation of kerogen.
- The kerogen elemental percentages of carbon, oxygen and hydrogen vary significantly during maturation. Carbon increases whereas hydrogen and oxygen decrease along with kerogen maturation.
- The hydrogen index drops significantly from 800 in the diagenesis stage to as low as 50 in the final stage of metagenesis.
- The fluorescence of kerogen under light of suitable wavelength gives a good indication of the thermal maturity of kerogen. At the end of metagenesis the kerogen does not emit fluorescent light.
- The relationship between vitrinite reflectance versus the H/C ratio of kerogen is shown in Figure 8.3. With maturity, the vitrinite reflectance of kerogen increases and the atomic H/C ratio decreases. Point P and other points upward in the figure represent the H/C ratio corresponding to immature kerogen. Any point below P represents the H/C ratio corresponding to mature kerogen. At lower point B all types of kerogen merge into each other with the same very low H/C ratio (0.3) and high vitrinite reflectance ($\%R_0$) of about 5.
- The vitrinite reflectance and thermal alteration are commonly used for the determination of the thermal maturity of kerogen. There is a definite relationship between vitrinite reflectance and TAI. The catagenesis (oil window) is represented by vitrinite reflectance between 0.8 and 2.0 which corresponds to a TAI value between 2.6 and 3.8. Both vitrinite reflectance and TAI reach values of 5 and 4 at the fully mature and stable stage of kerogen respectively at the end of metagenesis.

TABLE 8.3
Correlation of Kerogen Characteristics with Physical Parameter

Maturation Stage	Subsurface Temperature °C	Vitrinite Reflectance R_0 %	H/C Ratio %	TAI Number	Fluorescence		Pyrolysis T_{max}
					Color	λ max	
Diagenesis	50°C	0.2	1.7	1.0	Blue	470	400°C
		0.3	1.6	1.8	green	500 nm	
		0.4	1.5	2.2			
		0.6	1.4	2.5			
Catagenesis	150°C	0.8	1.3	2.6	Golden	540 nm	401°C
		1.0	1.1	3.0	Yellow	600 nm	440°C
		1.2	0.9	3.3	Dull yellow	640 nm	460°C
		1.5	0.8	3.6	Orange	660 nm	480°C
		2.0	0.7	3.8	Red	680 nm	500°C
Metagenesis	250°C	3.0	0.5	3.9	No fluorescence, not applicable.		501°C
		3.5	0.4	4.0			530°C
		4.0	0.3	4.0			550°C
		5.0	0.3	4.0			550°C
Remarks	Determines maturity.	Determines maturity.	Determines quality and maturity.	Determines maturity.	Determines quality and maturity.		Direct method.

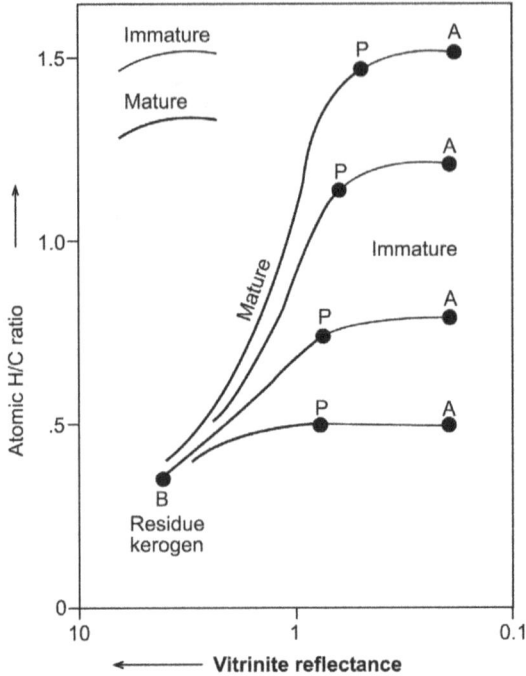

FIGURE 8.3 Correlation between vitrinite reflectance and atomic H/C ratio. Source: Modified from Douglas Waples, *Organic Geochemistry for Exploration Geologist*, Burgess Publication Company CEPCO Division, 7108 Ohms Lane. Minneapolis, Minnesota 55435, USA, 1981 (Figure 7.1, page 70).

8.24 BITUMEN EVALUATION

The next step after kerogen evaluation is bitumen analysis. Bitumen is a generic term, related to the hydrocarbons associated with and generated from the kerogen. Bitumen is a mixture of hydrocarbons of natural origin which may be solid, semi-solid, liquid or gas. Bitumen is completely soluble in carbon disulfide. In commercial practices the term bitumen is used for the semisolid or solid bitumen which includes asphalts, tars and pitches.

A bitumen molecule is composed of predominantly carbon and hydrogen together with hetero atoms nitrogen, sulfur and oxygen and trace amounts of metals, particularly vanadium and nickel. The physical properties and chemical composition vary considerably from one bitumen sample to another. The hydrocarbons derived from bitumen can be classified broadly into two groups, cyclic and non-cyclic compounds. Each series may be further subdivided into saturated and unsaturated compounds, as follows:

- Cyclic hydrocarbon compounds
  - Aromatic – semi saturated
  - Naphthene – saturated

- Non-cyclic straight chain hydrocarbon compounds
  - Paraffin – saturated
  - Isoparaffin – saturated
  - Olefin – unsaturated (not found in bitumen)

The non-hydrocarbon elements, called hetero atoms, in bitumen bear the same pattern as in other natural hydrocarbons. Sulfur, oxygen and nitrogen compounds are found mostly in ring structure. These hetero atoms are referred to as 'functional or polar groups'. Generally 'functionality' (presence of functional groups) relates to how the bitumen molecules interact with each other or with other molecules. Polarity refers to unbalanced electrochemical forces within the molecule, which produce dipoles. A dipole induces a molecular interaction that influences physical interaction. The hetero atoms impart functionality and polarity to the bitumen molecules, and they have a much greater effect on the properties of bitumen. A bitumen sample is analyzed by various methods.

- Solvent extraction and selective adsorption-desorption column chromatography
- Gas chromatography–mass spectrometry
- High-performance liquid chromatography

Solvent extraction along with the column chromatography method only separates the bitumen into five broad fractions (types). It does not give the individual component

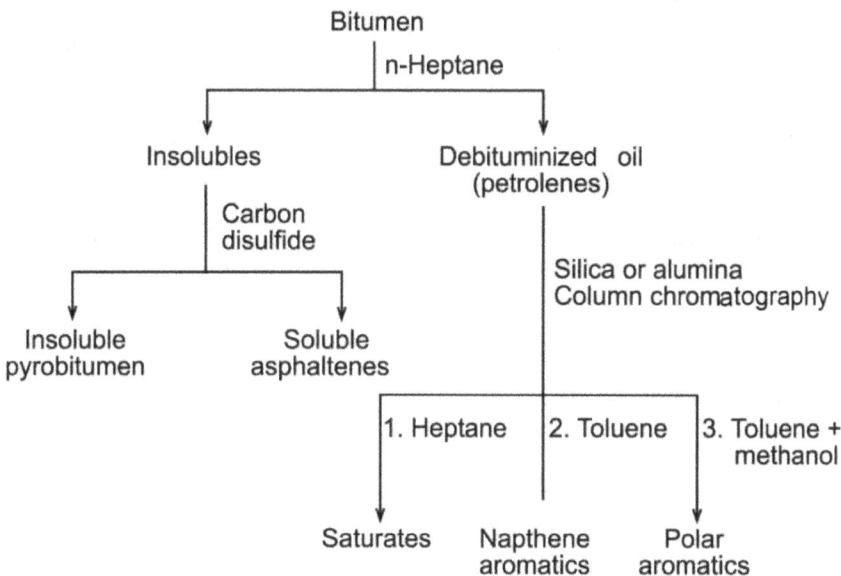

FIGURE 8.4 Principle bitumen components by solvent extraction and column adsorption-desorption chromatography method.

composition. To determine each component of the bitumen, a combination of the above mentioned methods are used. In the solvent extraction method, first bitumen is separated according to its solubility/insolubility in different solvents.

A solvent extraction scheme coupled with an adsorption-desorption column chromatography scheme is shown in Figure 8.4 using five different solvents of increasing polarity. The five solvents of increasing polarity are heptane, carbon disulfide, toluene and a toluene–methanol mixture. By this scheme, the following types of hydrocarbons are obtained.

8.24.1 PYRO-BITUMEN

The fraction of bitumen with the highest carbon content is known as pyro-bitumen. It is a black solid insoluble in carbon tetrachloride or carbon disulfide. It is a minor component of bitumen.

8.24.2 ASPHALTENE

Asphaltene is soluble in carbon tetrachloride and carbon disulfide but insoluble in a non-polar solvent such as heptane and naphtha. The heptane insoluble portion of bitumen is known as asphaltene. Asphaltenes are brown to black amorphous solids. They have a hydrogen/ carbon ratio of 0.81–1.00 which indicates that they are aromatic in nature. Their molecular weight range carbon s from 500 to 7000.

8.24.3 PETROLENE OR MALTENE (OIL)

The heptane soluble portion of bitumen is oil and is termed as petrolene or maltene. These are asphaltene and pyro-bitumen-free fractions of bitumen. Their hydrogen/ carbon ratio is in the range of 1.4-1.8 which indicates that their molecular structure consists of ring and saturated hydrocarbons. The ring compounds are naphthene, aromatic and aromatic containing hetero atoms (S, O and N). Saturated hydrocarbons are naphthene along with straight and branched chain hydrocarbons. The fraction may have appreciable wax content. Their molecular weights are in the range of 300–700. Petrolene is used to conduct column chromatography to get another three fractions of oil:

- Saturated compound. Normal paraffin, iso-paraffin and naphthene hydrocarbons.
- Saturated ring compound, naphthene-aromatic ring hydrocarbons without hetero atoms.
- Polar-aromatic hydrocarbons. They are aromatic compounds containing appreciable amounts of hetero atoms in a ring system.

In addition to the above-mentioned components the bitumen contains dissolved (associated) gases. The gases are analyzed by the gas chromatography method.

8.25   CONCLUSION

The earth as a whole is composed of chemicals. Petroleum geochemistry uses chemistry principles to explain the subsurface chemical transformation of organic matter that leads to the formation of oil/gas. Geology is concerned with solid state matter that changes with the environment and time. Solid material occurs in amorphous and crystalline states. Solid exhibit isomorphism, polymorphism, enantiotropy and monotropy. Among about 3000 solid minerals, silicate minerals are the most prominent. The earth's crust consists of about 92% silicate minerals. The silicates are formed by the fusion of oxides of metals and non-metals, particularly SiO_2, Al_2O_3, CaO, etc. The formation of silicates may have occurred in hot magma in conditions of high temperature and pressure. Other minerals are formed in aqueous medium facilitated by ionic potential, hydrogen ion concentration, oxidation-reduction, ionic potential and the colloidal behavior of minerals. Before the formation of sedimentary rock, the earth was covered by exposed igneous rock. The gradual cooling of igneous rock and its interaction with the atmosphere, hydrosphere and biosphere gradually developed the earth's crust to the present stage. The biosphere provided the necessary organic matter which through geological conditions and time transformed to oil/gas, in subsurface conditions. The main components of organic matter are proteins, carbohydrates, lipids (oils, fats, waxes), lignin, cellulose and the minor but important nucleic acid. The organic matter first converted to kerogen. The kerogen is evaluated for its maturity by geochemical methods for its potential for petroleum generation. The most effective and direct method is the pyrolysis of source rock/kerogen sample. In pyrolysis a known quantity of sample is cracked (pyrolysis) at an elevated programmed temperature range from 200 to 850°C, to yield different fractions of oil along with gases and inorganic residue. The oil fractions are correlated with the oil already generated and the potential oil that will be generated in future from the kerogen. Indirect methods for kerogen maturity determination are elemental analysis, H/C ratio, vitrinite reflectance, hydrogen index, TAI, and fluorescence. These are the specific physical/chemical properties of kerogen. Their value at a particular stage is related to the kerogen maturity and quality and quantity of the potential oil of the kerogen. The geochemical technique is directly applied for the analysis of oil/gas seepage from underground to the surface, in both visible and invisible leakages. But it has limited application in predicting the subsurface deposits. It works well for shallow depths but gives erroneous results for deeper reservoirs. The bitumen extracted from the kerogen is analyzed by several analytical techniques, namely solvent selective extraction, adsorption-desorption column chromatography, GCMS and HPLC. The selective solvent extraction coupled with adsorption-desorption column chromatography separates the bitumen into five fractions, two solids, and three oil fractions. The solids are pyro-bitumen and asphaltene. The three oil fractions are polar aromatics, naphthene-aromatics and saturate hydrocarbons.

References

TEXT BOOK

1. B.P. Tissot and D.H. Welte, *Petroleum Formation and Occurrence*, Springer Verlog, New York, 1984.
2. Norman J. Hyne, *Non-Technical Guide to Petroleum Geology, Exploration, Drilling and Production (e-book)*, Penn Well Book, Oklahoma, USA, 1995.
3. Norman J. Hyne, *Dictionary of Petroleum Exploration, Drilling and Production*, Penn Well Book, Tulsa, Oklahoma, USA, 1991.
4. Peter K. Link, *Basic Petroleum Geology*, OGCI and Petro Skills Publication, Tulsa, USA, 2014.
5. L.P. Dake, *Fundamental of Reservoir Engineering, Development in Petroleum Series 8*, Elsevier Publication, New York, 1995.
6. A.J. Dikkers, *Geology in Petroleum Production*, Elsevier Publication, New York, 1985.
7. Abhijit Y. Dandekar, *Petroleum Reservoir Rock and Fluid Properties*, Taylor & Francis, New York, 2007.
8. F.K. North, *Petroleum Geology*, George Allen & Unwin Publisher Ltd., 40 Museum Street, London WC1ASILU, United Kingdom, 1985.
9. John Milsom, *Field Geophysics* (third edition), John Wiley and Sons Ltd, The Atrium, South Gate, Chi-chester, West Sussex, England, 2003.
10. V. Sokolov, Victor Purto and L. Zellikoff, *Petroleum*, MIR Publishers, Moscow, 1972.
11. G.D. Hobson, *Modern Petroleum Technology*, Applied Science Publishers, London, 1975.
12. H.K. Abdel Aal, Mohammed Aggour and M.A. Fahim, *Petroleum and Gas Field Processing*, Marcel Dekker Inc., New York, 2003.
13. Frank Jahn, Mark Cook and Mark Graham, *Hydrocarbon Exploration & Production*, Development in Petroleum Science 46, TRACS International Ltd., Falcon House, Union Groove Lane Aberdeen, Elsevier, Oxford, UK, 2003.
14. Syed Iqbal Mohsin, *Introduction to oil and Gas Industry*, Department of Geology, University of Karachi, Pakistan (Unpublished).
15. Aphonsus Fagan, *An Introduction to the Petroleum Industry*, Department of Mines and Energy, Government of Newfoundland and Labrador, Canada, 1991.
16. Philip Kearey and Michael Brooks, *Geosciences Texts, An Introduction to Geophysical Exploration*, Black Well Scientific Publication, Oxford, 1991.
17. B. Durbin Milton, *An Introduction to Geophysical Prospecting*, 3rd Edition, McGraw Hill Book Company, London, 1976.
18. G. Henry, *Geophysics for Sedimentary Rock*, Technip Edition, Paris, 1997.
19. Mandouh R. Gadallah and Ray Fischer, *Exploration Geophysics an Introduction*, Springer Verlag, Berlin, 2009.
20. *Hand Book of Engineering Geophysics*, Bison Instrument Inc., 5780, West 36th Street, Minnesota 55416, USA, 1976.
21. John M. Hunt, *Petroleum Geochemistry and Geology*, W.H. Freeman and Company, New York, 1996.
22. Kevin McCarthy, Katherine Rojas, Martin Niemann, Daniel Palmowski, Kenneth Peters and Artur Stankiewicz, Basic Petroleum Geochemistry for Source Rock Evaluation, *Oilfield Review* Summer 2011:23, no.2, Schlumberger, 2011.

23. Douglas Waples, *Organic Geochemistry for Exploration Geologists*, Burgess Publishing Company, Minnesota, USA, 1981.
24. Brain H. Mason, *Principles of Geochemistry*, John Wiley and Sons, New York, 1982.
25. G.S. Sodhi, *Fundamental Concepts of Environmental Chemistry*, Narosa Publishing House, New Delhi, 2006.
26. Giovanni Martinelli, *Petroleum Geochemistry*, ARPA, Environmental Protection Agency, Reggio Emilia, Italy.
27. B.S. Bhal, Arun Bhal and G.D. Tuli, *Essential of Physical Chemistry*, Rajendra Ravindra Printers, Ram Nager, New Delhi, 2006.
28. Douglas M. Considine, *Energy Technology Hand Book*, McGraw-Hill Company, New York, 1973.
29. Michael Webber, *Fundamental of Petroleum, Energy Option and Policy*, 5th Edition, Edited by Debby Denney, University of Texas Publication, Texas, 2011.
30. Arun Bhal and B.S. Bhal, *Organic Chemistry*, Rajendra Ravendra Printers, Ram Nager New Delhi, India, 2006.
31. Johannes Karl Fink, *Petroleum Engineers Guide to Oil Field Chemicals and Fluids*, Elsevier Publication, New York, 2012.
32. S. Boyer and J.L. Mari, *Seismic Survey and Well Logging, Oil and Gas Exploration Tech., Institute Francais Du Petrole Publication*, Edition Technip, Paris, 1994.
33. Carl R. Noeler, *Text Book of Organic Chemistry*, Saunders Company, Philadelphia, USA, 1966.
34. Stephen Hawking, *A Brief History of Time*, Bantam Books Publisher, London, 1988.
35. Petroleum Technology. *Exploration, Production, and Refining*, Volume 1, Part 1, A John Wiley Inter Science Publication, New York, 2007.
36. Encyclopedia of Chemical Technology, Kirk Othmer (different volume & years), Wiley Inter Science Publication, New York.
37. Encyclopedia Britannia, Inc., William Benton Publisher, London, UK, 1973–74.

WEB SITE

1. Sedimentary Structure Chapter 3, Deposition environment chapter 6 &10, Basin Chapter 11, Sedimentary Rock Stratigraphy Chapter 8, 16 &17, ocw.mit.edu.
2. Sedimentary Rock and Sedimentary Structure, Geology 2 / Geosite.
3. Norman McLeod, Principles of Stratigraphy.
4. Bio-stratigraphy, Bio-chronology, Magneto-stratigraphy, Wikipedia.
5. Photosynthesis Process, School of Life Science.
6. Mahboob, Properties of Reservoir Rock.
7. Geological Time Scale, New World Encyclopedia.
8. Zvi Sofer, Stable isotopes in Petroleum, Amoco Production Company, Tulsa, Oklahoma.
9. Three Types of Faults: Normal, Reverse, and strike slips, Earth how, 2019.
10. What is Salt Dome? Geosciences News and Information, Geology .com.
11. Hassan Z. Harrz, Evaporite Salt Deposit.
12. Hedges J.I., Baldock J.A., Glens Y., Peterson M.L., Wickham S.G., the Biochemical and Elemental Composition of Marine Plankton,
13. Erik Kristensen. Organic matter diagenesis at oxic/anoxic interface in coastal marine Sediments.
14. Paleontology, Department of Paleontology, The natural History Museum, London.
15. Continental Drift, Paleontology and Geology Glossary, Enhanced Learning.
16. Bleil D.F., A Method of Geophysical Prospecting.
17. Stahl W.J., Carbon Isotope in Petroleum Geochemistry, Springer Link.

18. Continental Margin, Ut Dallas, edu. pujana/ocean.
19. Coring and core Analysis (Formation Evaluation), IPIMS e-learning.
20. Reservoir, AAPG, Wiki, March 2019.
21. Introduction to Geophysical Exploration, Introduction to Geophysical Exploration, Gravity, Magnetics, DC Resistivity & Refraction Seismology, kau/edu/sa.
22. John Milson and Asger Eriksen, Field Geophysics Hand Book, John Wiley and Sons.
23. Tom Boyd, Introduction to Geophysical Exploration, 1996 – 1999.
24. Mariila N.O., Gravity Method.
25. Dr. Laurent Marescot, Electromagnetic Surveying- TOMOQUEST,
26. Ground Penetration Radar, G.P.R. Ohio State.
27. Introduction to Electromagnetic Exploration Methods (D1), Geophysics 223, March 2009.
28. Introduction to Electro stratigraphy 3, Origin of Earth Geomagnetic Field.
29. Induced Polarization Method, GEOL 335.3, kau.edu.sa.
30. Ward S.H. AFMAG - Air Borne and Ground. Geophysics,
31. Mohammad Dawood, Stephen Hallinan, Rolf Herrman, Frank van Kleef, Near Surface Electromagnetic Surveying, Abdu Dhabi UAE.
32. Kevin McCarthy, Katherine Rojas, Martin, Niemann, Daniel Palmowski, Kenneth Peters and Arthur Stankiewicz, Basic Petroleum Geochemistry for Source Rock Evaluation, 2011.
33. Composition of Crust, Chemical Element, Minerals & Rock, sandatlas, Org.
34. Washington H.S. Composition of Earth Crust, Pubs.usgs/gov.
35. Components of living Cell and their function, your article library.
36. Dr. Marcela Helesicova, Chemical Composition of Living Organism.
37. Journal of Marine Chemistry, Vol. 78, 47-73 2002.
38. Origin of Life, Oxford Bibliographies.

Index

341

For Product Safety Concerns and Information please contact our EU
representative GPSR@taylorandfrancis.com
Taylor & Francis Verlag GmbH, Kaufingerstraße 24, 80331 München, Germany

www.ingramcontent.com/pod-product-compliance
Lightning Source LLC
Chambersburg PA
CBHW060757220326
41598CB00022B/2461